Modern Solid State Fermentation

Hongzhang Chen

Modern Solid State Fermentation

Theory and Practice

Springer

Hongzhang Chen
Institute of Process Engineering, CAS
Beijing, China, People's Republic

ISBN 978-94-007-6042-4 ISBN 978-94-007-6043-1 (eBook)
DOI 10.1007/978-94-007-6043-1
Springer Dordrecht Heidelberg New York London

Library of Congress Control Number: 2013933342

Printed on acid-free paper

Springer is part of Springer Science+Business Media (www.springer.com)

Preface

Solid-state fermentation originated in China and has a long history. Because of the water-saving, energy-saving, high-yield, and cleaning production advantages, solid-state fermentation has gradually received much attention by countries worldwide. After decades of rapid development, China has become the major power in the fermentation industry, and the proportion of the fermentation industry in the national industrial output value has gradually increased. However, because of insufficient understanding of the essence of solid-state fermentation itself and lag in research and development of fermentation equipment and related supporting technologies, many problems still need to be solved in large-scale application. At present, some related solid-state fermentation books mainly focused on the description of the fermentation process yet ignored basic solid-state fermentation principles. The main content of this book concerns new scientific research and the related industrial application of solid-state fermentation; for improvement of the technical level of the industry, the book systematically introduces the basic principles and applications of solid-state fermentation.

Based on the essence of solid-state fermentation and the natural solid-state fermentation process, this book examines the biological characteristics of related fermentation microorganisms; based on the principle of "three transports and one reaction" and the theory of porous media, the engineering process is clarified in detail. Basic theory is used to explain the process of solid-state fermentation and to highlight its unique advantages. Consequently, each chapter clarifies principles, technologies, processes, and applications. Technologies are derived from the basic theory principles; mature processes are the coupling of different technologies, and the modern solid-state fermentation technology platform is established by the organic integration of multiple process.

This book is a monograph that systematically discusses the basic principles of solid-state fermentation and its application. First, Chap. 1 introduces solid-state fermentation connotation and development. Chapters 2 and 3 bring out the essence of solid-state fermentation and related influencing factors, respectively, from biological and engineering perspectives. Chapters 4, 5, and 6 give the principles, the matching process, and the industrialization of aerobic solid-state fermentation, anaerobic solid-state fermentation, and adsorbed carrier solid-state fermentation, respectively. Finally, based on study and understanding of solid-state fermentation, several prospects for solid-state fermentation development are proposed.

My research is financially supported by the National Basic Research Program of China (973 Project, No. 2011CB707401); the National High Technology Research and Development Program of China (863 Program, SS2012AA022502); and the National Key Project of Scientific and Technical Supporting Program of China (No. 2011BAD22B02). In addition, my master's and doctoral programs were essential preconditions for publishing this book. Especially, Master Yingyi Duan, Dr. Zhiguo Zhang, Dr. Guanhua Wang, Dr. Qin He, Dr. Guanhua Li, Dr. Litong Ma, and Dr. Ning Wang participated in writing some chapters. This book cites many references of our predecessors and colleagues. I wish to express my sincere thanks to all of them.

Some errors may exist in this book. Please point out the mistakes and give directions to me whenever you see any weaknesses or shortages. I sincerely hope to receive correction and guidance letters from readers.

National Key Laboratory of Biochemical Engineering
Institute of Process Engineering
Chinese Academy of Sciences
Beijing, P.R. China

Hongzhang Chen

Contents

Chapter 1
Introduction

Abstract Since its inception, solid-state fermentation has provided many daily necessities for human beings. However, at present, solid-state fermentation is developing slowly because of an absence of understanding of its essence and the fermentation process. Therefore, solid-state fermentation only constitutes a small part of the fermentation industry as a whole. Compared to liquid fermentation, heat transfer efficiency in solid fermentation is low, the parameters are difficult to monitor and control, and the design and amplification of bioreactors are difficult. This chapter summarizes recent research on the principles and applications of solid-state fermentation. The purposes of this chapter are not only to help readers understand the application of solid-state fermentation but also to encourage further consideration of the subject. This chapter clarifies the connotation and the status quo of solid-state fermentation and emphasizes the basic theory of biology and the principles of regulation and the transfer process. The applications and advantages of solid-state fermentation are stated. Solid-state fermentation engineering is divided into four parts: upstream, midstream, downstream and auxiliary technology; all are introduced in detail.

Keywords Solid-state fermentation • Application • Regulation • Upstream engineering • Midstream engineering • Downstream engineering

1.1 Solid-State Fermentation

1.1.1 Solid-State Fermentation Connotation

After the 1940s, the huge demand for acetone, butanol, and penicillin caused the liquid fermentation industry to increase rapidly. The solid-state fermentation industry began to decline, and the advantages of solid-state fermentation

H. Chen, *Modern Solid State Fermentation: Theory and Practice*,
DOI 10.1007/978-94-007-6043-1_1, © Springer Science+Business Media Dordrecht 2013

were concealed by the rapid development of liquid fermentation. Consequently, solid-state fermentation only constitutes a small part of the fermentation industry as a whole. Currently, the problems of liquid fermentation, such as high energy consumption and serious pollution, are becoming increasingly prominent, significantly limiting the sustainable development of fermentation. People are again taking note of the advantages of solid-state fermentation, such as its water-saving, energy-saving, and low-cost properties. Solid-state fermentation has begun to play an important role in the chemical, pharmaceutical, and environmental fields, which points out a clear direction for the sustainable development of the entire biological and chemical industry. Thus, the principles and applications of solid-state fermentation have become a new research hot spot in recent years.

Fermentation is the process by which microorganisms catalyze nutrients, synthesize secondary metabolites, and complete other physiological activities under anaerobic or aerobic conditions. During the process, the desired microorganisms or microbial metabolites are accumulated. Therefore, there are three elements of fermentation research: the clear target product, the producing strain, and the desired training environment (nutrients, temperature, humidity, oxygen, etc.).

The unique feature of solid-state fermentation is that there is nearly no free water in the solid substrate (Chen and Xu 2004). Solid-state fermentation is a three-phase system consisting of the continuous gas phase, the liquid film, and the solid phase. It should be noted that there is no obvious relationship between the water content of the substrate and the content of free water. Because of the strong hydraulic holding ability of the solid substrate, such as that of the sugar beet plant material, even if the water content of the substrate is more than 80 %, there is seldom free water among the solid substrate. Consequently, the content of water cannot be defined as the only standard of solid fermentation.

The substrate can be divided into two categories based on its digestibility: nutritional carrier substrate or inert carrier substrate. The nutritional carrier substrate is crops (wheat bran, soybean meal, etc.) or agricultural and forestry wastes (straw, bagasse, sawdust, etc.). This substrate not only performs as a physical structure for the growth of microbes but also provides a carbon source, a nitrogen source, and growth factors for the microorganisms (Singhania et al. 2010). Nutritional food crops and agricultural and forestry wastes are the most commonly used substrates in actual production applications. The inert carrier substrate is a porous substrate that is chemically inert and difficult to be decomposed by microorganisms, such as polyurethane foam, macroporous resin, perlite, and vermiculite. These substrates only play a supporting role in the fermentation process; microorganisms can obtain nutrition from the fluid culture that is distributed in the porous media gap (Wu 2006).

Table 1.1 presents the development process of solid-state fermentation. From this, we can conclude that solid-state fermentation technology has provided many products for humans since the beginning of human civilization. In recent years, the application of solid-state fermentation technology has greatly expanded.

Table 1.1 Development of solid-state fermentation products

Time	Products
2000 B.C.	Bread, vinegar
1000 B.C.	Sauce, koji
550 B.C.	Kojic acid
7th century	Kojic acid was introduced to japan
Sixteenth century	Tea
Eighteenth century	Vinegar
1860–1900	Sewage treatment
1900–1920	Enzyme
1920–1940	Gluconic acid, citric acid
1940–1950	Penicillin
1950–1960	Steroid
1960–1980	Protein feed
1990	Bioremediation, biological detoxification, biotransformation, biopulping, aflatoxin, ochratoxin, endotoxin, gibberellic acid, zearalenone, cephamycin

1.1.2 The Difference Between Solid-State Fermentation and Liquid Fermentation

The water content of the solid substrate can be effectively maintained in the range of 12–80 %, mostly around 60 %. In contrast to solid-state fermentation, the typical water content of liquid fermentation is above 95 %. The current fermentation technology is liquid fermentation. Although application of and research for this technology have been long term, it still has many problems that need to be overcome. Detailed comparison between solid-state fermentation and liquid fermentation is shown in Table 1.2.

1.1.3 Advantages and Applications of Solid-State Fermentation

Compared to liquid fermentation, the main advantage of solid-state fermentation is a sufficient supply of oxygen. There is less organic wastewater and higher product yield in solid-state fermentation. The solid environment is more similar to the natural habitat of filamentous fungi. High value-added products could be produced by solid-state fermentation using low-cost industrial and agricultural residues as substrate. Consequently, solid-state fermentation is the most promising technology that can comprehensively utilize renewable resources.

At present, solid-state fermentation products mainly include traditional foods (vinegar, soy sauce, flavor spices); microbial cells (single-cell protein, spirulina, edible fungus, etc.); microbial enzymes (amylase, glucosidase, cellulase); and other microbial metabolites (nucleotides, lipids, vitamins, amino acids, etc.). Here, we summarize briefly.

Table 1.2 Detailed comparison of solid-state fermentation with liquid fermentation

Solid-state fermentation	Liquid fermentation
There is no free water, and the water content of substrate is low	Water is the main component of the culture
Microorganisms absorb nutrients from the wet solid substrate; a nutrient concentration gradient exists	Microorganisms absorb nutrients from the liquid culture; there is no nutrient concentration gradient
The culture system consists of three phases (gas, liquid, solid), and gas is the continuous phase	The culture system mainly consists of liquid; the liquid is the continuous phase
Inoculation size is large, more than 10 %	Inoculation size is small, less than 10 %
The required oxygen is from the gas phase; low energy consumption	The required oxygen is from dissolved oxygen; there is a larger amount of dissolved oxygen
Microorganisms adsorb on or penetrate into the solid substrate	The microorganisms uniformly distribute in the culture system
At the end of fermentation, the medium is a wet state substrate, and the concentrations of products are high	At the end of the fermentation, the medium is liquid, and the concentrations of products are low
High production rate and high product yield	Low production rate and low product yield
Mixing is difficult, and the growth of microorganisms is restricted by nutrient diffusion	Mixing is easy, and the growth of microorganisms is not restricted by nutrient diffusion
Removal of metabolic heat is difficult	Temperature control is easy
Heterogeneity	Homogeneity
The fermentation process is hard to detect and control online	The fermentation process can be detected and controlled online
Extraction process is simple and controllable; little waste organic water	Extraction process is usually complex; there is a large amount of waste organic water
Low water activity	High water activity
Simple fermentation container	Sealed fermentation container
Natural enrichment or artificial breeding strains	Pure strains
Energy consumption and equipment investment are high	Energy consumption and equipment investment are low
Low raw material cost	High raw material cost

Source: From Chen and Xu (2004)

1.1.3.1 Application of Solid-State Fermentation in the Food Industry

Solid-state fermentation originated in the field of traditional food production. Fermentation products such as wine, sauce, and vinegar are all daily necessities for people. For example, bread is a product that is fermented by yeast. Cheese is produced from mixed fermentation using *Lactococcus lactis* and *Streptococcus*. Soy sauce and miso involve the cultivation of *Aspergillus oryzae* on soybeans and wheat flour. Cultivation of lactic acid bacteria that transfer the alcohol into acetic acid creates vinegar. Traditional vinegar-brewing and acetic acid fermentation stages both use solid-state fermentation technology. “Red rice” is produced by cultivation of *Monascus purpureus* on cooked rice, which produces a dark red pigment. At the end of the fermentation, the red fermented rice is dried and ground, with the powder used as a coloring agent in cooking. Wine is also a daily product for human life; it is discussed in detail further in this book.

1.1.3.2 Solid-State Fermentation in Pharmacy and Chemistry

With the innovation of solid-state fermentation technology, applications have expanded and have shown good prospects in the pharmaceutical and chemical fields. Lactic and citric acid production had previously been accomplished by solid-state fermentation. Recently, solid-state fermentation technology has also been applied to the production of fumaric, oxalic, and linolenic acid. For example, lactic acid production involves the cultivation of filamentous fungi or bacteria using manioc, sugar beets, bagasse, and other agricultural waste as substrate. Soccol (Wu 2006) compared production of lactic acid under solid-state fermentation and liquid fermentation using *Rhizopus orzae*. The results showed that both the production levels and the production rates of solid-state fermentation were higher than for liquid fermentation.

Microbial secondary metabolites, such as amino acids, vitamins, and other biologically active substances, all have good value for medical and industrial applications. Recent studies showed that people have been successfully using solid-state fermentation technology to produce secondary metabolites such as antibiotics, bacterial toxins, auxin, immune drugs, and alkaloids. *Ceratocystis fimbriata* could grow on solid substrates such as cassava residue, apple pomace, soybeans, and coffee residue and can produce a variety of flavor compounds. Solid-state fermentation also has been applied in a variety of other fields, such as those for production of biosurfactants, glutamic acid, amine, pigments, vitamins, carotenoids, xanthan gum, and more.

1.1.3.3 Application of Solid-State Fermentation in the Energy and Environmental Protection Fields

Solid-state fermentation is important for solving the energy crisis and environment pollution (Chen and Qiu 2010; Chen and He 2012). Agricultural residues are often rich in nutrients, providing an ideal habitat for the growth of microbes. So, people tend to use agricultural residues to produce products with high value. Solid-state fermentation has been successfully applied for biofuels, biopesticides, biotransformation, biological detoxification, and bioremediation.

Fuel ethanol production from solid-state fermentation is the current research hot spot. Its advantages are briefly stated as follows: elimination of the sugar extraction process; cost savings; no wastewater discharge; and low energy consumption. Many scholars have studied ethanol production using solid-state fermentation and achieved good results.

Pest control methods using solid-state fermentation to cultivate insect pathogen and parasitic fungi are garnering increased attention. Production costs could be greatly reduced and virulence to insect pests greatly increased. Cultivation of fungi such as *Beauveria bassiana* or *Colletotrichum truncatum* with insecticidal ability is one of the most effective ways. The gas double dynamic solid-state fermentation technology invented by the Institute of Process Engineering, Chinese Academy of Sciences, has been successfully applied in industrial production of pesticides by cultivation of *Bacillus thuringiensis*, and the virulence could reach 10,000 IU/mg (Chen and He 2012).

Another important application of solid-state fermentation in environmental protection is the biotransformation of crops and waste to improve their nutritional value. The strains for biotransformation are commonly white rot fungi. For example, cassava is an important food for people in some regions of Africa, Asia, and South America, yet the protein, vitamin, and mineral contents are low. Several studies used solid-state fermentation to improve its nutritional value. Some residues also could be used to produce high-protein substances or single-cell protein; thus, solid-state fermentation could increase the protein content of products, which is more conducive to animal digestion (Chen and He 2012). Biotransformation of lignocellulosic substrate to animal feeds has broad application prospects, which will be discussed detailedly in the later chapter.

1.2 Principles and Regulations of Modern Solid-State Fermentation

Modern solid-state fermentation could use the study of the development of liquid fermentation to overcome the problems of traditional solid-state fermentation. Fermentation process control is an important way to achieve process optimization and is the main way to improve solid-state fermentation production efficiency. To meet the specific needs of production, researchers should fully understand the internal metabolic regulation of fermentation microorganisms under certain circumstances. The fermentation process control requires detailed understanding of dynamic biological characteristics, quantitative analysis of the metabolic network, and utilization of engineering knowledge to regulate environmental variables. Therefore, solid-state fermentation engineering control requires maximization of products, cognition and modification of metabolic genetic properties, and regulation and improvement of culture conditions as final targets.

1.2.1 Solid-State Fermentation Process Regulation Based on Biological Characteristics

Enhancing the potential production ability of the strain is the main way to improve fermentation efficiency. To achieve this goal, researchers should fully understand the laws of microbial metabolic regulation and fermentation and use engineering concepts to analyze and regulate fermentation processes.

In the early stage of solid-state fermentation development, fermentation strains were mainly obtained by natural selection and random mutagenesis measures. Combined with optimization of conditions and media components, the production yields were improved significantly. With the recent development of the fermentation industry, the mutagenesis screening method has been unable to meet industry fermentation requirements. Because of this, the biochemical mechanism of the synthesis should be explored extensively. Initially, researchers focused on the specific biochemical or transporting steps that seemed to limit microbial production. Subsequent researchers paid attention to the overall metabolic network because other channels would be influenced by its variation.

Because of the particularity of the solid substrate, solid-state fermentation metabolic expressions of fungi and bacteria are different from those in liquid fermentation. Chen Hongzhang studied the gas cycle stimulation effect on the growth of microorganisms; the perspective of membrane permeability and the activity of the key enzyme were the basis of the study (Duan and Chen 2012; Chen et al. 2002). The results showed that the 0–0.2-MPa air pressure pulsations of the normal operating range would not affect cell membrane permeability. Yet, the activity of the key enzymes hexokinase and adenosine triphosphatase (ATPase) were affected by the gas double dynamic condition. At present, fermentation substrates are selected based on microorganism physiological and metabolic primary properties in the solid-state environment. People usually use a uniform design, response surface methodology, and other measures to optimize the fermentation process.

1.2.2 *Solid-State Fermentation Process Regulation Based on Substrate Characteristics*

1.2.2.1 Process Characteristics and Issues

The characteristics of the solid-state fermentation nutritional carrier substrate can be recognized from both macro and micro aspects. Macroscopically, the dry weights and air permeability of the substrate are dynamic during the fermentation process. Microscopic aspects mainly include the growth of microbes, the adhesion on the surface of the substrate, as well as decomposition by the microbe.

The main characteristic of nutritional carrier substrate fermentation is the substrate layer pressure drop, which typically determined the cell growth trend previously. For example, in a fermentor with a height of 6.5 cm and diameter of 2 cm, the pressure drop changes of an inert macroporous resin carrier in the fermentation process are 0.21–0.65 cm (H_2O)/cm (stromal bed). In the process in the same fermentor, the maximum bagasse pressure drop changes are up to 2.17 cm (H_2O)/cm (stromal bed); at the beginning of fermentation, the value could reach 0.45 cm (H_2O)/cm (stromal bed). In a fermentor with a height of 15 cm and diameter of

4 cm, the wheat bran pressure drop change is 0.12 cm (H_2O)/cm (stromal bed). The initial substrate pressure drop changes are determined by the packing type, yet the substrate pressure drop changes in the fermentation process are determined by microbial growth.

The pore of the substrate will be contracted and blocked by the growth process of the microbe, which would result in channeling within the layer. At the same time, the substrate would contract, crack, and be blocked by the winding and utilization of filamentous fungi in the fermentation process. The gas first goes through the fissures, but not inside the substrate, which results in heat and mass transfer difficulties.

From the microscopic point of microbe growth, fungi growth in the solid substrate is limited by the surface tension of water, so microbial growth mainly concentrates in the pore and the edge of the liquid film. A large pore size is suitable for an adequate oxygen supply, yet at the same time, the microbe and nutrient contact would be impeded by the large steric hindrance. The small particles have small steric hindrance, which is suitable for the full contact of microbe and nutrient, but the diffusion of oxygen is affected. Therefore, only a suitable particle size could satisfy both mycelial growth and demand for oxygen and nutrients. However, previous studies showed that even if the mycelia had spread to all of the solid particles, only 34 % of the pores were occupied. Consequently, the gas phase was also continuous among the particles.

For nutrient supply, sugar and other small molecules can only be moved by the diffusion of water; the nutrient must be dissolved in the solid substrate water film before it can be utilized by microbes. Consequently, solid-state fermentation nutritional accessibility is worse than liquid fermentation because of the small amount of free water. From the perspective of the regulation of the environmental parameters (temperature, humidity, etc.), the thermal conductivity of air is much lower than for water; therefore, when using gas as the continuous phase, it is difficult to achieve high heat transfer efficiency. In addition, the hyphae often intertwine with substrate, which hinders migration and causes mass transfer difficulty. In the logarithmic period of microorganism growth, the temperature gradient difference caused by the accumulation of metabolic heat is up to 3 °C, which causes water evaporation from the substrate. Therefore, the heat and mass transfer difficulties caused by structure modification of the substrate may be the root cause of restricted development of solid-state fermentation.

The heterogeneity of the substrate is another important factor that affects the solid-state fermentation process. Because of substrate particle size heterogeneity, heat conductivity, and so on, the growth of the microbe, concentration of products, temperature, and pH are difficult to monitor.

1.2.2.2 Solid-State Fermentation Process Modeling and Regulation

Bioreactor amplification needs step-by-step and multilevel tests, which are time consuming and costly. Sometimes, the results may not be good. Since the 1950s, mathematical modeling has provided scientific and effective methods that have become routine with a wide application range. Mathematical modeling has played an active role in the solid-state fermentation amplification process at both macro and micro levels of research and application (Duan et al. 2012). On one hand, researchers improved the common solid-state fermentation reactor (tray, packed bed, rotating drum) model (Wang et al. 2010; Mitchell et al. 2010; Fernández-Fernández and Pérez-Correa 2007) and modified the model based on its operation (forced ventilation or not, stirred or not). On the other hand, at the micro level, researchers proposed mathematical modeling of substrate particle digestion, microbial growth, and enzymatic kinetics to explain the fermentation process microscopic mechanism. Whether static (tray, packed bed type) or dynamic solid-state fermentation (rotating drum, stirring), the heat, momentum, and mass transfer laws within the stromal layer are similar (Mitchell et al. 2006). Although three transfer principles are different because of the different types of substrate (wood cellulose, starch, inert carrier matrix), microorganism, and fermentation process, they all have the following common features: The fermentation substrate is solid, the gas is used as a continuous phase, and nearly no free water is present in the substrate. Thus, solid-state fermentation is a solid-liquid-gas three-phase system (Fig. 1.1).

Compared with other porous substrates (e.g., soil, rock), the growth of mycelia in the nutritional carrier substrate has a significant impact on the structure of the substrate and the transfer properties. The substrate structure always changes throughout the fermentation process, and the porosity and the mass and heat transfer properties of the substrate also change (Fig. 1.2). For the nutritional carrier substrate, the metabolic heat generated in the cell growth process, moisture, oxygen, and carbon dioxide all will affect solid-state fermentation.

Microscopic Model and Regulation of Substrate

The macro descriptive models of moisture, temperature, and oxygen gradient are always differential equations that require complicated algorithms (Mitchell et al. 2006). Consequently, microscopic models for microbial growth are simple empirical equations.

The reported growth kinetics models include linear, exponential, logistic, and other models (Lenz et al. 2004). These nonlinear regression models do not involve a specific physical meaning. Mitchell developed a segmented model that considered the parameters of the factual physical meaning of the growth of the microbes (Viccini et al. 2001). The model proposed that the growth rate of microbes would decline after the logarithmic growth phase. Therefore, the logistic model can only be applied to the logarithmic phase.

Dalsenter (Dalsenter et al. 2005) added the functional relationship between the physiological state of microbes and the changes of temperature and water activity to the logistic model description, but they did not indicate how to characterize the physiological state.

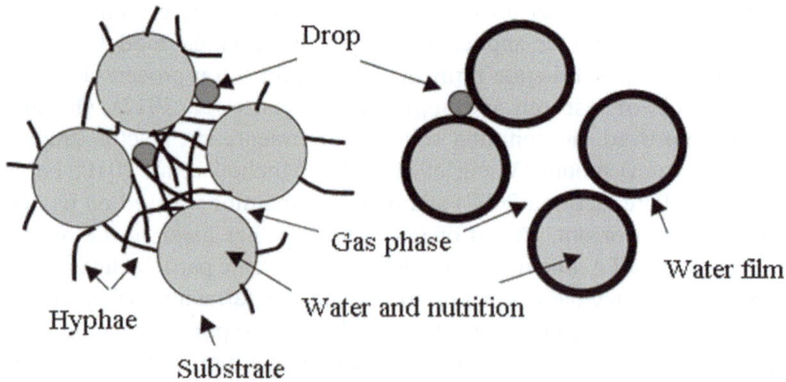

Fig. 1.1 Internal structure of the nutritional carrier substrate (Liu et al. 2006; Zambra et al. 2011)

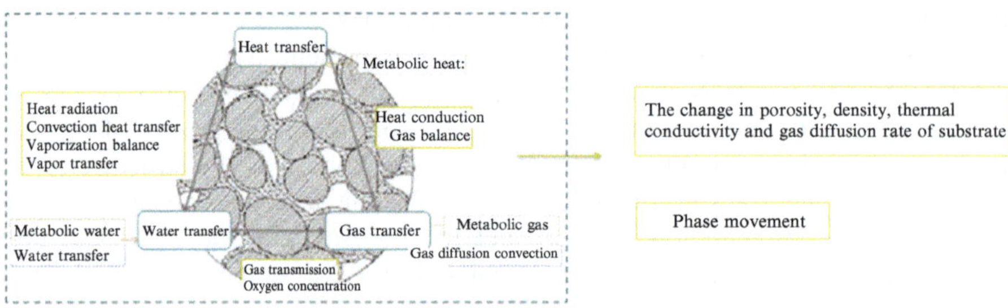

Fig. 1.2 Summary of the heat and mass transfer within the fermentation substrate

Researchers also were concerned about the interaction between the growth of microbes and the surrounding condition and established a functional relationship between microbial metabolism and the water, particle length, oxygen concentration within the substrate, carbon dioxide concentration, and heat transfer (Bovill et al. 2000). These functions all included the growth and the metabolism of microbes, yet the model parameters were hard to determine. Therefore, estimating the impact of microbial growth on the surrounding environment from the overall change of the substrate was preferable.

In addition to the microbial growth model, researchers studied the transfer of oxygen, enzymes, and carbon dioxide on the substrate surface (Rajagopalan et al. 1997). Mitchell studied the glucoamylase diffusion law on the agar substrate and assumed that the substrate was infinitely large and the pelotons lacked internal structure. Although many studies revealed the metabolite law of diffusion, this model was not promoted because of the irrationality of the assumptions and the neglect of the porous substrate structure. Rajagopalan established a model that could avoid the unreasonable assumptions; the results showed that the concentration of oxygen within the liquid film was more important than the water content of the substrate (Couto et al. 2002; Mitchell et al. 2004). However, taking into account the complexity of filamentous fungi growth in the substrate particles, the model results were often too simplified and failed to describe the actual situation.

Macro Model and Regulation of the Substrate

The solid fermentation modeling object is to study the quality of heat transfer between the substrate bed and its surrounding environment on the macro level. It is mainly used to assess the applicability of the bioreactor and to optimize the manipulated process (ventilation rate, temperature, etc.).

Compared to the experience of microscopic modeling and its randomness, macroeconomic modeling mainly emphasizes the physical meaning of the model. Transfer items in the model are substantially the same, including the conduction and convection heat in the gap between the substrate, the diffusion convection of air, the generation and diffusion of water vapor, as well as the heat conduction and convection between the bioreactor wall surface and the surrounding environment (Couto et al. 2002). These transfer forms are suitable for all the macro models of solid-state fermentation, and when the operations (such as forced ventilation, stirring, etc.) are varied and the bioreactors are different, the different substrate values, coefficients, and physical parameters may differ. At present, solid-state fermentation model improvement is mainly concerned with physical parameter and transfer item details.

In the late 1980s and early 1990s, the models were mainly concerned with the tray and rotating drum fermentation bioreactors (Mitchell et al. 2003). Differential equations analyzed the variation of heat, gas, and biomass, but did not consider the heterogeneity in the axial and radial directions of the substrate. At the same time, the first modeling mainly revealed the cooling effect of evaporation and ventilation and initially investigated the conductivity of the substrate. In the mid-1990s, the tray solid-state fermentation bioreactor model began to use partial differential equations (Sargantanis et al. 1993), which revealed the variations of mass and heat transfer in solid-state fermentation. The model studied not only the coupling of the mass and heat transfer but also the impact on the substrate size, porosity, and diffusion efficiency of oxygen when the microbes lived on the surface of the substrate. Researchers could understand the transfer process of the different parts of the stromal bed using partial differential equations, so an increase in stroma bed height attempted to obtain a greater packing coefficient for increasing solid-state fermentation capacity. As a result, the packed bed bioreactor began to replace the tray bioreactor and became a research hot topic. In the late 1990s, many packed bed bioreactor macro models appeared. The packed bed model did not adequately describe the actual transfer process in the packed bed; therefore, it cannot be regarded as a success. Although there are few references related to these models, the results still reflected a relatively correct trend.

Until the late 1990s, the appearance of Smith's model allayed those problems. This model still assumed that the forms of substrate were regular, but the diffusion of water vapor and oxygen and the water content were coupled with the growth of microbes and heat transfer; at the same time, the microbial growth model was improved, which could reflect the decline of the microbe. It is worth mentioning that the three phases of substrate and porosity were also embodied in the model (Ashley et al. 1999; Mitchell et al. 1999; Hasan et al. 1998). Therefore, Smith's model not only was more in line with the overall actual situation but also was a qualitative breakthrough when compared to conventional solid-state fermentation models.

At the beginning of this century, people gradually recognized that the increases in the pressure drop and the difference of axial temperature in the packed bed solid fermentation process were the main reasons that limited the scale-up process (Smits et al. 1999). Experimental and model results showed that these two issues should be solved to promote solid-state fermentation development. Since the beginning of this century, studies of the solid-state fermentation macroscopic model have mainly focused on two reactors: a packed bed reactor with stirring or forced ventilation operation and a rotating drum bioreactor. The rotating drum bioreactor model was mainly concerned with the shear force damaging and mixing effects on the substrate. The temperature and oxygen were distributed uniformly in the substrate because of mixing by drum rotation and the stirring blades. Consequently, the macro qualitative heat transfer model of the rotating drum bioreactor was represented by ordinary differential equations, which included a variety of delivery items.

Recently, studies of the forced ventilation transfer process and the stirring packed bed bioreactor were generally from the model established by Smith (Hasan et al. 1998; von Meien and Mitchell 2002), which emphasized the porosity and the mass and heat transfer in the gas, liquid, and solid phases. The transfer theory of solid-state fermentation has been further improved. However, the follow-up models ignored the decline of microbes, as well as the dynamic changes of the porosity and the permeability of substrate in fermentation.

Despite the development of the macro and micro models of solid fermentation, rare reports related the two together. The micro-level research results were difficult to apply to improve the macro-level model. At present, the reported macro model still ignores the change of the nutritional carrier substrate physical properties caused by microbial degradation, which cannot truly serve the modification of the reactor and process.

In summary, macro modeling of solid-state fermentation neglects variation of heat and gas transfer caused by the structural changes of substrate and ignores heat and mass transfer coupling. Substrate particles are usually assumed regular, ignoring a high degree of substrate heterogeneity and porosity.

1.3 Development of Solid-State Fermentation Engineering

Modern fermentation engineering technology research can be divided into four parts: upstream, midstream, and downstream engineering and auxiliary technology (Fig. 1.3). Broadly, the difference between modern solid-state fermentation engineering and traditional solid-state fermentation is whether the various components of fermentation engineering have achieved intensification, mechanization, and automation. Narrowly, researchers deem the large-scale intensive purebred fermentation as the important difference.

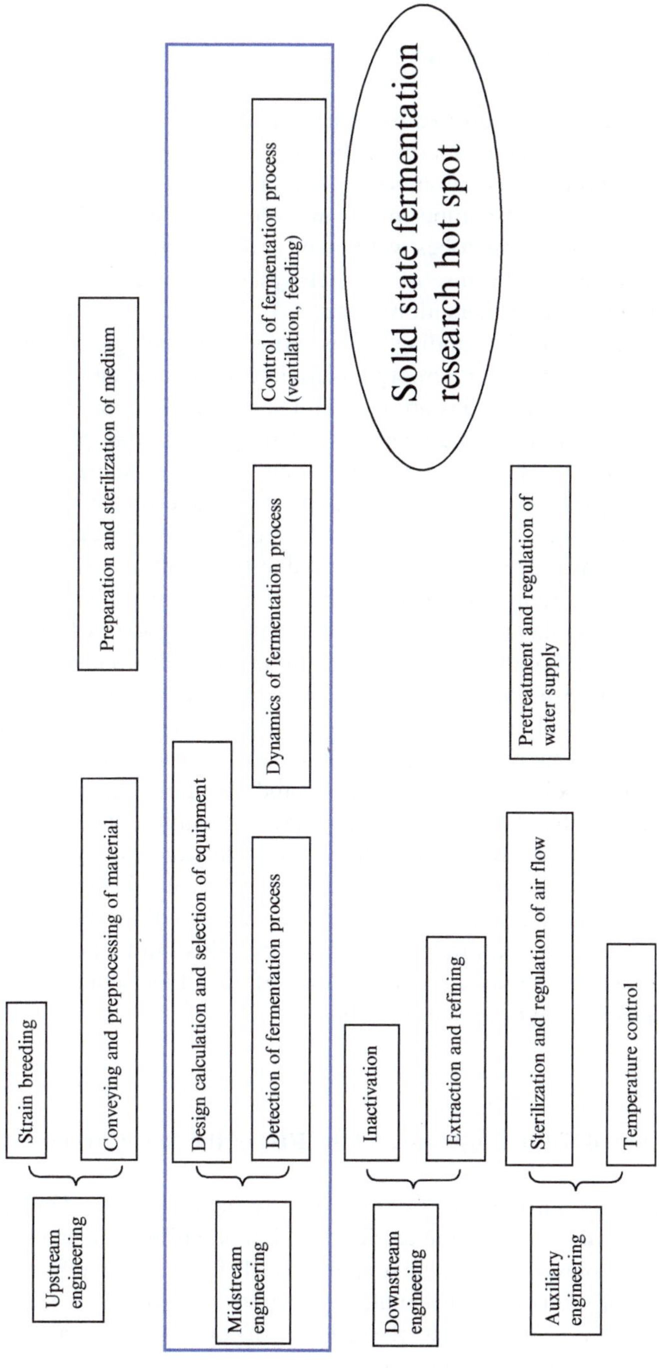

Fig. 1.3 Fermentation engineering research topics

1.3.1 Upstream Engineering of Solid-State Fermentation

1.3.1.1 Pure Culture Technology

The main focus of research on upstream engineering of solid-state fermentation was the breeding and cultivation of microbes. In the early development of solid-state fermentation (e.g., koji fermentation, sauce fermentation, etc.), microbes used in the food industry were often natural mixed microbes or limited mixed microbes. These fermentation process microbes were mainly from the local environment, not from artificial inoculation. Because of the lack system knowledge, there was less human intervention in the fermentation process.

With the increased understanding of microbial physiological metabolism, scholars divided fermentation microorganisms into dominant and characteristic microbes. The dominant microbes are the solid-state fermentation process determining factors. Characteristic microbes are microorganisms that play a decisive role in fermentation product quality. Dominant microbes sometimes can be replaced by other microbes or microbial enzymes; characteristic microbes cannot be replaced. Based on this, researchers began to regulate the fermentation process actively to strengthen the effect of the natural mixed culture and to meet the products' need. Early studies began from biogas and liquor production. Process regulation also plays a role in the modern field of oil recovery and hydrometallurgy.

Since the second half of the last century, with the rapid development of ecology and metabolic engineering, researchers used molecular technology to understand the complex physiological metabolism of mixed microbes. This makes it possible to use the limited mixed microbes in fermentation. Artificial regulation has a more significant impact on solid-state fermentation production efficiency. Mixed solid fermentation is adjustable, but there is no in-depth recognition of mutual relations and the limited mixed microbes. Therefore, the selection and combination of synergistic microbes are random without effective theoretical guidance. Researchers cannot effectively regulate strain variation in the mixed culture system applied, hindering mixed fermentation production efficiency. This limits the industrialization process. Therefore, the most important issue of mixed fermentation is solving the synergistic mechanism of mixed microbes.

1.3.1.2 Selection of Solid Substrate and the Rise of the New Carrier Substrate

Substrate is an important element of solid-state fermentation. As described, the solid substrate not only serves as the carrier for microorganism attachment but also provides a carbon source, a nitrogen source, growth factors, and other nutrients. As the main body of solid-state fermentation, substrates determine the fermentation process transfer effect and affect the choice of production strains, process conditions, and fermentation equipment.

Solid substrate should be chosen based on the principle of maximization of target products. In the traditional fermentation industry, the substrates mainly include cereals and pulses. Cereal raw materials include wheat, rice, millet, corn, and sorghum; nitrogen sources are mainly raw soybeans, peas, soybean meal, wheat bran, and corn steep liquor. Solid substrates used in the energy and environment fields are usually from agricultural waste, the main components of which are hemicellulose or lignocellulose. Feed, edible fungus, and the production of fuel ethanol all have a relationship to bran, wheat bran, and straws.

For solid-state fermentation, microorganisms can only use small-molecule sugar (such as glucose, xylose, etc.) and cannot directly transport macromolecules into cells. Therefore, the fermentation material must go through the process of crushing, cooking, and steam explosion so the macromolecular substrates can be hydrolyzed by the microorganisms or enzymes into the soluble unit.

The nutritional carrier substrate pretreatment process is more complicated; caking occurs easily during the fermentation process. In the 1960s, researchers began to utilize inert carrier substrate (vermiculite, perlite, plastic foam, etc.) to overcome these difficulties. The structure of these substrates can be maintained unchanged because of their chemical inertness, which could provide a more stable environment for the growth of microorganisms. Because microbes could obtain nutrition from the liquid culture distributed within the substrate pores, the fermentation process is easily monitored. It can be concluded that inert carrier substrate solid-state fermentation has good prospects because of the water- and energy-saving advantages as well as ease of control.

1.3.2 Solid-State Fermentation Midstream Engineering

1.3.2.1 Midstream Engineering Development

A good fermentation process requires not only strains with good traits and appropriate culture medium but also optimal environmental conditions. Solid-state fermentation bioreactors can provide suitable environmental conditions and space for the growth of microorganisms.

As shown in Table 1.3, Pandey's monograph (Pandey and Larroche 2008) stated the details of the fermentation process development, with a focus on the evolution of solid-state fermentation technology.

1.3.2.2 Solid-State Fermentation Process Control Parameters

The solid-state fermentation control process parameters are closely related to the metabolic regulation of microorganisms (Chen and Li 1998). Based on the metabolic needs of the fermentation microorganisms, the control of water activity, oxygen content, temperature, and pH are the main solid-state fermentation

Table 1.3 Evolution of solid-state fermentation technology

Type		Technology development	Production scale
Static		Cellars, water tanks	8–12 m^3
		Tray	0.15–0.25 m
Dynamic	Ventilation	Tray	100 × 10 × 6 cm
		Packed bed bioreactor	100–130 L
	Stirring	Rotating drum bioreactor	13 m^3
		Rotating drum bioreactor	200 L
	Fluidization	Gas-solid fluidized bed	8,000 L
	Double dynamic solid-state fermentation technology	Gas double dynamic solid-state fermentation technology	50,000 L

Source: From Pandey and Larroche (2008)

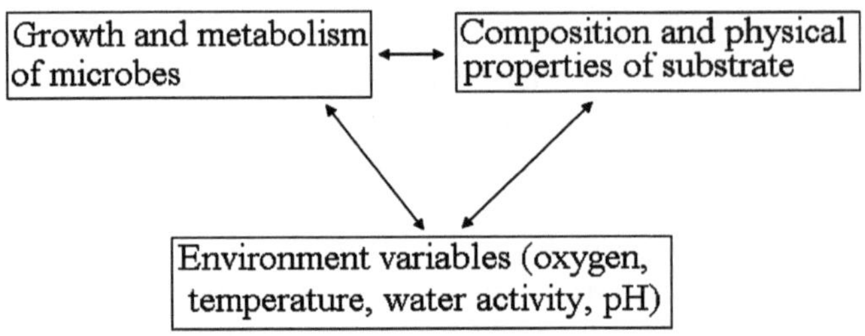

Fig. 1.4 Relationship of the three variables: microbial metabolism, physical properties of substrate, and fermentation parameters

parameters. In the solid-state fermentation process, the water, gas, and heat caused by the growth microbes are the dominant factors that determine the environmental changes. The growth and metabolism of the microorganisms are affected by the mass and heat transfer properties of the substrate itself; at the same time, the growth of microorganisms and the utilization of nutritional carrier substrate could change the structure and physical properties of the substrate (Fig. 1.4).

Substrate Water Content and Water Activity

Whether microorganisms can grow on a substrate depends on the water activity of the substrate (α_W). Generally, bacteria need an α_W of 0.90–0.99; for yeast, the value is 0.80–0.90, and fungi need 0.60–0.70. For aerobic fungi, because the rate of oxygen diffusion in the water is only 1/200,000 of that in air, the water film tension becomes the main limiting factor affecting mycelial extension. The increase in water content will hinder mycelial stretching in the pores. In addition to meeting the microbial physiological requirements, the water content plays a decisive role in the variation of the three-phase structure related to water retention, permeability, and thermal conductivity. Actually, researchers always use sterile and humidified air to improve the α_W of the condition to ensure normal microbial growth.

Fermentation Temperature

The fermentation temperature affects microbial growth, metabolism, and spore germination. Heat buildup is the typical effect of temperature on solid-state fermentation.

Because of poor heat conductivity and accumulation of metabolic heat in the material combined with substrate shrinkage and decreased porosity, gas convection is seriously hindered. Previous studies showed that the major resistance to heat transfer in solid-state fermentation was low conduction efficiency; stromal thermal conductivity was determined by its water content. Therefore, moisturizing is a common measure of temperature control. In addition, routine operations (e.g., forced ventilation, jacket cooling) all can solve these problems. Evaporative cooling is one of the main solid-state fermentation temperature control measures. It can take away 60–80 % of calories from the substrate. The evaporative cooling rate can be adjusted by regulating the forced ventilation airflow rate and the water content of the medium. In general, increasing the forced ventilation airflow rate could reduce the temperature gradient of the medium. From present research, in a large-scale solid-state fermentation system it is difficult to maintain the temperature at an ideal range. So, coupling of ventilation, temperature, and humidity is usually a control measure used in large-scale solid-state fermentation.

Substrate Oxygen Concentration

Oxygen is a key factor affecting solid-state fermentation. Oxygen consumption rate (OUR) and carbon dioxide production (CDPR) can be used to assess the state of the solid-state fermentation process, but different microorganisms cause these assessments to vary.

The k_La (transfer coefficient of oxygen) can also be applied in liquid fermentation to measure ventilation efficiency.

Ghildyal et al. (1994) studied the impact of the gas concentration gradient on product yield in a tray solid-state fermentation bioreactor. The results showed that the variations of O_2 and CO_2 concentration gradients were obvious, which seriously affected product yield. With gradient increases, the yield decreased. Gowthaman et al. (1993) studied the impact of gas concentration gradient on the product in a packing bed bioreactor. The results showed that the gas concentration gradient can be eliminated and the ability of mass transfer can be enhanced by forced ventilation, which resulted in increases in enzyme activity.

pH Value

It is difficult to measure and control the pH values in the solid-state fermentation system. At present, there is little research on the changes and regulation of pH in the solid-state fermentation process. Commonly, if the initial pH value of the medium is adjusted, the variations of pH value during the solid-state fermentation process need not be considered. However, in many fermentation processes, the pH values change characteristically. Because of the low water content of the substrate, the pH values are difficult to determine by conventional detection. Materials with a buffer capacity are often used as a substrate to eliminate the adverse effects of the changing pH values. Sometimes, nitrogen-containing inorganic salts (such as urea) are used as nitrogen sources to offset the fermentation process pH variation.

New Process Control Developments

Traditional methods of bioreactor control are empirical. The pH value, dissolved oxygen, and concentration of residual substrate are controlled as isolated points. The regulation fails to reflect the biological dynamic characteristics of the fermentation process, which has many limitations. To optimize and regulate the fermentation process, researchers should understand the microbial dynamic growth state (growth rate, morphology, concentration, etc.); the rate of oxygen demand; as well as the influence of a variety of fermentation conditions on these dynamic variables. Consequently, researchers should establish fermentation models that involve acquisition, processing, integrated computing, and parameter estimation of data and real-time processes.

For a long time, there were few measurement parameters for industrial solid-state fermentation that determined that industrial solid-state fermentation process control can only be represented by extensive and empirical formulas. Process condition optimizations were often based on the purpose of the product after fermentation, neglecting process consideration. Recently, our group developed a new solid-state fermentation bioreactor, the gas double dynamic solid-state fermentation bioreactor, which has been successfully applied at the industrial level. Many parameters (e.g., temperature, humidity, circulating air, light, oxygen) could be controlled online by a line computer monitoring system.

1.3.3 Solid-State Fermentation Downstream Engineering and Auxiliary Technology

Solid-state fermentation downstream technology includes the refining, sterilization, and purification of the fermentation product. In fact, the problems faced by solid-state fermentation refining and the refined product are similar to those for liquid fermentation, which puts forward higher requirements for modern separation and purification technology.

For mixed fermentation (such as compost, silage, etc.), the products do not need to be sterilized and can be used directly. For food, pharmaceutical, chemical, and fermentation products, the products must be sterilized. For different fermentation substrates and fermentation products, the sterilization techniques also vary.

Table 1.4 Solid-state fermentation contamination

Process or places	Possible reasons
Various stages of pure culture	Contamination of bevel or culture; incomplete sterilization of medium; aseptic work is nonstandard; lapse of clean bench filters; poor sterile room hygiene
Media contamination	Medium is itself seriously contaminated; failure of sterilization equipment; sterilization time and temperature failed to meet the standard; sterilization is not thorough; medium caking
Air system contamination	Incompetent air sterilization; supply duct contamination; pipeline leakage
Fermentation contamination	Incompetent fermentation equipment cleaning; equipment fails to maintain a positive pressure

1.3.3.1 Solid-State Fermentation Contamination

There are many reasons for contamination during solid-state fermentation. Every process should be examined carefully to identify the contamination source and find the appropriate measures to ensure safe fermentation. The reasons for contamination of the solid-state fermentation process are summarized in Table 1.4.

In the process of expanding culture, researchers should examine whether the medium is contaminated to avoid greater losses. Common sterile test methods are broth culture, slant culture, and disk incubation. Two common methods are phenol red broth culture and two-disk culture.

1.3.3.2 Contamination Control

For solid-state fermentation, contamination is often inevitable. As long as the operation is proper, the contamination can be controlled so it is harmless. Common practices include (1) control of substrate water activity; (2) increase in the inoculation amount; and (3) adjustment of pH values (addition of acetic or lactic acid to the medium may also enhance the antibacterial capability of the culture). (4) Cooling can be used. In summer, the temperature is high; media are often easily contaminated. So, the culture temperature is controlled at a lower temperature range, which can greatly reduce bacterial contamination. (5) Salt addition (commonly about 15–18 % concentration) will have an increased inhibitory effect when the salt concentration is higher, but high concentrations will inhibit enzyme activity.

1.3.3.3 Sampling and Testing of Fermentation Products

Sample detection for solid-state fermentation includes the detection of raw materials and semifinished and finished products. Traditional physical and chemical tests of solid-state fermentation materials include those for water content, water activity, acidity, ash, temperature, nutrient content, and other conventional items.

For direct consumption products, the detection process should also include the number of microbes, the concentration of toxins produced by microbial metabolism, as well as the heavy metal content. Because of the heterogeneity of the solid-state fermentation process, even with the same strains and production process, the quality of products in the different batches and the production processes will be very different; even in the same fermentation vessel, the product is not the same. Consequently, the sampling process must be uniform based on the material characteristics of solid-state fermentation.

1.3.3.4 Posttreatment of Fermentation Products

At the end of solid-state fermentation, according to requirements, products usually need a drying treatment. Drying is one of the oldest methods for anticorrosion and saving the products that still plays an important role in some modern industries (Hu and Xu 2009). According to the different products, the drying methods can be summarized as follows: (1) natural drying (air dried) (2) hot air drying, (3) spray drying, (4) vacuum drying, and (5) freeze drying. There are three main factors that affect the drying rate: material water content, drying medium, and drying equipment. Additional important influencing factors are as follows: (1) the nature and shape of the material (physical structure, chemical composition, and the mutual binding mode of materials); (2) the temperature of the material itself; (3) the initial water content of the material; (4) the temperature, humidity, and velocity flow of drying air; and (5) dryer structure.

References

Ashley VM, Mitchell DA, Howes T. Evaluating strategies for overcoming overheating problems during solid-state fermentation in packed bed bioreactors. Biochem Eng J. 1999;3:141–50.

Bovill R, Bew J, Cook N, D'agostino M, Wilkinson N, Baranyi N. Predictions of growth for *Listeria monocytogenes* and *Salmonella* during fluctuating temperature. Int J Food Microbiol. 2000;59:157–65.

Chen HZ, He Q. Value-added bioconversion of biomass by solid-state fermentation. J Chem Technol Biotechnol. 2012;87(12):1619–25. doi:10.1002/jctb.3901.

Chen HZ, Li ZH. Bioreactor engineering. Chin J Process Biotechnol. 1998;18:46–9.

Chen HZ, Qiu WH. Key technologies for bioethanol production from lignocellulose. Biotechnol Adv. 2010;28:556–62.

Chen HZ, Xu J. Modern solid state fermentation: theory and practice. Beijing: Chemical Industry Press; 2004.

Chen HZ, Xu FJ, Tian ZH, Li ZH. A novel industrial-level reactor with two dynamic changes of air for solid-state fermentation. J Biosci Bioeng. 2002;93:211–4.

Couto SR, Gundin M, Lorenzo M, Sanroman M. Screening of supports and inducers for laccase production by Trametes versicolor in semi-solid-state conditions. Process Biochem. 2002;38:249–55.

Dalsenter FDH, Viccini G, Barga MC, Mitchell DA, Krieger N. A mathematical model describing the effect of temperature variations on the kinetics of microbial growth in solid-state culture. Process Biochem. 2005;40:801–7.

Duan YY, Chen HZ. Effect of three-phase structure of solid-state fermentation substrates on its transfer properties. Chin J Chem Ind Eng. 2012;63:1204–10.

Duan YY, Wang L, Chen HZ. Digital image analysis and fractal-based kinetic modelling for fungal biomass determination in solid-state fermentation. Biochem Eng J. 2012;67:60–7.

Fernández-Fernández M, Pérez-Correa JR. Realistic model of a solid substrate fermentation packed-bed pilot bioreactor. Process Biochem. 2007;42:224–34.

Ghildyal N, Gowthaman M, Raghava Rao K, Karanth N. Interaction of transport resistances with biochemical reaction in packed-bed solid-state fermentors: effect of temperature gradients. Enzyme Microb Technol. 1994;16:253–7.

Gowthaman M, Ghildyal N, Rao K, Karanth N. Interaction of transport resistances with biochemical reaction in packed bed solid state fermenters: the effect of gaseous concentration gradients. J Chem Technol Biotechnol. 1993;56:233–9.

Hasan SDM, Costa JAV, Sanzo AVL. Heat transfer simulation of solid state fermentation in a packed-bed bioreactor. Biotechnol Technol. 1998;12:787–91.

Hu WF, Xu GR. Solid-state fermentation principle, devices and applications. Beijing: Chemical Industry Press; 2009.

Lenz J, Höfer M, Krasenbrink JB, Hölker U. A survey of computational and physical methods applied to solid-state fermentation. Appl Microbiol Biotechnol. 2004;65:9–17.

Liu W, Fan AW, Huang XM. Heat transfer qualitative theory and application of porous media. Beijing: Science Press; 2006.

Mitchell DA, Pandey A, Sangsurasak P, Krieger N. Scale-up strategies for packed-bed bioreactors for solid-state fermentation. Process Biochem. 1999;35:167–78.

Mitchell DA, von Meien OF, Krieger N. Recent developments in modeling of solid-state fermentation: heat and mass transfer in bioreactors. Biochem Eng J. 2003;13:137–47.

Mitchell DA, von Meien OF, Krieger N, Dalsenter FDH. A review of recent developments in modeling of microbial growth kinetics and intraparticle phenomena in solid-state fermentation. Biochem Eng J. 2004;17:15–26.

Mitchell DA, Krieger N, Berovic M. Solid-state fermentation bioreactors: fundamentals of design and operation. Berlin: Springer; 2006.

Mitchell DA, Cunha LEN, Machado AVL. A model-based investigation of the potential advantages of multi-layer packed beds in solid-state fermentation. Biochem Eng J. 2010;48:195–203.

Pandey A, Larroche C. Current developments in solid-state fermentation. Berlin: Springer; 2008.

Rajagopalan S, Rockstraw DA, Munson-McGee SH. Modeling substrate particle degradation by *Bacillus coagulans* biofilm. Bioresour Technol. 1997;61:175–83.

Sargantanis J, Karim M, Murphy V, Ryoo D, Tengerdy R. Effect of operating conditions on solid substrate fermentation. Biochem Eng J. 1993;42:149–58.

Singhania RR, Sukumaran RK, Patel AK, Larroche C, Pandey A. Advancement and comparative profiles in the production technologies using solid-state and submerged fermentation for microbial cellulases. Enzyme Microb Technol. 2010;46:541–9.

Smits J, Van Sonsbeek H, Tramper J, Knol W, Geelhoed W, Peeters M, et al. Modelling fungal solid-state fermentation: the role of inactivation kinetics. Bioprocess Biosyst Eng. 1999;20:391–404.

Viccini G, Mitchell DA, Boit SD, Gern JC, da Rosa AS, et al. Analysis of growth kinetic profiles in solid-state fermentation. Food Technol Biotechnol. 2001;39:271–94.

von Meien OF, Mitchell DA. A two-phase model for water and heat transfer within an intermittently-mixed solid-state fermentation bioreactor with forced aeration. Biotechnol Bioeng. 2002;79:416–28.

Wang EQ, Li SZ, Tao L, Geng X, Li TC. Modeling of rotating drum bioreactor for anaerobic solid-state fermentation. Appl Energy. 2010;87:2839–45.

Wu ZQ. The technology and application of solid state fermentation. Beijing: Chemical Industry Press; 2006.

Zambra C, Moraga N, Escudey M. Heat and mass transfer in unsaturated porous media: moisture effects in compost piles self-heating. Int J Heat Mass Transf. 2011;54:2801–10.

Chapter 2
Biotechnology Principles of Solid State Fermentation

Abstract As discussed in Chap. 1, solid-state fermentation is an important bioprocess. In this process, microorganisms are the most important participant. Because of the unique characteristics of the solid matrix and the rich interface environment it forms, microorganisms in solid-state fermentation show some different features compared with liquid culture. This chapter mainly discusses the relationship of solid-state fermentation and the solid matrix, the physiological metabolism and growth characteristics of microorganisms in the solid matrix, and the interactions between the microorganisms and the solid matrix. The aseptic techniques and inoculation techniques for large-scale solid-state fermentation are discussed in the last part of this chapter. Note that the dynamics model of fractal dimension established in my laboratory can quantitatively characterize the variation law of morphology with the microbial growth in solid-state fermentation and can be used as indicators for biomass yield in the solid-state fermentation process, which provides a new way to automate process control for solid-state fermentation.

Keywords Microbial metabolism • Biofilm • Growth dynamics • Numerical simulation • Pretreatment • Aseptic technique • Reaction kinetics

2.1 Overview of the Microbial Physiology of Solid-State Fermentation

2.1.1 Microbial Growth and Metabolic Characteristics

Because microbes are the major participants in the solid-state fermentation process, this chapter mainly discusses the physiological metabolism and growth characteristics of microorganisms in the solid matrix and the interactions between the microorganisms and the solid matrix. Based on the unique microbial metabolic processes and living environment, the mechanism of solid-state fermentation is explored to

H. Chen, *Modern Solid State Fermentation: Theory and Practice*,
DOI 10.1007/978-94-007-6043-1_2, © Springer Science+Business Media Dordrecht 2013

obtain better understanding of the complicated process and to use microorganisms to serve humankind.

Solid-state fermentation can be roughly divided into two categories according to for microorganism requirements: a pure or a nonpure culture process. A nonpure culture process (composting, silage, koji making, etc.) manages the microbial populations by control of the environmental conditions, such as oxygen, temperature, and humidity. However, multimicrobial community structures are invariably formed in the solid matrix, so different species assume a different role. Therefore, it is difficult to require a definite characteristics of substrate for specific microorganisms. In pure cultures, all of the biological reactions are finished by a single kind of microbe. Therefore, these organisms often require a variety of desirable traits to adapt to the demands for industrial production, especially for the production of enzymes, organic acids, antibiotics, and other metabolites. Filamentous fungi usually grow on solid substrate without free water (such as plant tissue, moist soil, etc.); some of them can also obtain higher protein expression, mainly because these fungi are not only able to degrade insoluble nutrients to ensure their own growth but also enter into the internal matrix through growth. Similar to the filamentous fungi, many actinomycetes have the same ability. In general, microorganisms for solid-state fermentation should have certain characteristics, such as the ability to use polysaccharides, especially some insoluble nonstarch polysaccharides, as a carbon source; a complete enzyme system; depth to the material layer or even enter the matrix; rapid growth rate; strong resistance to infectious microbes; and growth under low-water conditions.

2.1.1.1 Microbial Metabolism in the Solid-State Matrix

Whether in a solid or liquid culture, microorganisms show common characteristics of metabolism. However, the solid matrix interface environment provides microorganisms some different characteristics compared with liquid culture. The most typical characteristics are coexistence of multimetabolic processes of homologous microbes and the synergistic effect of heterogeneous microbes induced by the solid interface environment. For example, the biofilms formed by the microbes growing on solid surfaces are not homogeneous microenvironments for the microbes. For microbes growing in the bottom that belong to the anaerobic layer, their anaerobic enzymes are induced, and the activities of enzymes related to the tricarboxylic acid cycle drop to a very low level. Microbes in the layer that contains little oxygen have inhibited anaerobic enzymes; the final electron transport chain includes cytochrome, and cytochrome oxidase activity increases sharply to capture oxygen molecules effectively. However, the tricarboxylic acid cycle enzymes continue to have less activity. As for the layer with a saturated oxygen concentration, the enzyme activities related to the tricarboxylic acid cycle increase because the metabolism-limiting factor is not an electron acceptor. The microbial cells in the outermost layer show a poisoned state, and both their growth rates and their enzyme activities are reduced because of the high oxygen concentration. Apart from the oxygen gradient, there are other gradients in the actual biofilms, such as of carbon and nitrogen source distribution. Excess glucose in the biofilm surface causes repression of the metabolites.

The synergism of different kinds of microbes is another characteristic of solid matrix metabolism. The solid matrix interface environment makes the symbiosis of different microbes possible, which is fundamentally different from most liquid cultures. Moreover, there are interactions among the different microbial species (Xin et al. 2010). For example, in the microbial community on the surface of a tooth, *Fusobacterium nucleatum* and *Prevotella bryantii* grow under pH 5.0–7.0, and in their growth process, they are able to keep the biofilm neutral by ammonia and organic acids, which are produced by glutamate and aspartate fermentation (Takahashi 2003). The neutral microenvironment allows acid-sensitive microbes such as *Porphyromonas gingivalis* to survive successfully. Another example is acid resistance bacteria *Streptococcus* mutants; they are able to grow in an acidic environment, and the lactic acid produced by sugar metabolism inhibits the growth of *Streptococcus* (*Streptococcus sanguinis*) to increase the composition of *Streptococcus* mutants in the biofilms. However, the *Streptococcus* can also produce hydrogen peroxide, which shows a strong antagonism to *Streptococcus* mutants and inhibits a variety of other anaerobic bacteria (Kreth et al. 2008).

2.1.1.2 Uniqueness of Extreme Microbial Metabolism

There are many extreme environments in nature, such as those with high temperature, low temperature, high salt, high alkali, high acid, high acid and heat, drought, high pressure, or high radiation. Most organisms usually cannot exist in such an environment; however, some microorganisms can adapt to these environmental conditions because of long-term natural selection, by which a unique metabolism and genes evolve to answer the strong environmental limiting factor. So, the in-depth study of the microbial metabolic processes in the natural environment, especially in extreme environments, has significance for the development and utilization of microbial resources and solid-state fermentation. The study of microbes in extreme environments has begun in Japan, the United States, and Europe and shows significant progress in revealing the mystery of extreme life forms and utilization of special mechanisms and special products.

Thermophilic Microorganisms

From the temperature of microbial growth, Williams defined thermophilic microorganisms as those with a highest growth temperature above 60 °C, and the optimal growth temperature should be higher than 50 °C. The highest growth temperature of extreme thermophilic microorganisms should exceed 90 °C, and their optimal growth temperature is higher than 65 °C. There are thermophilic microbes in nature, such as near a volcano, hot springs, compost, coal pile, solar radiation, desert, and so on. An artificial high-temperature environment (such as found in power plants, high-temperature industrial process wastewater) can also be an ideal habitat for thermophilic microorganisms. Thermophilic microorganisms may be present in fungi, actinomycetes, and archaea, which have been popular topics for recent research.

Thermophilic microorganisms can grow and reproduce in a high-temperature environment, depending on their unique mechanism of adaptation. First, the membrane has a high degree of saturation of a fatty acid to form a hydrophobic bond with great strength that contains a large composite resin special structure in the cell membrane lipid bilayer. For example, the cell membrane of archaebacteria from glycerol and isoamyl alcohols (C20 and C40 explant alkanols) by an ether bond formed by combining pairs of phytane diglycerol ethers are by explants alkanol glycerol diethers tail-to-tail connected to form both sides of a hydrophilic group monolayer lipid maintaining the integrity of a hydrophobic inner layer, and variability separated by the bilayer film at high temperatures is avoided. Second, the thermophilic microorganisms could produce several particular protein, such as heat-stable protein, that is rich in leucine, proline, and glutamate. The thermal stability of this proten is significantly higher than ordinary bacterial homologous proteins. They also produce unique thermal stability promotion factors, and Ca^{2+}, Mg^{2+}, and Zn^{2+} can stabilize the protein structure. *Pyrodictium occultum* produces molecular chaperone proteins that stabilize the other cell proteins by protecting and refolding. Finally, the genetic material of the thermophilic microorganism also has a unique form, such as higher levels of DNA G-C, which have more hydrogen bonds to form a more stable double-helix DNA structure.

The solid-state fermentation of thermophilic microorganisms is important (Wakelin et al. 2012). In the food and fermentation industry, high-temperature α-amylase is produced by *Bacillus stearothermophilus* to complete the conversion of starch to dextrose, and β-glucanase can improve the speed of wort separation to reduce beer turbidity and gelatinous precipitate.

Acidophilic Microorganisms

Acidophilic microorganisms include prokaryotic and eukaryotic categories; they mainly are those microorganisms that can only grow when the environmental pH is less than 3.0, and their optimum growth pH is between 1.0 and 2.5. The acidophilic microorganisms usually are distributed in metallothionein deposit acidic mine water, bioleaching, coal deposit acidic mine water, as well as sulfur hot springs and soil. Most current studies are focused on the prokaryotic acidophilic microorganisms. Based on their growth temperature range, prokaryotic acidophilic microorganisms can be divided into three groups: normal temperature, medium temperature, and high temperature. Room temperature acidophilic prokaryotic microorganisms mostly are autotrophic, such as *Thiobacillus ferooxidans*, followed by polyculture and heterotrophic acidophilic bacteria, which can either use sulfur as energy for autotrophic growth or take advantage of organics to grow heterotrophically. Most acidophilic microorganisms are separated from the natural environment, such as *Acidothiobacillus ferrooxidans,* which is enriched and separated from acid mine drainage.

The hypotheses regarding the mechanism by which acidophilus bacteria maintain a nearly neutral pH intracellular environment and adapt to an external acidic environment can be grouped into three categories: pump theory, shield theory, and Donnan equilibrium. Studies of the transmembrane H^+ gradient and the potential difference of thermophilic acid from Achaea have shown that the permeability of a plasma membrane proton is indirectly determined by tetramer located in the membrane lipid. Such a transmembrane tetramer can form a solid layer to keep the proton impervious at the growth pH range.

Acidophilic microorganisms receive widespread attention because of important application prospects in the bioleaching of low-grade ore and coal desulfurization. Especially, inorganic autotrophic bacteria are advantageous for potential application in bioleaching of low-grade ore, recovery of precious metals, raw coal desulfurization, and environmental protection.

2.1.1.3 Biofilm System

The cell membrane, also known as the plasma membrane or endometrial cell membrane, is semipermeable and surrounds the soft and fragile cytoplasm. It is composed of protein and phospholipid and is about 800 nm thick. Two layers of phospholipid molecules are symmetrically arranged to form the main cell membrane component; the polar head of phospholipid molecules faces the outer part of the cell, and the nonpolar end is arranged toward the inside. Biofilm is the basic form of the cell and has a variety of vital functions, such as material transport, energy conversion, cell recognition, information delivery, and metabolic regulation. Study of the biofilm system is significant for exploring the metabolism of microbes in solid-state fermentation and for realizing the regulation of microbial life activities.

For microorganisms, the C3 position polar head of the glycerol molecule has different *R* groups, such as phosphatidic acid, phosphatidyl ethanolamine, phosphatidyl glycerol, phosphatidyl choline, phosphatidyl serine, and phosphatidyl inositol, that play a special physiological role. Recent studies have found that the archaeal cell membrane has unique characteristics, such as isoprene repeat units of the hydrophobic tail, the ether bond of the hydrophilic head and hydrophobic tail, alternating odd and even molecular layers, and unique lipids in some archaea.

The following is a brief introduction to rhodopsin and the light-mediated synthesis system of halophilic bacteria. Under anaerobic conditions, the rhodopsin produced by halophilic bacteria is embedded in the cell membrane to constitute the purple membrane, accounting for about 50 % of the entire membrane. Rhodopsin has a strong absorption peak at 570 nm, and its chromophore is usually present as full-*trans* structures in the internal membrane and temporarily converted into the *cis* state by light incitation. Through this process, the H^+ protons are transferred to the outer membrane. Bacterial rhodopsin is converted into the more stable *trans* isomer after absorption of a proton from the internal cell. Such a cycle forms a proton gradient and electrochemical potential to provide the energy for the synthesis of adenosine triphosphate (ATP), which is necessary for life activities (Shen et al. 2009).

2.1.2 Characteristics of the Solid-State Fermentation Interface and Their Impacts on Microbial Metabolism

Solid-state fermentation is an important interface biological reaction process; it is fundamentally different from submerged fermentation. The interface usually shows different physicochemical properties from the main phase. After the microbes settle on the interface and become biofilms, the microorganism biofilm structure and physiological activity form the microenvironment, which results in different environmental conditions inside and outside the membrane. For example, in the direction perpendicular to the interface, an oxygen concentration gradient is observed, as is gradients of pH value and other substances (Donlan 2002), which indicates significant impacts on microbial physiological activities, such as gene expression. At the same time, some community activities can be accomplished because of the microbial community formed in the biofilms, which seems to be impossible for the free cells in liquid fermentation.

2.1.2.1 Solid-Liquid Interface and Its Features

The interface is the transition zone between two phases in close contact. The thin transition zone is generated by the uneven force in interface layer, which displays different physical and chemical properties from the main phase. For example, the surface tension, interfacial adsorption, and polar molecules on the interface show completely different distribution rules compared with the main phase (Zhu and Zhao 1996). When the interface area is small, the role of the interfacial phenomena is often negligible; however, these effects must be taken into account for a solid-state fermentation system that has a porous matrix with a stable interface.

The interfacial tension, which is caused by cohesion among liquid molecules, is mutual attraction in the surface layer when the two phases are in contact and the specific force keeps the tendency to contract of the main phase (Zhu and Zhao 1996). As the molecules in the surface layer are sparser than those in the internal phase, they suffer from internal cohesion in the direction of the internal phase, which results in the liquid surface layer contraction trend. At a certain volume, globularity shows the smallest surface area; the droplets always tend to form spherically under the surface tension effect.

Interfacial adsorption is also an important interfacial phenomenon. There are adsorptive effects on the solid-liquid and gas-solid surface. The fundamental reason for solid-liquid interface adsorption is its spontaneous reduction tendency. When pure liquid contacts a solid surface, the liquid molecules are often aggregated at the solid-liquid interface to lower the interfacial energy. Because of the charge on the interface, the interface shows much electrical behavior, such as double layer, electrophoresis, electro-osmosis, and electrosettlement. These electrical phenomena originate from three aspects: ionization, adsorption, and ion dissolution of amino and carboxyl groups. Amino and carboxyl groups are common macromolecular substances, such as polysaccharides and protein, which show positive and negative charges, respectively. Their proportions can determine the electrical properties of the interface; some charged particles are always adsorbed to the interface to make the interface charged, and the dissolution of the ions may also charge the interface. The easier hydration of cations compared to anions causes the interface to show a negative charge in the majority after contact with the aqueous medium.

The interface is three dimensional. As mentioned, the interface should actually be considered as an interface layer, which is a transition region of mutual penetration formed by the penetration and diffusion of the molecules in the two-phase boundary because of the potential energy difference and the chemical potential imbalance of the two phases and intermolecular gravity. It is not a geometrical plane, but a three-dimensional structure, which means that the interface has a certain thickness. The thin layer exists as a gradient variation of the composition and structure. The interface is basically formed by the molecules of the two phases. However, the molecular distribution of the two phases in the interface region is not uniform, which has a relationship to the distance of a single phase state and shows an obvious gradient variation of properties.

The microorganism is able to survive in a natural environment with poor nutritional status mainly because of the role of the interface. The physical and chemical properties of the interface are different from any one phase of the two phases. This has important implications for the distribution of ions, macromolecules, and colloids in the interface. The interface has a surface energy and thus can adsorb certain biological macromolecules to reduce the surface energy; macromolecules with adsorption roles generated in the metabolic process of microorganism growth can help the microorganisms grow at the interface.

2.1.2.2 Solid Matrix Characteristics

Porosity and Surface Area

As aerobic solid-state fermentation is a reaction system with the gas phase as the continuous phase, the porosity and specific surface area of the solid particles are important factors that affect oxygen transfer in solid-state fermentation and the degree of difficulty for substrate utilization. Some scholars compared the production of amylase using wheat and grinding grains as the matrix. They found that specific surface area increased, and the seed coat that prevented microbial invasion had been destroyed after grinding the grains. Because of the increasing porosity and surface area, aerial hyphae increased. For these reasons, increased matrix porosity and specific surface area accelerated the fermentation process and increased enzyme production (Rahardjo et al. 2005).

Water Activity

Water activity is an important for organisms to complete physical activity. Although little or no free water exists in solid-state fermentation, the water attached to the solid particles still determines whether the biological reaction can proceed successfully. The important factor in determining whether microbes can grow on a solid matrix is water activity α_w. The water activity is defined as the ratio of the equilibrium vapor pressure of the water in the material to that of pure water at the same temperature (Chen and Xu 2004).

$$a_w = \frac{P_W}{P_W^0} = \gamma_W X_W \tag{2.1}$$

P_w = partial pressure of water;
P_w^0 = equilibrium vapor pressure of pure water;
X_W = water proportion;
γ_w = activity coefficient.

The water activity value represents the amount of unconjugated water in the microbial environment, which is related to the water content of the matrix. The water activity not only has a relationship with the insoluble substance that dissolved in water but also has a relationship with soluble substance. Water activity is important in the solid-state fermentation of filamentous fungi. It relates to the germination of the spores, mycelial extension, spore formation, generation of other metabolites, and many other aspects. Studies by Gervais and Molin (2003) showed that different fungi have different suitable water activity values for growth, and water is an important factor for inducing fungal spore generation.

Different microorganisms have different water activity requirements, and suitable water activity for general bacteria is 0.9–0.99; for yeast, it is 0.8–0.9, and for fungi and a few yeasts it is 0.6–0.7. Thus, the solid-state fermentation process needs to adjust the water activity value of the material according to the characteristics of different microorganisms (Chen and Xu 2004).

2.1.2.3 Effects of Interface on Microorganism Growth

The difference between solid-state fermentation and submerged liquid fermentation is the bioreaction on the biological interface. The reaction substrate is present in solid form; the transfer process in the reaction system is extremely complex, including gas-solid, gas-liquid, and liquid-solid. Gas is its main flow medium. In the growth process, the microorganisms adsorb on the interface through various mechanisms, such as the secretion of polysaccharides and glycoproteins to adsorb on the surface for many bacteria; fungi use rhizoids or haustoria to fix on the substrate surface. Hyphal tips lead to many enzymatic reactions on the solid matrix surface by the fixed adsorption structure, which facilitates the penetration of the matrix particles in the interior and completes the mass transfer of the biological reaction. Compared with the inside liquid phase, the interface can provide a good microenvironment for microbes. Whether nutrition-type material or an inert carrier, the surface and internal voids can form the interface for free growth of microorganisms, and their growth state often shows significantly different performance compared with submerged liquid fermentation.

Gradient Effect in the Solid Surface Biofilm

Microorganisms grow on the interface as biofilm in solid-state fermentation. The microorganism physiological and biochemical characteristics after the formation of biofilms are usually different from those of free culture, which are the response to environmental changes and the foundation of biological reactions for many industrial processes. The interfacial microbial physiological changes are mainly functions of film thickness, and the physiological phenomenon is mainly caused by a gradient formed by the penetration of oxygen and other substances (Costerton et al. 1995).

Ecological Effects

In nature, microorganisms often grow on the interface as biofilm. Particularly for a solid surface, it may be formed at rich interface conditions because of its complexity. This creates conditions for a variety of microorganisms to grow together. When various microorganisms coexist, they form a microscopic ecosystem on the interface, and the microorganisms grown on the interface often reflect significant ecological effects. The typical characteristics of these effects are that all kinds of microorganisms coexist and complete biochemical reactions collaboratively (Table 2.1), such as in rumen biofilm.

Rumen microorganisms include two categories, bacteria and ciliates. These microorganisms are present in rumen fluid (20–30 %), on the surface of the solid particles (70–80 %), and the stomach wall, where the two types form biofilms.

Table 2.1 Microorganisms in teeth

Types	Gram positive	Gram negative
Cocci	*Streptococcus*	*Neisseria*
	Peptostreptococcus	*Veillonella*
Rod	*Actinomyces*	*Campylobacter*
	Bifidobacterium	*Capnocytophaga*
	Corynebacterium	*Eikenella*
	Eubacterium	*Fusoacterium*
	Lactobacillus	*Haemophilus*
	Propionibacterium	*Leptotrichia*
	Rothia	*Pevotella*
		Porpyromonas
		Selenomonas
		Treponema

Source: From Gervais and Molin (2003)

The microbes attached to the solid surface degrade the insoluble material (cellulose, starch) to obtain sugars and organic acid or methane after subsequent degradation. Microorganisms absorbed on the stomach wall play a role in preventing adsorption of pathogenic microorganisms (Costerton et al. 1987). This effect is particularly evident in solid-state fermentation. Solid-state fermentation using a single-strain biofilm to build a system is relatively infrequent; most systems in which biofilms play a crucial role have a variety of microbial symbiotic systems.

The composting process is a typical interface biological response. Whether composting is aerobic or anaerobic, a variety of microbial interactions exists (Table 2.2). The microorganisms in the aerobic microbial composting process include bacteria, fungi, and actinomycetes; even yeasts and protozoa and their species are in large numbers. The process is the corporate result of multiple microbial communities with a population that experiences nonstop succession. Single microorganisms, no matter how high their water activities, cannot compare with the community role of various kinds of microorganisms.

In the early stage of composting, material temperature is moderate, and many microorganisms are able to adapt it. When entering the high-temperature phase, mesophilic microorganisms are replaced by thermophilic microorganisms. At 50 °C, activity is mainly from thermophilic fungi and actinomycetes. When the temperature rises to 60 °C, the fungi stop their activity, and only thermophilic actinomycetes and bacteria are still active. When the temperature reaches 70 °C, most microorganisms are dead or dormant except spores (Yu et al. 2012).

The microorganisms' relationship in anaerobic composting is more complex. The anaerobic digestion process is actually a series of biochemical coupling reactions of a variety of microorganisms (the hydrolysis and fermentation flora, flora that produce hydrogen and acetic acid, homoacetogenic flora, and methanogenic flora) (Yu et al. 2012). The hydrolysis fermentation flora hydrolyze complex organic compounds into organic acids, alcohol, and other substances; hydrogen- and acetate-producing flora degrade the organic acids and alcohols into acetic acid, CO_2, and H_2; homoacetogenic bacteria flora convert CO_2 and H_2 to acetic acid; finally, the

Table 2.2 Biological species and amount in composting

Microorganisms	Number of microorganisms/(g/kg)
Bacteria	—
Actinomycetes	10^8–10^9
Fungi	10^5–10^8
Algae	10^4–10^6
Virus	$<10^4$
Protozoa	10^4–10^5

Source: From Li and Zhang (2000)

methanogens convert formic acid, acetic acid, methanol, and methylamine into H_2 and CO_2. From the perspective of carbon source metabolism, the relationships of the microorganisms in the composting process include the relationship between nonmethanogens and methanogens, the relationship within the nonmethanogens or methanogens, and the relationship between nonmethane bacteria and methane bacteria more important. First, because the carbon sources for methane bacteria use are rare, nonmethanogens must degrade substances such as carbohydrates and proteins into H_2, CO_2, acetic acid, and other small molecules for methanogens to grow and produce methane. The nonmethanogens can eliminate the trace amount of oxygen in the system to reduce the oxidation reduction potential and remove methanogen inhibitors, such as benzene, phenol, cyanide, and heavy metal ions. The significance of methanogens to nonmethanogens is to decompose organic acid and relieve product inhibition. In addition, the two types of microorganisms work together to maintain the system pH. However, there is a competitive relationship between these two microorganism types, mainly the use of H_2.

2.1.3 Filamentous Microorganisms on the Solid Matrix

Filamentous fungi are the major sources of bacteria in the present solid-state fermentation process. In the life of filamentous fungi, the mycelium not only acts as a nutritional body to absorb nutrients and excrete metabolic waste but also acts as a propagule to complete the life cycle. In the entire solid-state fermentation process, mycelial growth often changes the solid matrix characteristics, which are conducive or not conducive to microbial growth, so studies of mycelial change and growth are important for exploring the growth and metabolism of filamentous fungi in solid-state fermentation.

2.1.3.1 Tip Growth Mechanism

The most accepted theory for filamentous fungi growth is the vesicle theory of tip growth proposed by Grove et al. in 1970, which has been supported by experimental observation (Xin and Li 1999). The top region of filamentous fungi is approximately elliptical or hemispherical. The brown spitzenkorper region seen under the microscope is actually the core vesicles surrounded by the subvesicles.

The spitzenkorper plays an important role in mycelial elongation; the generation of the new branch of the mycelium is accompanied by a new spitzenkorper. When growth stops, the spitzenkorper disappears; when the mycelium changes elongation direction, the spitzenkorper is relocated in the top of the mycelium. The cytoplasm vesicles originate from the Golgi or endoplasmic reticulum (ER). These vesicle membranes contain hydrolases and polysaccharide synthesis enzymes. Cytoplasmic vesicles are transferred as a vesicular form to the Golgi from the ER membrane. Then, they are concentrated and processed in the Golgi and converted from the ER vesicle membrane into the plasma membrane. After that, the vesicles are released from the Golgi, transferred to the hyphal tip, and fused with the protoplasm membrane; then, they release their contents into the cell wall. This fusion process makes the vesicle contents enter the interspace of the cell walls to synthesize the cell wall; the vesicle membrane is also incorporated into the plasma membrane, increases the plasma membrane area, and leads to hyphal tip growth. In addition to vesicles, in the tip of mycelium there is the chitosome, which is a small vesicle surrounded by membrane. The chitosome transports chitinase to the cell surface.

Besides tip growth, filamentous fungi develop the mycelial branch, which is actually a new top in the mature mycelial wall. The mechanism of new tip generation is similar to tip growth, including vesicle aggregation. The mycelial branch growth is an altered form of top growth. Filamentous fungi growth in solid medium not only extends and branches hyphae; when fungal mycelia or spores grow on a solid medium, they always generate mycelia and spread. In this process, a branch occurs at almost any point along the mycelium, and it will form a circular colony in the plane. In the late phase of bacterial colony development, partial degradation at the contact point of the mycelium induces the anastomosis phenomenon and makes the colonies form a complete network structure. The anastomosis phenomenon is common in the higher fungi but rarely occurs in the low fungal hyphae. The colonies may be different for the same fungi grown on different compositions of media. Colonies under fixed conditions, such as media components, temperature, incubation time (typically 3–5 days), show constant characteristics, such as shape, size, color, and decoration, which can be used as a basis for microbial classification and identification. The shape of the fungal colonies can be summarized as loose, tight, flat, and smooth. Some colony growth shows concentric circles or a radiation pattern. There are various colony colors, and it is always difficult to describe common colors. The colony size varies significantly, and after a certain incubation time, colonies can extend to the whole dish or have limited growth, only 1–2 cm in diameter. Most fungal hyphae are transparent; some produce pigment to make the mycelium dark brown and black or a bright color (Fig. 2.1).

2.1.3.2 Biological Role of Mycelia

Filamentous fungi growth often spreads on the matrix. For example, *Mucorales* often form extended hyphae, and when spreading a certain distance, generate root-like mycelial rhizoids on the substrate surface and then move forward to form new

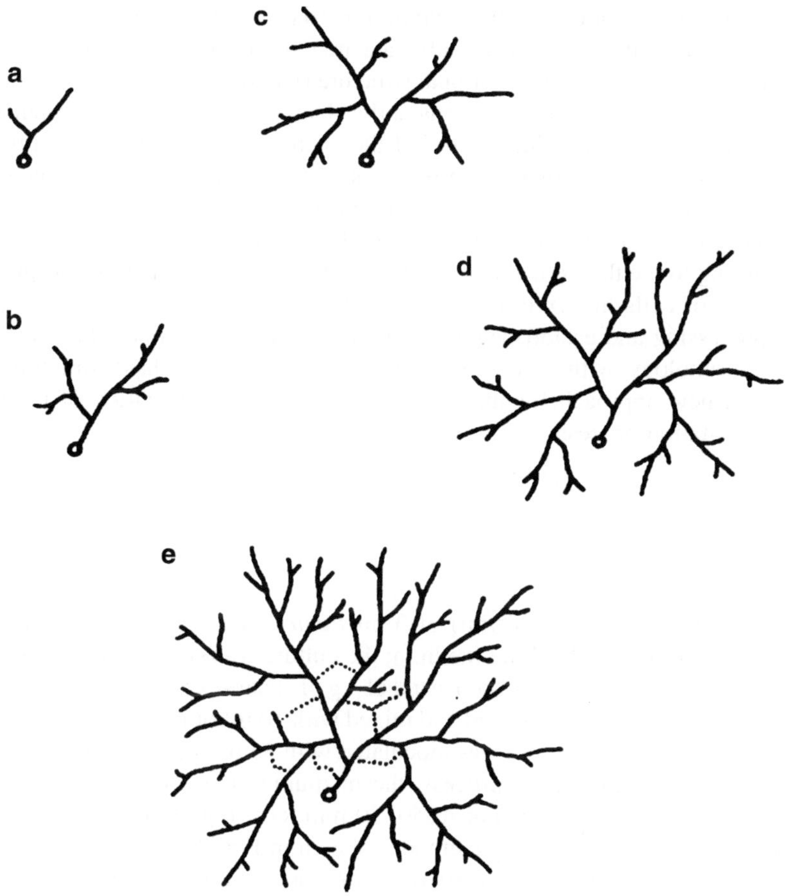

Fig. 2.1 Process from spore germination to colony formation (Xin and Li 1999)

creeping hyphae. *Rhizopus* and *Absidia* are representative of the typical prostrate hyphae and rhizoids that contact the substrates as nutrient absorption organs. Some fungi generate a root-like nonnuclear mycelium of single cells or single cells that function in nutrition and reproduction.

2.1.3.3 Filamentous Microbial Growth Dynamics on a Solid Matrix

The three-phase composition of the solid matrix affects cell growth metabolism by the influence of its transfer properties. In the actual fermentation process, the process of bacterial growth and metabolism has an impact on the fermentation substrate. Both the cell growth and the composition of solid matrix are in a state of common variation and mutual influence, which is significantly reflected in solid-state fermentation using nutritional food and straw as the carrier matrix. As the mainstream current solid-state fermentation matrix, the main characteristics of the nutritional solid-state fermentation substrate are that the chemical composition can be used by microbes and organizational structure (length, porosity) can be changed, so the nutritional matrix is in a constantly changing state during fermentation. Although previous studies indicated that the changes in the matrix structure will affect cell growth metabolism, most matrix structure variation in the fermentation process is qualitatively described by electron microscopy; the fermentation substrate unevenness causes poor representation by electron microscopy.

According to the theoretical basis of fractal geometry, I studied the morphological changes of cells-matrix in the bacterial fermentation process using the digital image processing and established a kinetic model. They examined the changes of matrix morphology with cell growth under different three-phase structures and provided a new approach for the characterization and optimization of the solid-state fermentation process.

Materials and Methods

A spore suspension (5.2×10^8/ml) of 3 ml *Penicillium decumbens* JU-A10 was seeded in 50 ml of liquid culture medium; the culture was shaken for 24 h at 30 °C and 170 rpm. The solid medium consisted of 4 g of steam-exploded wheat straw and 1 g wheat bran. Then, it was stirred and mixed with 10 ml of inorganic salt solution. After sterilization, the medium was inoculated with 3 ml seed liquid and cultivated at 30 °C. In the fermentation process, the medium was removed every 24 h for pictures with a Cannon Powershot A650 (50-mm lens; Fig. 2.2).

The digital image-processing method is shown in Fig. 2.3. First, the RGB (red-green-blue) color image was converted to grayscale and the background adjusted. Histogram equalization and a low-pass filter were used to enhance the grayscale. Then, the improved image was converted into a binary image using the Ostu threshold segmentation method (Mitchell et al. 2004; Rahardjo et al. 2006). In addition, erosion operation enhanced image quality. Finally, to extract the image outline, a candy operator was used for the subsequent calculation of the fractal dimension. These steps were achieved using Matlab 7.1 programming.

The samples were accurately weighed after drying at 60 °C for 24 h, and the water content of the samples was obtained. Culture medium biomass was measured by the glucosamine method. The fractal dimension was calculated by the methods of Otsu et al. (Otsu 1975).

Calculation was as follows: The obtained image was divided into a grid side length of 1–1,024 pixels. When the images were measured with a different scale, the following relationship was obtained:

$$N(s) = cs^{-D_{\mathrm{B}}} \tag{2.2}$$

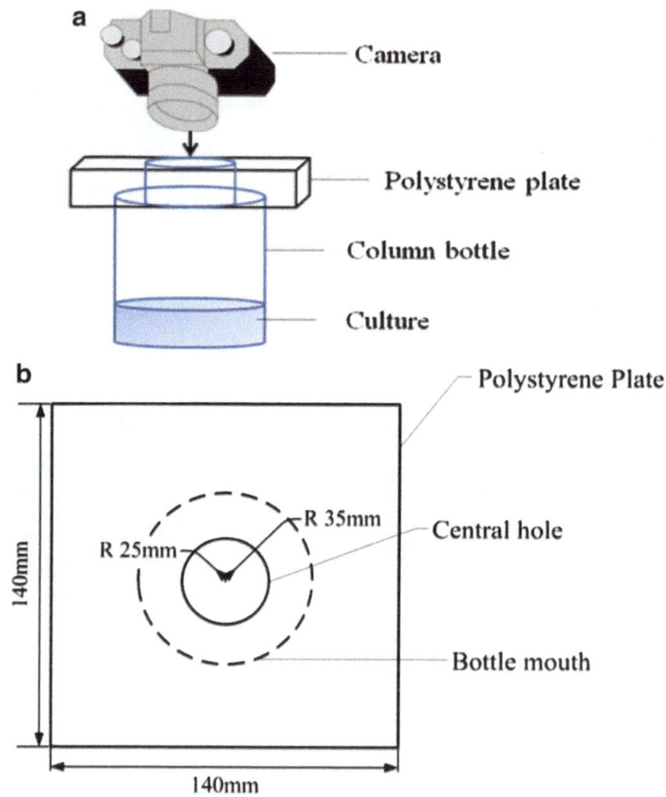

Fig. 2.2 Schematic diagram of image acquisition device: (**a**) Three-dimensional diagram of device; (**b**) *top view* of device

D_B is the fractal dimension, $N(s)$ is the number of grids covering the test area, and c is a constant. By taking the logarithm on both sides of this equation, using the regression method and fitting the equation of log $N(s)$ and log s, $-D_B$ was obtained.

$$\log N(s) = \log c + (-D_B) \log s \tag{2.3}$$

Results and Discussion

Relationship Between the Solid-State Fermentation Fractal Dimension and Mycelial Growth

To investigate the effect of mycelial growth on lignocellulose matrix morphology, photos were taken every 24 h of bacteria-matrix mixture samples in the fermentation process; the images were processed to calculate the fractal dimension. The results are shown in Figs. 2.4 and 2.5; with the growth of the mycelia, the structure of the matrix fiber was damaged, fibers became slender, and the entire mycelia-matrix mixture structure became more complicated.

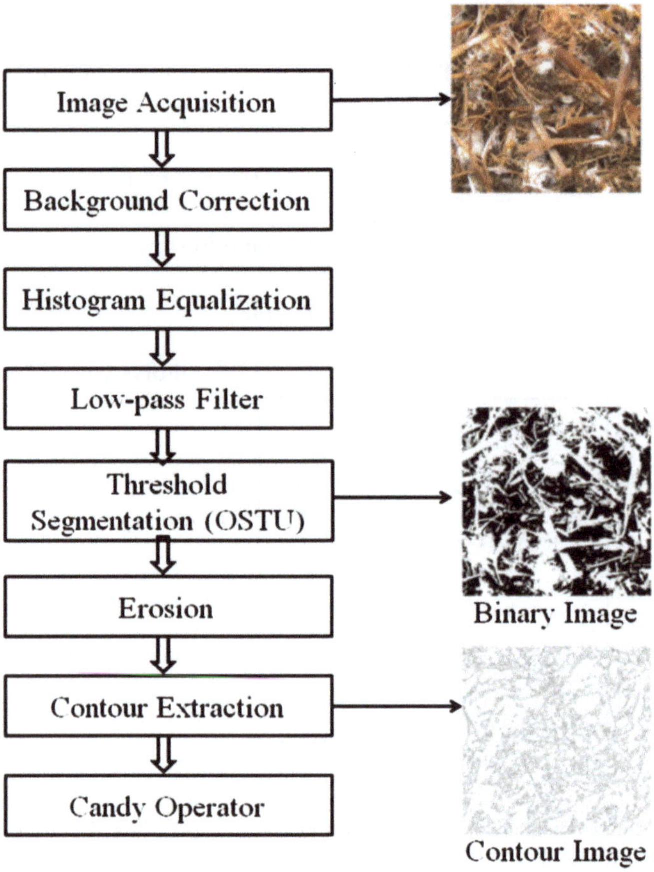

Fig. 2.3 Numerical image-processing step (OSTU is a method of image segmentation)

According to the meter box dimension principle, the fractal dimension of the matrix-bacterial mixture demonstrated that the outline of the matrix changed from complicated to simple and to complicated again; the pore structure of the matrix also experienced the same changes. Combined with previous research, the changes can be explained based on the growth of the microbial cells in the porous matrix (Fig. 2.6). At the beginning of cultivation, only a small amount of bacterial cell existed in the pore of the porous cellulose matrix; the pore morphology was intact, and the overall structure of matrix was also not significantly affected. With the continuous growth of bacterial cells, the internal pore of the matrix was occupied by bacterial cells. This resulted in porosity reduction and decreasing fractal dimension, which happens in the log phase and the earlier stage of the stable phase. When the cell growth accessed the stable phase, further matrix utilization destroyed the overall structure, and the matrix debris increased the porosity and the fractal dimension. Late in the stable phase, the bacterial cells secreted large amounts of enzymes. The matrix was further degraded into smaller fragments, which further increased porosity and the fractal dimension.

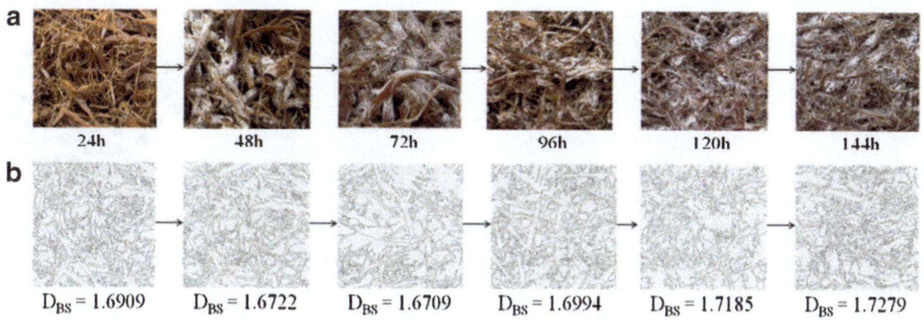

Fig. 2.4 (**a**) Morphological changes of matrix-bacterial mixtures in solid-state fermentation; (**b**) Matrix-bacterial contours after digital image processing

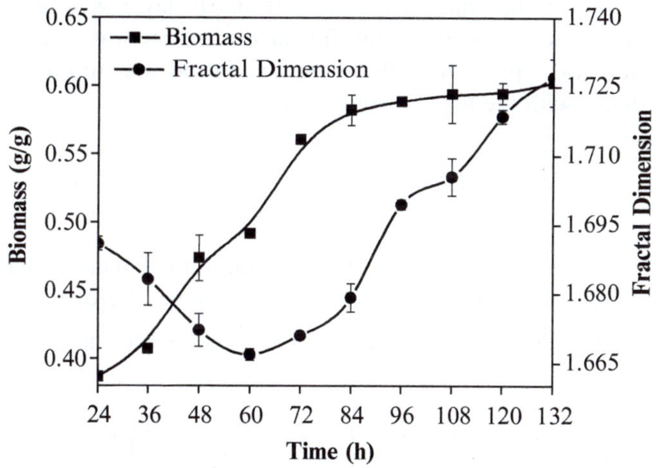

Fig. 2.5 Changes of matrix fractal dimension with cell growth in solid-state fermentation

Establishment of the Bacterial Growth and Fractal Dynamics Model of Bacteria-Matrix Mixture

The fractal dynamics model of a matrix-bacteria mixture was established based on these results. A logistic equation is the best simulation model for microbial growth, so the model was applied to simulate the growth of *Penicillium decumbens*.

The model was as follows:

$$\frac{dX}{dt} = \mu_M X\left(1 - \frac{X}{X_M}\right) \tag{2.4}$$

X = the biomass in the solid matrix, g/g;
X_M = the maximum biomass, g/g;
μ_M = the maximum specific growth rate, 1/h.

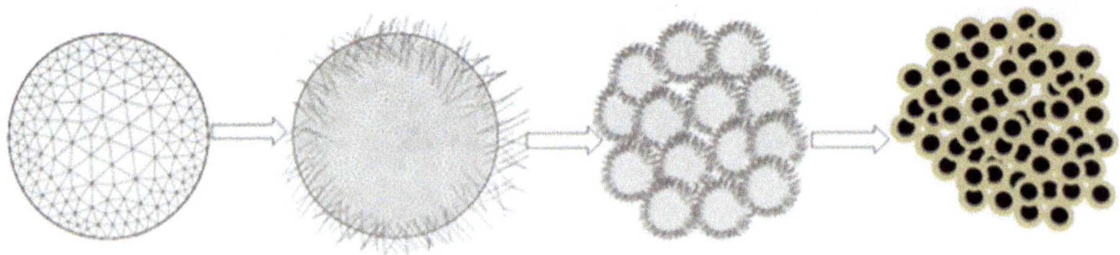

Fig. 2.6 Physical model of morphological evolution of matrix-bacterial complex in solid-state fermentation

The temperature of the matrix and water activity were constant in the experiments, so the effects of environmental factor variation on microbial growth were ignored in the model. Based on the analysis, it was assumed that an equal fractal dimension for the whole bacterial cell-matrix layer was reasonable. The biomass at the inflection point of the fractal dimension was termed X_g. When $X < X_g$, the relationship of D_B with biomass variation was approximately linear and in accordance with the logistic equation as follows:

$$\frac{dD_B}{dt} = \alpha_M D_B\left(1 - \frac{D_B}{D_{Bm}}\right) = k_1\alpha_M X\left(1 - \frac{X}{X_M}\right) \tag{2.5}$$

When $X > X_g$, because the bacterial cells were in the balanced state of growth and apoptosis, the variation of D_B was affected by bacterial growth and death, and the kinetic equation contained a decreasing bacterial coefficient $\left(\frac{X}{X_g} - 1\right)$:

$$\frac{dD_B}{dt} = \beta_M D_B\left(1 - \frac{D_B}{D_{BM}}\right) = k_2\beta_M\left(\frac{X}{X_g} - 1\right)\left(1 - \frac{X}{X_M}\right) \tag{2.6}$$

In the equation, α_M and β_M represent the maximum specific reduction rate and the maximum specific growth rate, respectively; D_{BM} is the minimum fractal dimension at the inflection point; D_{BM} is the maximum fractal dimension of the fermentation process; and k_1 and k_2 are equivalence factors of α_M and β_M, respectively. Derived by the following formula to obtain the equation, these two equations directly reflect the variation of fractal dimension to bacterial growth, where δ and η represent the equivalent reduction rate and equivalent increase rate of D_B respectively.

$$\begin{cases} \dfrac{dD_B}{dt} = \dfrac{dX}{dt}\dfrac{k_1\alpha_M}{\mu_M} = \dfrac{dX}{dt}\dfrac{\delta}{\mu_M}, & D_B > D_{Bm} \\[2ex] \dfrac{dD_B}{dt} = \left[\dfrac{dX}{dt}\left(\dfrac{1}{X_g} - \dfrac{1}{X}\right)\right]\dfrac{k_2\beta_M}{\mu_M} = \left[\dfrac{dX}{dt}\left(\dfrac{1}{X_g} - \dfrac{1}{X}\right)\right]\dfrac{\eta}{\mu_M}, & D_{Bm} < D_B < D_{BM} \end{cases} \tag{2.7}$$

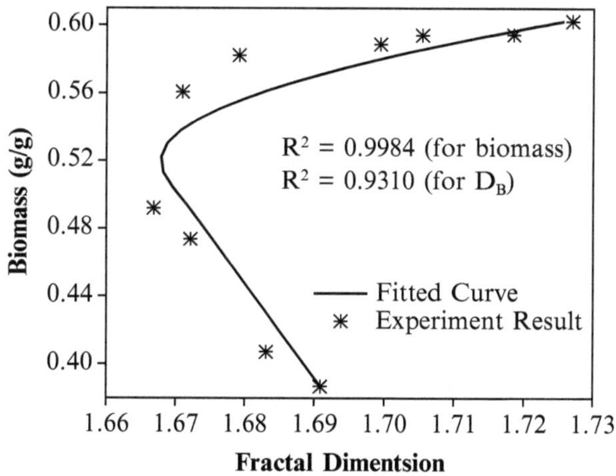

Fig. 2.7 Contrast of real and fitted values of matrix fractal dimension and biomass

The present study used a minimum error method to calculate the model parameters δ, η, and μ_M; the initial values of X_M, X_0, X_g, and D_B were needed for the calculation. The calculation process was accomplished using Matlab 7.1. In this algorithm, the approximate ranges of δ, η, and μ_M were confirmed first. The optimal parameters were calculated by repeating the cycles of the known range, and under these optimal parameters, the values of X and D_B showed the smallest deviation compared with the experimental values. For the steam-exploded wheat straw-bran matrix in this research, the optimal δ, η and μ_M were −0.006, 0.17, and 0.0333, respectively. The errors of X and D_B obtained from the optimal parameters and the experimental values were 0.776 and 0.0932 %, respectively. The relatively good fit showed that the model established in this study was suitable for prediction of microbial growth and fractal dimension variation of bacteria-matrix mixtures in the steam-exploded wheat straw-bran fermentation system (Fig. 2.7).

Structural Variation of Matrix-Bacterial Morphology Under Different Three-Phase Structure Indexes

To detect the sensitivity of the model parameters and the applicability of the model, this study used the fermentation substrate of steam-exploded straw-bran with different water contents and particle sizes. We investigated the impact of different microbial growth on the morphology and simulated the experimental results using the established model to calculate the corresponding δ, η, and μ_M. As shown in Fig. 2.8, both the changing trends of fractal dimension and microbial growth were reduced first and increased later under different matrix conditions. Specific fractal dimension changes were closely related to the growth of the mycelia and showed a high degree of specificity.

The calculation results (Table 2.3) were in accordance with the experimental values; the error ranges of biomass X and fractal dimension D_B were 0.5405–5.22 % and 0.4544–3.8847 ‰, respectively.

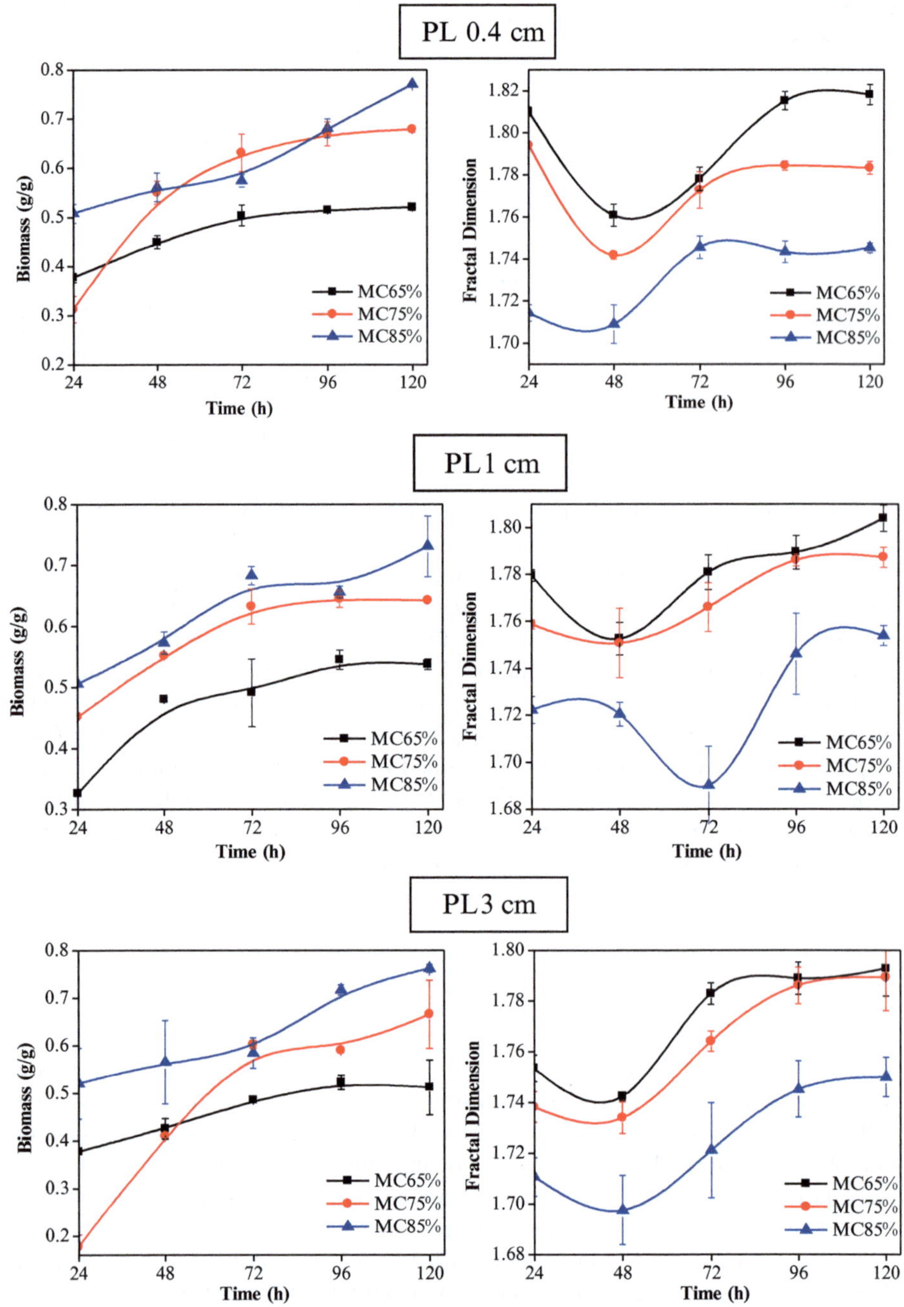

Fig. 2.8 Fractal dimension matrix changes with different water contents and particle sizes. *PL* particle length MC, moisture content

Table 2.3 Fractal dynamics model fitting results

Sample	δ ($\times(-10^{-3})$/h)	η ($\times10^{-2}$/h)	μ_M ($\times10^{-2}$/h)	RE of D_B ($\times10^{-3}$)	RE of biomass ($\times10^{-2}$)
A_1B_1	26	19.4	3.82	1.55	3.6292
A_1B_2	14.7	9.9	6.82	0.4544	3.2566
A_1B_3	1.7	4.3	1.52	0.6512	2.7616
A_2B_1	8.7	30.5	5.91	1.6811	5.2203
A_2B_2	3.9	12.8	4.96	1.1957	1.8694
A_2B_3	2.3	32	3.11	1.396	1.5456
A_3B_1	5.4	9.1	3.87	3.8847	2.1083
A_3B_2	1.2	2.3	6.21	1.6141	0.5405
A_3B_3	2.2	5.7	2.73	4.908	3.9831

A_1, A_2, and A_3 are the size of particles 0.4, 1, and 3.5 cm, respectively; B_1, B_2, and B_3 are the initial water contents of 65, 75, and 85 %, respectively

When the moisture content of the matrix was lower than 85 % (w/w) in the $X < X_g$ stage, the decreasing fractal dimension rate of the small-particle matrix was higher than that of big particles; when the moisture content was above 85 %, there was no significant difference in the fractal dimension of the different particles. This was mainly because the growth of mycelia between the particles was affected by the water film tension on the particle surface (Fig. 2.9).

Because the rate of oxygen in the water was 1/200,000 of that in the air, the water film tension was the main limitation for mycelial extension, and the increasing water content would stop the mycelia from stretching in the pore and further decrease the fractal dimension variation rate. In addition, the large pore diameter was an important factor. Previous studies showed that, regardless of the matrix particle size, the aerial hyphae were mainly concentrated in 50 μm of particle surface. Because the influence of the same size bacterial film on the small particles was greater than that on the big particles, the impact of microbial growth on the fractal dimension of small particles was more apparent. In addition, it is noteworthy that the relation of δ and μ_M was not correlated linearly, which can be explained by the equation that their relation was influenced by both D_{B0} and X_0.

Overall, the fractal dimension can reflect the influence of microbial growth on the matrix structure specifically and the cell growth under different matrixes. The previous study also showed that fractal dimension could indicate. Application of image processing and a fractal dynamics model can characterize the microbial growth state of a porous fermentation substrate, and this method is more universal.

The dynamics model of fractal dimension established in this study can quantitatively characterize the variation law of nutrient matrix morphology with the microbial growth in solid-state fermentation, can reflect the changes of matrix structure because of microbial utilization under different matrix conditions, and can be used as an indicator for biomass yield in solid-state fermentation. In solid-state fermentation, the nutritional carrier matrix intertwines with bacteria to form a mixture difficult to separate; the morphology and structure changes of the mixture are

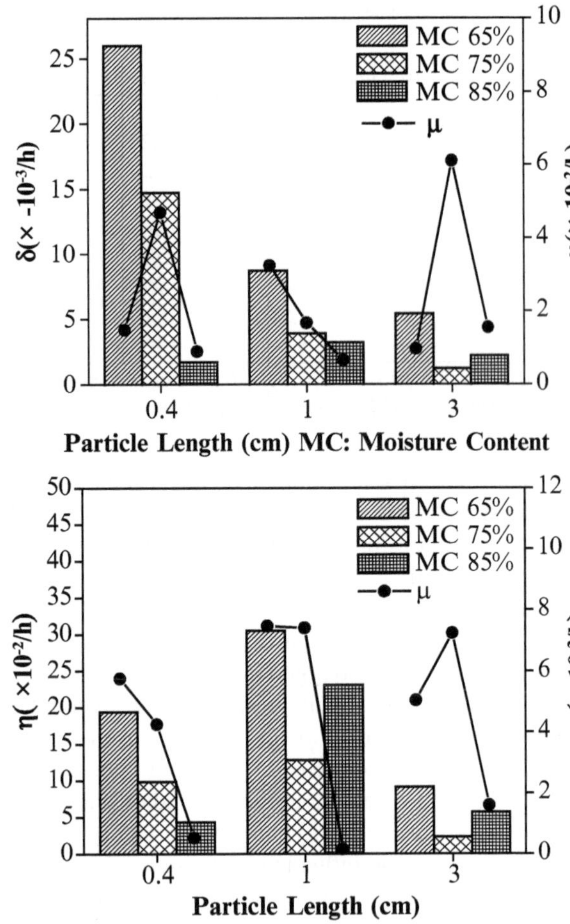

Fig. 2.9 Relationship of matrix fractal dimension change and rate of cell growth

closely related to microbial growth. In the early stage of cell growth, the internal matrix space is occupied by the bacteria to reduce the fractal dimension, and with the disintegration of the matrix because of microbial utilization, the fractal dimension of the matrix increases. In accordance with this law, a fractal dynamics model that reflects the matrix structure changes with cell growth was established in the study, and the model had high reliability and wide usability. The variation of fractal dimension with microbial growth under different water contents and fiber lengths demonstrated that the model had high reliability, and the variation rate of the fractal dimension showed high specificity to specific growth rates of microbes. Thus, measuring the changes of the fractal dimension can ascertain internal cell growth. Based on the specific quantitative relationship, this study established an online monitoring method for the fermentation process using a digital image-processing and fractal dynamics model. Specifically, to obtain the fractal dimension by digital image processing of the fermentation, photos captured by frequency cameras, and combination with the proposed model and model parameters δ and η, the microbial specific growth rate can be calculated, and the microbial biomass can be quantified according to the inoculation amount. Compared with other automated measuring techniques, this method has the advantages of low cost, rapid results, and accuracy. Moreover, by adjusting the focal length, the method can be used for the determination of the large-area matrix. For a porous nutritional carrier fermentation substrate, the specific quantitative relationship between cell growth and matrix morphology structural changes can be used for online monitoring in solid-state fermentation and provide a new way to automate process control.

2.1.4 Bacterial Growth on the Solid Matrix

Bacteria are single-cell prokaryotes, and their size is generally 0.5–2 μm. Under natural or artificial conditions, there are adsorption phenomena on the solid surface. An adsorption effect is mainly determined by bacterial surface structure and physicochemical properties and the nature of the solid surface. Bacteria adsorbed on the solid surface can even change their adsorption state according to the surrounding environment. This adsorption is more typical in solid-state fermentation and always forms a biofilm.

2.1.4.1 Bacterial Surface Structure

The bacterial surface structure includes the cell wall and cell wall appendages (pili and flagella) and capsule, which determine the physical and chemical properties of the bacterial surface. The bacterial cell wall is located outside the cell membrane, and its main function is to maintain cell integrity. According to the different cell wall components, bacteria are divided into two categories: the gram-positive and gram-negative bacteria.

The cell wall structure of gram-positive bacteria is relatively simple and relatively thick (about 20–80 nm); it contains many cross-linked peptidoglycans. These peptidoglycans consist of *N*-acetyl-glucosamine-*N*-acetyl muramic acid units. Four peptides are connected to the *N*-acetyl muramic acid molecule; these peptide chains are then linked by a peptide bridge chain to form the stable peptidoglycan network structure. The cell walls of gram-negative bacteria are thin (about 10–15 nm), but their structure is relatively complex. In addition to peptidoglycans, the main components include the outer membrane, located outside the peptidoglycan layer. The outer membrane is composed of lipopolysaccharide, a phospholipid bilayer, and lipoproteins. The inside of the outer membrane is lipoproteins, which connect the phospholipid bilayer and peptidoglycan.

The capsule is a mucus-like wrapped substance outside the cell wall, generally with a water content above 95 %. Its main component is a polysaccharide (there are also polypeptide, lipid, or lipid-protein complexes for a few bacterial capsule). The polysaccharide structure may be the same or different types, and the molecule may be linear or branched. Part of the bacteria has filamentous flagella and fimbriae on the surface of the cell wall. The main role of the flagella is cell movement. Fimbriae are mainly found in gram-negative bacteria. The main role of fimbriae is to bond cells (red blood cells, epithelial cells) and settle in a variety of cell surfaces, which is related to the pathogenic characteristics of the pathogen.

2.1.4.2 Bacterial Surface Physicochemical Properties

The electrical property of bacteria and a solid surface an important factors in the adsorption of bacteria on a solid surface. The surface of both gram-positive and gram-negative bacteria is generally negatively charged. Therefore, when the solid surface is positively charged, the bacteria can absorb on it because of the interaction of the charge (Loh and Hubbard 2002). The hydrophobic nature of the bacterial surface is also an important factor affecting bacteria adsorption. Microorganisms can secrete many specific substances, including surface-associated protein and polysaccharide, to change the surface hydrophobicity. Because of surface hydrophobicity changes, microorganisms may be fixed or moved on the interface, especially for some bacteria without the ability to move (Neufeld et al. 1980).

2.1.4.3 Adsorption on a Solid Surface

The adsorption of bacteria to the interface of a solid surface from the free state to the biofilm form can be roughly divided into four steps (Kolter and Greenberg 2006; Van Loosdrecht et al. 1990).

First, the microbes move from the liquid phase to the solid-liquid surface, including movement by free diffusion, convection transfer, and voluntary movement. There are many reasons for free diffusion of microorganisms, such as Brownian motion caused by their tiny size. Microorganism deposition under static conditions is also important. The speed of free diffusion is generally slow, so the main motion mode in the case of a nonstatic state of the aqueous phase is achieved by the transfer of flow. A third microbial movement type is voluntary. Some microorganisms can move on the interface by random movement or chemotaxis because of the gradient concentration of certain chemical substances.

The initial adsorption process is the second step. This adsorption process is a physical process mainly affected by physical properties, such as the electrical properties of the interface and the microbial surface and hydrophobic conductivity. Initial adsorption is both reversible and irreversible. Reversible adsorption can exhibit the nature of Brownian motion and be removed from the interface by a mild shearing action. Irreversible adsorption is more secure. The adsorbed microorganisms do not do Brownian motion and cannot be detached from the interface by slight shearing action such as in the stirring operation process.

The third step is the adhesion process. After adsorption of microorganisms at the interface, there will be some specific substance or bacterial structure (e.g., pili, surface-specific protein) that adheres the microorganism firmly to the solid surface. Some scholars emphasized that polysaccharides play an important role in the microbial membrane and a decreased role in the adhesion process.

The fourth step is colonization of microorganisms in the solid surface. After firm adhesion to the interface, microorganisms begin to grow and form biofilms on the interface.

2.1.5 Yeast Growth on a Solid Matrix

Different from the tip growth of filamentous fungi, yeast growth is an increased number of cells. Yeast breeding includes asexual and sexual reproduction. Sexual reproduction often requires specific nutritional and environmental conditions, and asexual reproduction is the main means of yeast growth, which includes budding and schizont formation.

Yeast budding occurs in a variety of ways, depending on the species, such as multipolar budding, double-polar budding, and single-polar budding. Multipolar budding refers to the generation of buds from different parts of yeast; a typical example is *Saccharomycodes cerevisiae*. Many types of yeast can form pseudohyphae, but the pseudohyphae often occur in solid culture; it is difficult to observe pseudohyphae in liquid fermentation.

The budding process is actually a cell cycle. This cell cycle is similar to the cycle of plant cells, which includes the G1, S, G2, and M phases. The preparations for DNA synthesis (chromosome despiralization into chromatin, the synthesis of various RNAs, proteins, and enzymes) are finished in the G1 phase, and the S phase completes DNA replication and the nuclear split. The G2 phase finishes preparations for mitosis. In the M phase, the replicative chromosome completes the distribution and gradually forms two cells. Although the yeast cell cycle is similar to that of plant cells, its speed of propagation is much faster. The completion of the cycle only takes about 1.5 h (plants often need more than a dozen hours to complete one cell cycle).

The yeast budding mechanism is relatively clear. Many protoplasm vesicles gather in the budding site before budding emerges. These vesicles contain the enzymes required for cell wall growth. The microtubules are also arranged in this region; their role is to flow the vesicles to the budding site. Yeast bud growth is accomplished by tip growth and equatorial expansion. Once bud growth is finished, the diaphragm is formed between the bud and maternal cell. Budding yeast breeding also contains schizont reproduction. However, the yeast schizont is not the same as that of prokaryotes such as *Schizosaccharomyces pombe*.

2.1.6 *Numerical Simulation of Solid-State Fermentation by Nutritional Carrier Matrix Based on Segmentation and Integration*

Because of matrix particle decomposition by bacterial utilization, the morphology and volume of the matrix layer are in a state of change, which is the essential difference between solid-state and liquid fermentation. I used the finite element method to solve single-field and multifield partial differential equations to study the simulation of real physical phenomena.

By multiple comparison statistical analysis of the matrix properties of different fermentation stages, the fermentation process was decomposed into certain significantly different phases. The fermentation effect was simulated based on the transfer theory of solid-state fermentation and spliced to a more realistic fermentation process. In this approach, a new numerical simulation of solid-state fermentation was established.

2.1.6.1 Thermokinetics Model of *Penicillium decumbens* Solid-State Fermentation

Penicillium decumbens Growth Thermokinetics

The growth curves of *P. decumbens* under different temperature conditions are shown in Fig. 2.10. From 25 to 35 °C, the cells showed a good growth trend, and the rate of cell growth increased with increasing temperature. The growth of bacterial cells stopped when the temperature was higher than 40 °C.

Using the logistic equation to fit its growth process, the relationship between the maximum specific growth rate and temperature is shown in Fig. 2.11. Based on the theoretical model, the functional relationship between the maximum specific growth rate of *P. decumbens* and temperature can be obtained:

$$\mu_M(T) = 6.598(T - T_{\min})(1 - \exp(0.000157 \bullet (T - T_{\max})))(R^2 = 0.927) \quad (2.8)$$

Dynamics of Cellulase Production by *Penicillium decumbens*

As shown in Fig. 2.12, the enzyme production of *Penicillium decumbens* and cell growth revealed a similar trend: The bacterial enzyme was hardly produced above 40 °C; in the range of 25–30 °C, the bacterial enzyme production rate increased as the temperature increased. Therefore, it can be inferred that the cellulase produced by *Penicillium decumbens* was a growth-associated type. The relationship between the rate of enzyme generation and cell growth was as follows:

$$\frac{dC}{dt} = \alpha \frac{dX}{dt} + \beta X \quad (2.9)$$

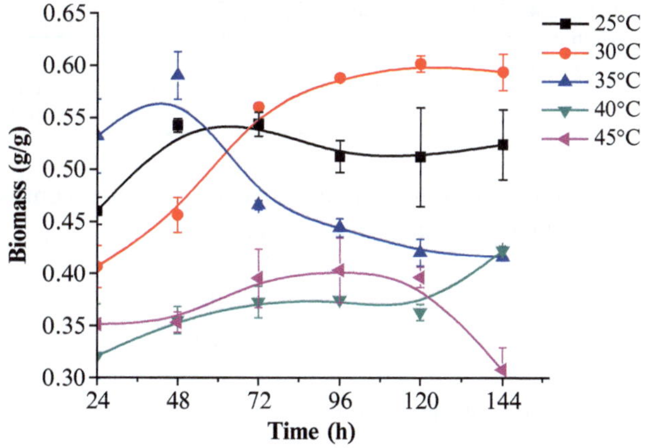

Fig. 2.10 Variations of fungal growth with temperature change

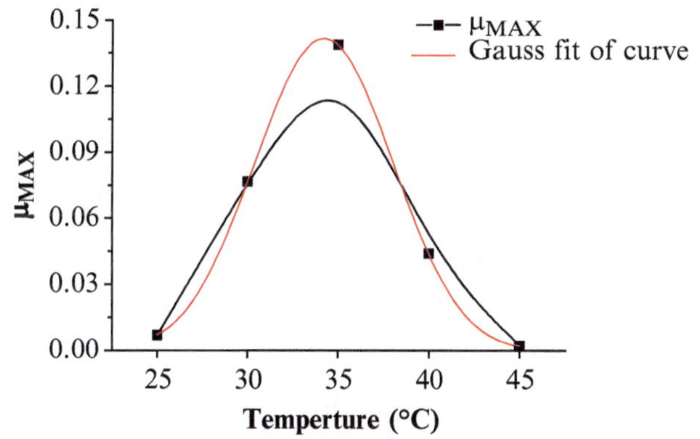

Fig. 2.11 Relationship between maximum specific growth rate and ambient temperature

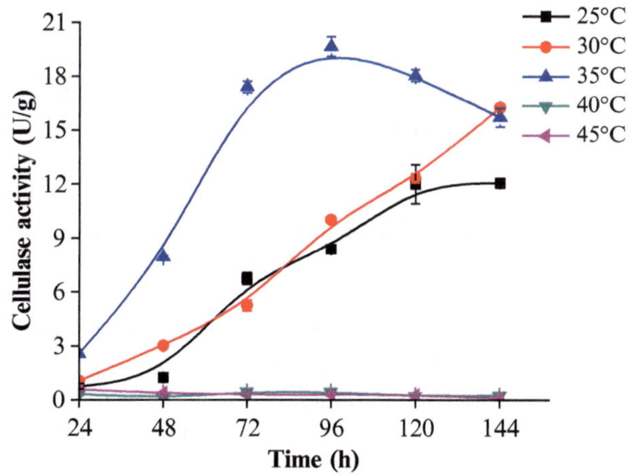

Fig. 2.12 Cellulase production of *Penicillium decumbens* in solid-state fermentation under different temperatures

Table 2.4 Dynamics model parameters for cellulase production

Parameter	C_{25}	C_{30}	C_{35}	C_{40}	C_{45}
α	380.6	−26.67	−41.98	0.370	−23.98
β	−0.029	0.230	0.331	0.008	0.003
R^2	0.945	0.992	0.813	0.817	0.891

Both sides of the equation are divided by the biomass X, and then the relationship of enzyme generation and cell growth is

$$\frac{\mathrm{d}C}{X\mathrm{d}t} = \alpha\mu + \beta \tag{2.10}$$

Integrating the equations:

$$C = X_{\max} \bullet \left[\frac{\alpha}{(X_{\max}/X_0 - 1)\exp(-\mu_{\mathrm{m}}t) + 1} - \frac{\alpha}{X_{\max}/X_0} + \frac{\beta}{\mu_{\mathrm{m}}}\ln\left(1 + \frac{\exp(\mu_{\mathrm{m}}t)}{X_{\max}/X_0 - 1}\right)\right] \tag{2.11}$$

C is FPA (filter paper activity) (IU/g); α and β are the model coefficients associated with the cell growth rate and cell mass respectively. The fitting results of this equation are shown in Table 2.4; at different temperatures, the difference in cell growth characteristics led to the different model coefficients.

To establish the model of quantitative relationship further among α, β, and $\mu_{\max}$, it was found that the model was in accordance with the Gaussian equation:

$$\alpha = \frac{752.94}{\sqrt{\pi/2}} \bullet e^{-8(\lg\mu_{\mathrm{M}}+1.94)^2} - 36.16 \quad (R^2 = 0.995) \tag{2.12}$$

$$\beta = 0.331 - \frac{0.51}{\sqrt{\pi/2}} \bullet e^{-473.71(\mu_{\mathrm{M}}-0.022)^2} \quad (R^2 = 0.999) \tag{2.13}$$

The process model of cellulase production by solid-state fermentation of *P. decumbens* can be obtained by the integration of the models.

2.1.6.2 Source Heat Calculation of *Penicillium decumbens* Solid-State Fermentation

As shown in Fig. 2.13, bacterial heat production in the adiabatic fermentation process was highest at about 48 h. With cell growth into the regression phase, the matrix internal temperature was gradually reduced, which was also affected by matrix water content and its increasing specific heat capacity. The temperature

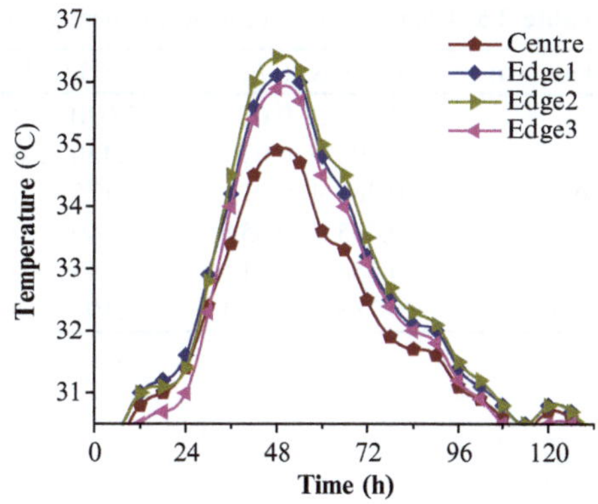

Fig. 2.13 Temperature changes of each matrix point in solid-state process

difference of the matrix at different fermentation process times was 6.8 °C, and the temperature difference between the central and around of fermentation reactors was ±1 °C. Therefore, the temperature of the substrate layer at the same time point was almost equal in the solid-state fermentation process of a small apparatus. The heat radiation and heat conduction of the matrix did not generate an obvious temperature difference in different parts.

The variation of matrix volume and specific heat capacity in the fermentation process is shown in Table 2.5. Heat production by the growth of microorganisms in solid-state fermentation can be calculated according to the calorie formula. From calculated results, with the microbial access into the logarithmic phase in adiabatic fermentation, the heat generated per unit of bacteria increased rapidly. With cell growth into recession, the heat production per unit of bacteria decreased. This is because of the decreased microbial metabolism and the increasing heat capacity of the whole matrix layer caused by the water produced from the metabolism of microbes.

2.1.6.3 Numerical Simulation of the Fermentation Process Based on Segmentation and Integration

The physical model of the matrix was divided into six periods: 0–12, 12–24, 24–48, 48–72, 72–96, and 96–120 h. The height difference among the time periods was significant.

As shown in Fig. 2.14, the simulated physical model was divided by six in accordance with the time period. The other physical properties of the matrix (0–120 h) changed continuously throughout the fermentation process. Eventually, numerical simulation results for the various physical models at the different time periods were spliced to obtain simulation results for the entire fermentation process.

Table 2.5 Changes of specific heat capacity and matrix density with cell growth

Time (h)	Biomass (g/g)	VSH (mJ/m^3/K)	Volume ($\times 10^{-4}$ m^3)	Q (mJ/g)
12	0.288 ± 0.011	2.030 ± 0.128	2.154 ± 0.072	16.451
24	0.290 ± 0.020	2.430 ± 0.067	2.054 ± 0.091	21.464
48	0.384 ± 0.012	2.175 ± 0.122	1.981 ± 0.055	36.684
72	0.495 ± 0.054	2.660 ± 0.163	1.834 ± 0.01	19.293
96	0.556 ± 0.011	2.448 ± 0.119	1.686 ± 0.011	11.581
120	0.783 ± 0.020	3.100 ± 0.119	1.539 ± 0.012	5.7263

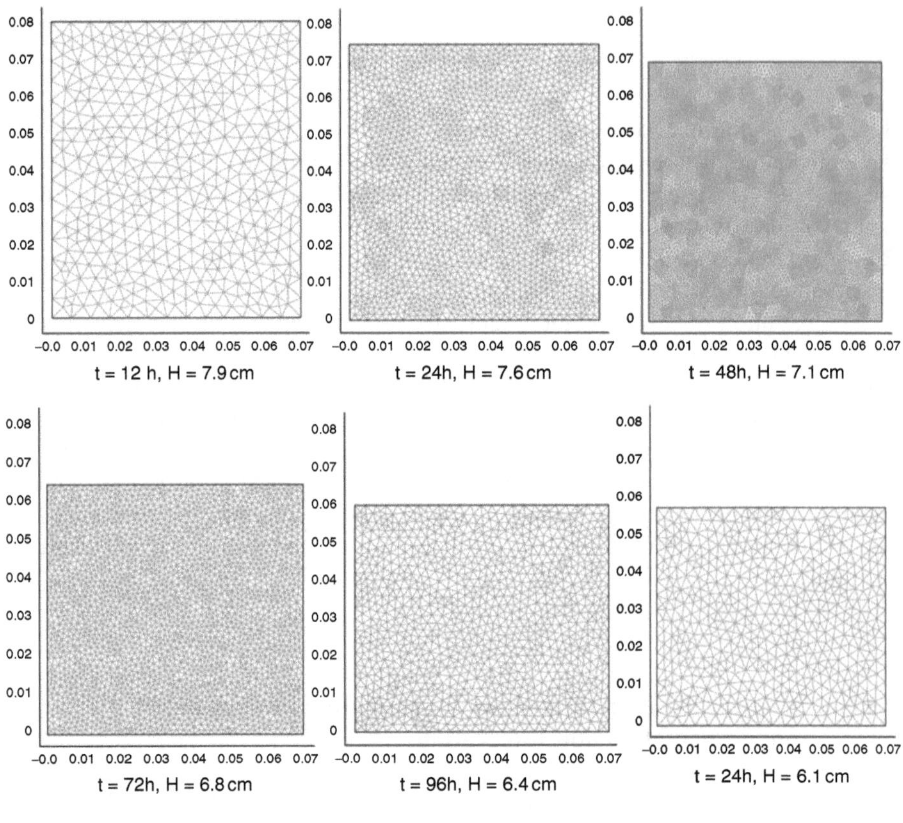

Fig. 2.14 Physical model of solid matrix

2.1.6.4 Numerical Simulation of Cellulase Production in *Penicillium decumbens* Solid-State Fermentation

Control Equations and Boundary Conditions

The physical model for static solid-state fermentation is shown in Fig. 2.15. The size of the physical area of the cross section is the same as the experimental matrix,

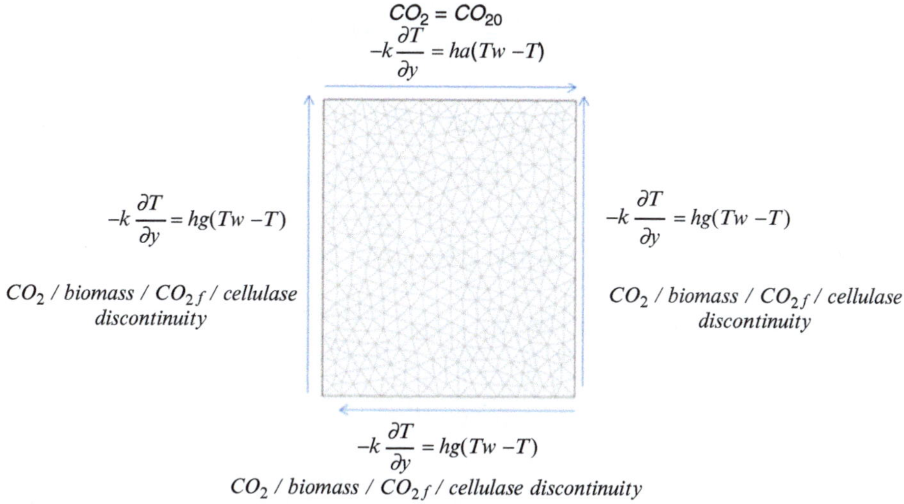

Fig. 2.15 Physical model and boundary conditions of nutritional carrier matrix

with a height of 8 cm and a bottom diameter of 7 cm. Under the assumed conditions, the heat transfer in the solid-state fermentation interior mainly included the following:

1. Oxygen diffusion within the matrix:

$$\frac{V_g \partial C_{O_2}}{\partial t} = D_{O_2}\frac{\partial^2 C_{O_2}}{\partial y^2} - K_{gas}\left(C_{O_2} - HC^f_{O_2}\right) \tag{2.14}$$

$$\rho_s Y_{O_2/X}\frac{dX}{dt} = K_{gas}\left(C_{O_2} - HC^f_{O_2}\right) \tag{2.15}$$

$$D_{o_2} = -0.002X^3 + 0.0005X^2 - 0.0004X + 0.00009 \tag{2.16}$$

$$\begin{cases} \frac{dK}{dt} = 4.981\frac{dX}{dt}, & K > 1.00083 \times 10^{-10} \\ \frac{dK}{dt} = 3.840\left[\frac{dX}{dt}\left(\frac{1}{0.414} - \frac{1}{X}\right)\right], & 1.00083 \times 10^{-10} < 1.58608 \times 10^{-10} \end{cases} \tag{2.17}$$

$$K = \frac{d^2 {V_g}^3}{180(1 - V_g)^2} \tag{2.18}$$

2. Heat generation and conduction in the matrix:

$$\frac{\partial}{\partial t}[(\rho c)_m T] + C_{P,l}T[v_l \bullet \nabla(\rho V_l)] + C_{P,g}T[v_g \bullet \nabla(\rho V_g)] + V_l C_{P,l}[v_l \bullet \nabla(\rho_l T)] + V_g C_{P,l}[v_g \bullet \nabla(\rho_g T)] = \nabla \bullet (\lambda \Delta T) + \theta \tag{2.19}$$

$$\rho_s C_p \frac{V_s \partial T}{\partial t} = k \frac{\partial^2 T}{\partial y^2} + \Delta H * \frac{dX}{dt} \tag{2.20}$$

$$k = 0.828 - 0.717 e^{(-X/0.949)}$$

$$C_p = 3.175 - 3.389 e^{(-X/0.641)}$$

$$\rho = 358.044 - 239.132 e^{(-X/0.995)}$$

3. Cell growth in the matrix:

$$\frac{dX}{dt}\frac{1}{X} = \mu(t) \tag{2.21}$$

$$\frac{dX(t)}{dt} = \mu_{max} X(t)\left(1 - \frac{X(t)}{X_{max}}\right) \tag{2.22}$$

$$\mu_{max}(T) = 6.598(T - 6.598) \bullet (1 - \exp(0.000157 \bullet (T - 318.15))) \tag{2.23}$$

$$\frac{dX}{dt} = \mu_m \left(\frac{C^f_{O_2}}{Ko_2 + C^f_{O_2}}\right) X \left(1 - \frac{X}{X_{max}}\right) \tag{2.24}$$

$$y = a\frac{dx}{dt} + B \tag{2.25}$$

For solid-state fermentation in a flask, the boundary conditions depend on the heat exchange with the outside world. The training environment temperature was 30 °C. The culture temperature was set at 30 °C. Heat exchange took place in the top layer of the matrix because of the connection with the air, and the oxygen concentration was consistent with the external environment. Heat exchange for the other surface of the matrix layer was achieved by the bottle wall, and mass transfer did not occur. Therefore, the boundary and initial conditions of the model were seen in Table 2.6.

1. When $t = 0$, $T = T_w$; $CO_2 = CO_{20}$; $CO_{2f} = CO_{2f0}$; biomass = $biomass_0$; Ccellulase = $Ccellulase_0$.
2. When $t > 0$, because of the different in container materials and air control, the heat exchanger sidewall efficiency was consistent with that of the bottom and different from the upper surface.

Numerical Simulation of the Sectional Fermentation Process

As shown in Fig. 2.16, the simulation results revealed that the temperature of different parts of the matrix layer at 8 cm was close at the same time (±0.05 °C).

Table 2.6 Constants in model

Constants	Value
Ko_2 (mol/m^3)	8e-3
Biomass$_{12}$ (kg/m^3)	8.269
K_{gas} (1/s)	1.7778e-3 (Rahardjo et al. 2006)
H	46 (Ooijkaas et al. 2000)
CO_2 (mol/m^3)	8.75
$Y_{O_2}X$ (kg/kg)	9.375
CO_{2f0} (mol/m^3)	0.19
h_y (m)	0.058–0.08
Ccellulase$_0$ (mol/m^3)	0.05
Biomass$_0$ (mol/m^3)	7.5
T_w (K)	303
ha (W/(m^2*K))	2e-2
hg (W/(m^2*K))	2.5e-2

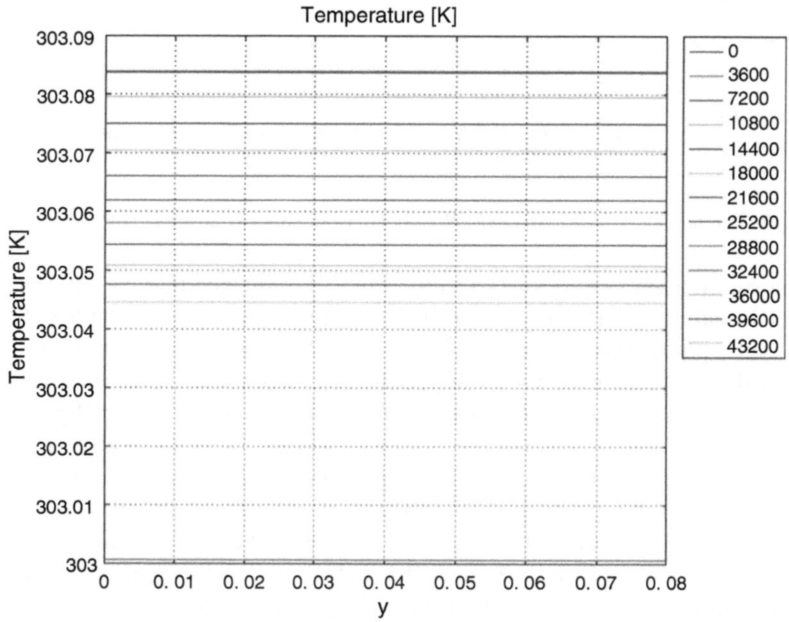

Fig. 2.16 Change of matrix temperature at 0–12 h

This is because, on one hand, the temperature of each part of the matrix depends on the exothermic capacity of the heat source, and a relatively uniform growth of the microbial cells in the various parts of the matrix results in the same rate of temperature change in the various parts of the matrix. Previous studies showed that the thermal conductivity and specific heat capacity of the matrix were mainly affected by the matrix water content under identical chemical matrix compositions. So, because of the lack of significant water content changes in the matrix, the relatively uniform thermal conductivity of the matrix makes the temperature distribute evenly.

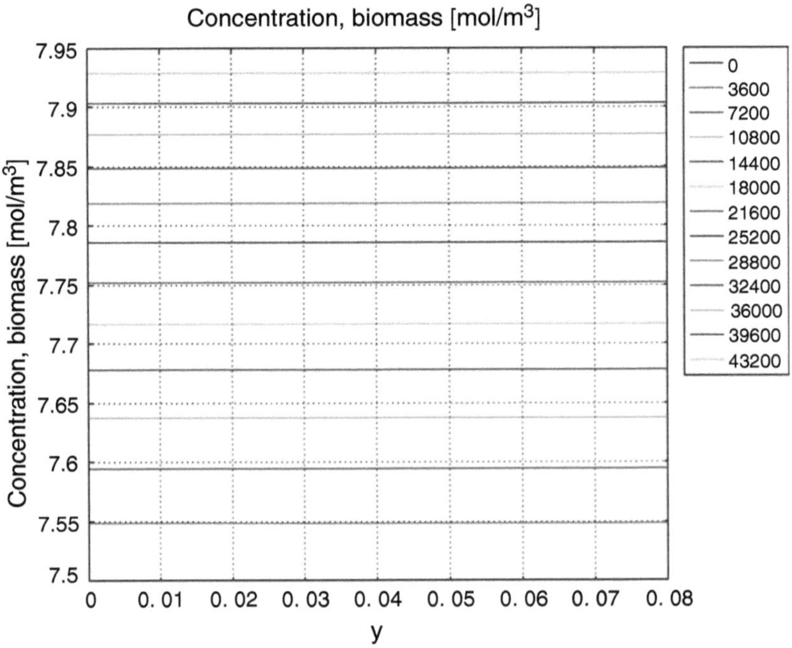

Fig. 2.17 Fungal biomass variation in SEWS (steam explosion wheat straw) substrates for 0–12 h

The simulation of cell growth results (Fig. 2.17) showed that the differences of bacterial quality in different sites at the same time was only ±0.0005 kg/m^3 and decreased with decreasing matrix depth. Although the oxygen content in the matrix decreased with decreasing matrix depth, the oxygen content of the internal matrix was high because of the porous matrix, so this factor did not limit cell growth.

At the same time, because of the sufficient heat exchange of small fermentation devices, the bacteria in the matrix could grow more evenly. As cellulase is associated with cell growth, the production distribution of cellulase in the matrix was more uniform (Fig. 2.18). In the remaining period, the distribution of the microbial cells in the matrix was relatively uniform; at different time points, the temperature, the cell, and enzyme activity of each point of the matrix were different.

Numerical Simulation of the Entire Fermentation Process

As shown in Figs. 2.19, 2.20, and 2.21, the simulation of cell growth, temperature change, and cellulase production was in good agreement with the experimental results. The simulation of cell growth commonly used logistic equation in previous reports, and results were comparatively idealistic and did not reflect the real growth process with fluctuations such as secondary growth. Section fitting can more truly reflect the current situation, which was closer to the practical fermentation process.

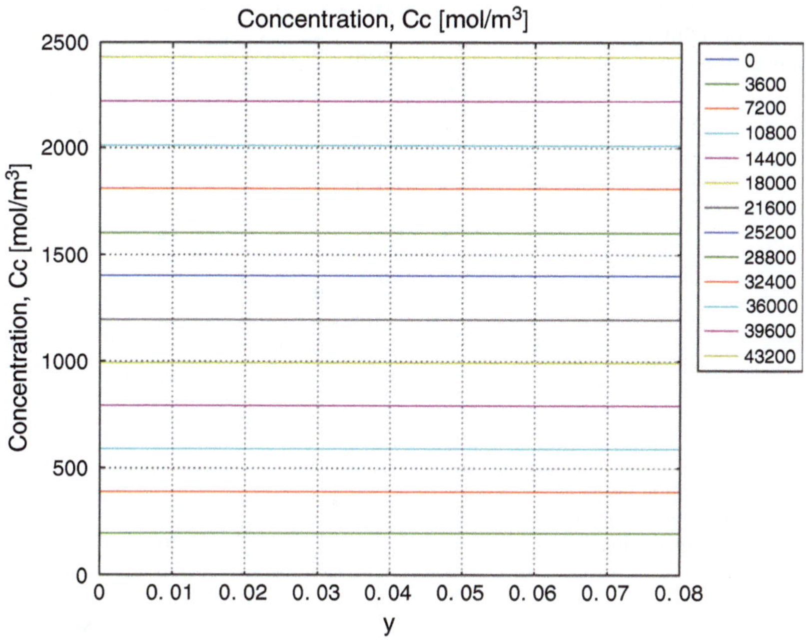

Fig. 2.18 FPA (filter paper activity) variation in SEWS-bran substrates for 0–12 h

Fig. 2.19 Numerical simulation of cell growth based on segmentation and integration

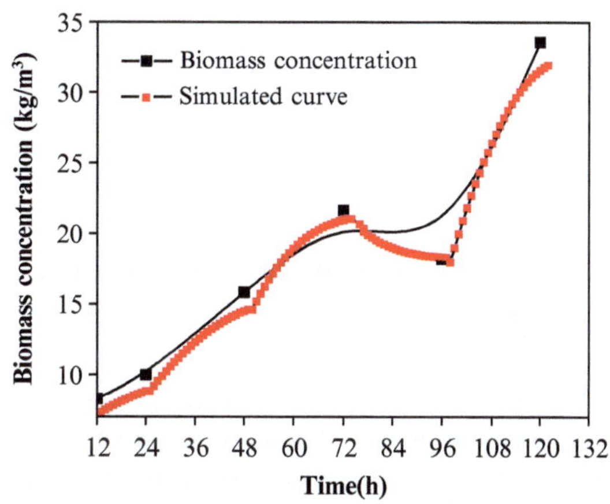

Simulation of the temperature changes in the fermentation process was also closer to the actual trend (Figs. 2.20 and 2.21) and reflected the complexity of the temperature change, which cannot be achieved by conventional simulation. However, the discontinuity of the temperature simulation results also reflected the inherent defects of the method: The chosen numerical method UMFPACK unsymmetric multifrontal sparse LU factorization package is a linear unsteady method and causes an unexpected outcome when it is used to solve nonlinear multiphysics coupling problems, especially fluctuations.

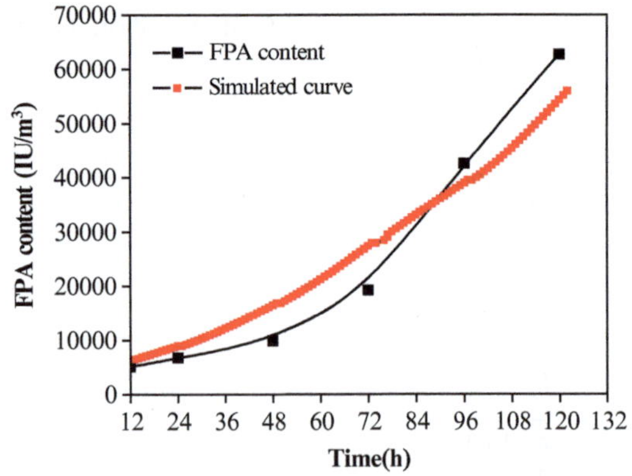

Fig. 2.20 Numerical simulation of cellulase fermentation

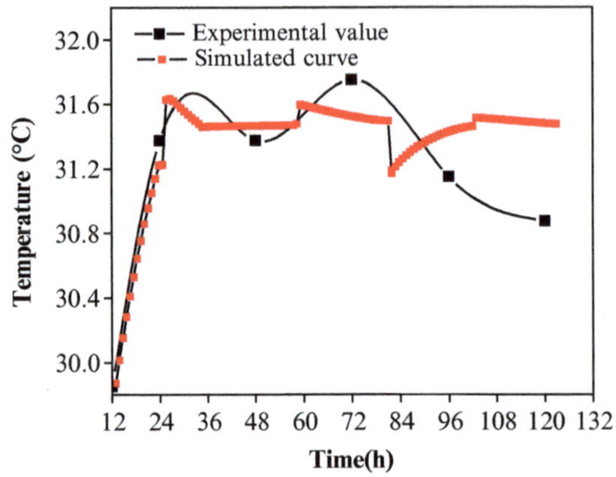

Fig. 2.21 Simulation result of matrix temperature change in solid-state fermentation

2.2 Properties of the Solid Matrix in Solid-State Fermentation

Compared with traditional liquid fermentation, in which water is the main matrix in the fermentation process, the substrate in solid-state fermentation is more diverse, has more extensive sources, and even contains the unique components required for microbial metabolism. Many cheap raw materials can be used as carbon and nitrogen sources, reducing production costs. On the other hand, the composition of the fermentation substrate is more complex; some of the ingredients cannot be directly used as a substrate for microbial fermentation. This section details the specific substrate in solid-state fermentation as well as substrate pretreatment technology, especially the unique steam explosion pretreatment technique developed in our laboratory.

2.2.1 Types of Solid Matrix Suitable for Solid-State Fermentation

The solid matrix for solid-state fermentation mainly comes from various solid biomass materials, mostly from the products or waste of agriculture and forestry. These can be broadly divided into several categories, including sugar-rich materials, lignocellulosic raw materials, grain, and inert carriers.

2.2.1.1 Sugar-Rich Raw Materials

Sugar-rich raw materials for solid-state fermentation are mainly sweet sorghum and sugarcane. The carbon source is the sugar in the raw materials. Sweet sorghum, also known as milo, is a mutant of ordinary sorghum. This crop is harvested not only for sorghum grain but also for the rich sugar in the stalk. The sugar content of sweet sorghum stalks is generally 15–23 %, and it can be divided into the sugar crystal and syrup types. The sugar crystal type of sweet sorghum mainly contains sucrose and is used for the production of crystallized sugar; the syrup type contains glucose and is used mostly in the production of sweet sorghum syrup. Sweet sorghum production is relatively high, up to 100,000 kg/ha (Xin and Li 1999). Sweet sorghum has not only a high sugar content but also characteristics tolerant to drought, waterlogging, and salinity. Sweet sorghum shows excellent adaptability to soil pH and is able to grow well in the pH range of 5.0–8.5. These traits give sweet sorghum an extremely wide cultivation range (Li 2002).

Botanical characteristics of sweet sorghum are similar to those of sorghum. The appropriate period of sweet sorghum harvest can be determined according to law and the purpose for sugar use. The sweet sorghum harvest period is the time the stalk has its highest sugar content for sugar production (Table 2.7). For forage purposes, whether the grain is mature or not, it can be harvested and immediately preserved by silage fermentation.

Because of the high yield and tolerance to drought and salinity, sweet sorghum can be used for ethanol production. The Brazilian government started ethanol production from sweet sorghum in 1975; similar studies have been conducted in America since 1978. Sweet sorghum has been listed as one of the major crops for alcohol preparation in America; Europe has also developed research for sweet sorghum since the 1980s. Enhancing the comprehensive development and utilization of sweet sorghum stalks has an important and far-reaching significance for alleviating China's energy shortage, improving the ecological environment, and promoting national economic stability and sustained development (Gnansounou et al. 2005).

Liquid fermentation of sweet sorghum is used for an ethanol product. However, this causes serious wastewater pollution. Solid-state fermentation is conducted in a state with no or almost no free water flow. The crushed sweet sorghum stalks are directly smashed for solid-state fermentation, saving juicing costs. Solid-state fermentation also has some other advantages, such as low operating costs and generation of less wastewater.

Table 2.7 Chemical composition of sweet sorghum stalks

Components	Whole sorghum	Pith	Bark
Cellulose (%)	12.4	8.7	19.2
Hemicellulose (%)	10.2	6.3	17.5
Lignin (%)	4.8	0.6	8.8
Sucrose (%)	55.0	67.4	32.2
Glucose (%)	3.2	3.7	2.4
Ash (%)	0.3	0.2	0.5

Some scholars have studied the solid-state fermentation of sweet sorghum stalks for ethanol production. Song et al. (2007) used active yeast tolerant to high temperature to produce ethanol by solid-state fermentation, obtaining optimal fermentation. The theoretical yield of ethanol was 0.332 g ethanol/g dry sweet sorghum stalks; the actual yield was 0.298 g ethanol/g dry sweet sorghum stalks (89.8 % of the theoretical yield), which means only about 3.01 t dry stalks were required to produce 1 t ethanol.

2.2.1.2 Lignocellulosic Materials

Lignocellulose is the largest biomass feedstock on Earth and one of the most concerned raw materials. These types of materials include wood, crop straw, bagasse, and other waste, such as corncobs. The lignocellulosic feedstock generally consists of cellulose, hemicellulose, and lignin. In plant materials, these components constitute the supportive skeleton of the plant body. The cellulose forms fine fibers to constitute the cell wall network skeleton; the hemicellulose and lignin are the filler and binder among the fibers.

Cellulose is the chain polymer compound formed by the β-1,4-glycosidic bond of D-glucose, which can be represented by the formula $(C_6H_{10}O_5)_n$ (n represents the number of glucose units, and the value of n is generally from hundreds to thousands). Cellulose is not soluble in water, dilute alkali, and acid at room temperature. In plant cell walls (straw, wood), the cellulose content generally accounts for 35–50 % (Weber et al. 1999); in cotton, the cellulose content can reach more than 99 %. Natural cellulose has a more complex form, including the interleaving presence of the crystallization and noncrystalline regions. Cellulose molecules in the crystalline region are arranged in order, and their density is relatively higher. However, the molecules in the noncrystalline regions show disarranged order and slightly lower density. The hydroxyl groups of the noncrystalline region are in a free state, and the hydroxyl groups of the crystalline region often form numerous intramolecular and intermolecular hydrogen bonds. Because of the presence of these bonds, natural cellulose presents a relatively dense structure. It is now generally believed five cellulose crystals exist: types I, II, III, IV, and X. Type I is the natural crystalline form; the others are artificial polymorphs obtained by manual handling.

Hemicellulose is another important component of lignocellulosic feedstock. It was once misunderstood that the term *hemicellulose* came about because of semifinished cellulose synthesis in the plant. Hemicellulos is a plant-derived polysaccharide containing the basis chain of D-xylose, D-mannose, D-glucose, or D-galactose and another sugar group as a branched chain connected to the base chain. Unlike cellulose, which is a uniform polysaccharide, hemicellulose is a polymer that consists of several sugar units. The contents of hemicellulose in various plants and their structures are different. Its main chain can consist of a single sugar unit or two or more kinds of sugar units.

Lignin is another important component in plant cell walls and is the unique chemical composition of gymnosperms and angiosperms. Lignin is a complex compound with a nonlinear and random connection of the phenylpropane unit. Currently, lignin is deemed to contain three monomers: coumaryl alcohol, coniferyl alcohol, and sinapyl alcohol. Different lignin structural units are divided into three types: syringyl alcohol (S), guaiacyl alcohol (G), and *p*-coumaryl alcohol (H). Lignin in plants exhibits a high degree of heterogeneity; different species and different growth phases have different structures.

The inter- and inner chain bond of cellulose and hemicellulose is connected via hydrogen bonds; lignin forms a stable carbohydrate complex with hemicellulose in addition to the internal hydrogen bonds. In the cell wall structure, cellulose molecular chains are regularly arranged to form the pro-fine fibers and microfibers subsequently. Lignin and hemicellulose wrap the microfine fibers. The cellulose forms the skeleton of the cell wall in the form of fine fibers, and the lignin and hemicellulose covalently cross-link to form a three-dimensional frame structure. The fine fiber bundle is embedded inside the frame structure. Lignin in the intercellular layer outside the cell wall bonds the two cells together. Thus, cellulose, hemicellulose, and lignin cross-link to form a complex cell wall structure that is difficult to degrade (Chen 2005). Such raw materials are mainly used to produce cellulase and xylanase. Compost, silage, and biopulping are also involved in the cellulose matrix solid-state fermentation process.

2.2.1.3 Grain

Grain is the raw material most commonly used for solid-state fermentation and many food-brewing processes; the production of organic acids and biopesticide products uses grain as a raw material. Compared with cellulose, the advantage of food is that the starch contained in the matrix can be used rapidly by many microorganisms. Many filamentous fungi have a strong ability to utilize starch, so a dedicated saccharification process is not required. Yeast and many bacteria do not have such an ability, and in the fermentation process, the degradation of starch is accomplished by the addition of koji or glucoamylase. These raw material types include cereals (rice, sorghum, corn, wheat, and oats), potatoes, and so on.

Sorghum is an important raw material for wine making and is widely cultivated in China, especially in the northeast. According to the characteristics, sorghum is

Table 2.8 Quality percentage of different parts of wheat

			Percentage	Percentage
Cortex	Outer cortex	Epidermis	0.5	13.5
		Epicarp	1.0	
		Endocarp	1.5	
	Inner cortex	Episperm	2.0	
		Nucellar layer	3.0	
		Aleurone layer	5.5	
Endosperm			84	
Embryo	Plumule		1.0	2.5
	Scutellum		1.5	

widely used in food, wine brewing, sugar making, and feed. Wheat is also an important raw material. Wheat cultivation has a history of more than 10,000 years; currently, wheat is still the basic food crop for most countries in the world. Wheat accounts for about 26 % of the total food cultivation area, and the output of wheat accounts for about 22 % of the total output. More than a third of the population around the world uses wheat as the main cereal. Although the morphology of cereal grains is different, their composition and structure are similar and broadly include the cortex, endosperm, and embryo. The internal structure of the wheat grain is delineated in Table 2.8. Wheat cortex is also called bran. The main wheat starch is in the endosperm. Wheat grinding obtains ground endosperm and isolates the bran and embryo. The endosperm contains about 70 % starch, 13 % water, and 12 % protein (Yang 2001).

In addition to the sorghum and wheat, potatoes, such as cassava, are used for solid-state fermentation. The main problem for cassava fermentation is the cyano glycoside contained in the epidermis. It can be solved by discarding the skin; at the same time, the amount of nitrogen and other trace elements will be reduced, so some inorganic salts and nitrogen sources are required for the growth of microorganisms. Other grain materials, such as corn, banana powder, sweet potato residue, buckwheat, and rye powder, can be used for solid-state fermentation. In addition, wheat bran is applied in the solid-state fermentation process. Although bran does not belong to food, the main composition of bran is also starch. So, bran can be used to ferment various products such as α-amylase.

The most important ingredients in the grain raw materials suitable for microbial utilization are starches. Compared with cellulose, starch has a similar chemical structure, but the chemical nature is dramatically different. The starch actually consists of two related polymer compositions: amylose and amylopectin. Amylose is a glucose polymer connected by α-1,4-glycosidic bonds. The amylose has a degree of polymerization of about 500–5,000. The binding mode of amylopectin includes α-1,4-glycosidic bonds and 5–6 % of α-1,6-glycosidic bonds at the branch point, which lead to the amylopectin branch. The average length of the amylopectin branch is about 24–30 glucose residues. In different crops, the proportions of amylose and amylopectin are different; for example, the ratio of amylopectin is about 69–77 % in corn, potato, and wheat starch. The starch is present in the form of particles in nature, and its size is generally 5–100 μm. The shape of the starch granules is also different because of the different species and is mainly decided by the amylopectin.

The starch-degrading enzymes include α-amylase, glucoamylase, pullulanase, and isoamylase; α-amylase and glucoamylase play the most important role. α-Amylase is starch endonuclease that randomly hydrolyzes α-1,4-glycosidic bonds to reduce the molecular weight of the starch rapidly. α-Amylase can convert amylose completely into maltose and maltotriose. α-Amylase cannot hydrolyze α-1,6-glycosidic bonds. Glucoamylase is an exonuclease that can act on the α-1,4-glycosidic and α-1,6-glycosidic bonds. However, the hydrolysis rate of glucoamylase on α-1,4-glycosidic bonds is much higher than that on α-1,6-glycosidic bonds. Both amylopectin and amylose are converted into glucose. Glucoamylase mainly exists in fungi, including *Aspergillus* and *Rhizopus,* which are important for solid-state fermentation. A number of yeast and bacteria can also produce glucoamylase. The synergistic effect of α-amylase and glucoamylase can greatly enhance the degradation of the native starch.

2.2.1.4 Inert Adsorption Solid Substrate Carrier

The inert carrier is a special class of substrate for solid-state fermentation, mainly used for inert carrier solid-state fermentation (Ooijkaas et al. 2000). The inert carrier is not a nutrient source for microorganism growth and production, but only a support for the adsorption of liquid medium. Commonly used inert carriers are usually porous materials, including natural minerals such as vermiculite and perlite and some synthetic materials such as polyurethane foam.

The large surface area of these porous materials not only helps absorb the liquid medium but also provides a space for microorganism growth. Under aerobic fermentation conditions, the porous structure is also conducive to oxygen transfer from the gas phase to the liquid phase. Inert carriers commonly used in solid-state fermentation include lignocellulosic materials, inorganic materials, synthetic polymer materials, and ion exchange resin.

Inorganic materials (mostly with porous structures) have long been used as carriers for solid-state fermentation; such inorganic materials are clay microgranules, perlite, vermiculite, and pozzolan. Inorganic materials as carriers are used mostly in the production of biopesticides (Ooijkaas et al. 2000). Such products often do not require separation and can be used directly with the carrier. Except for use as an inert carrier substrate for biopesticides, little other use has been reported, mainly because of the fixed physical properties of these substrates, which are difficult to adjust to different fermentation conditions.

The lignocellulosic inert carriers include bagasse, hemp, *Opuntia imbricate,* and others. The best features of this type of matrix are the low price and wide variety of sources. This type of material theoretically has better biocompatibility. This type of matrix has been used for the production of biological pesticides (Desgranges et al.

1993) and agricultural antibiotics (Weber et al. 1999). In addition to the low price, the degradation of the materials may be an important reason for use.

The organic polymer carriers include polyurethane, polystyrene, nylon sponge, and others. Judging from the number of recent reports, such materials are more popular at present. Polyurethane material has the advantages of a larger and uniform distribution pore size and good strength. Because polyurethane is a sophisticated material, it can be selected and customized based on experimental needs. Other advantages over other materials are good flexibility and relative ease of separation so the fermentation liquid can be easily separated from the carrier.

In addition to the commonly used carriers described, ion exchange resin can be used as an inert carrier material. However, there are relatively few such reports; Gelmi tried to use this carrier to produce the agricultural antibiotic gibberellic acid (Dominguez et al. 2001), and Gutierrez-Rojas attempted to use ion exchange resin to produce citric acid (Gelmi et al. 2002).

2.2.2 *Pretreatment of Solid Matrix for Solid-State Fermentation*

2.2.2.1 Pretreatment of Grain Raw Materials

Grain is the main raw material for wine- and spice-brewing processes. The different processes have different requirements according to the physical state of the raw materials; therefore, the raw materials are pretreated to adapt to the demands of the fermentation. There are many pretreatment means and procedures; and the relatively important common processes include crushing, screening, and cooking. Crushing and cooking are discussed here.

Crushing is the process of breaking the chunks of material into suitable size pieces by external mechanical forces. The purpose of crushing is to increase the uniformity of the solid particles and the specific surface area. The increase in specific surface area is beneficial for the dissolution of the soluble component and the precipitation of insoluble components; it improves raw material utilization. However, the degree of material crushing is not a matter of smaller size being better. If the raw materials are too small, the porosity of the particles decreases, which can reduce the gas volume in the matrix and limit the transfer of oxygen in the matrix. There are many types of crushing. Breaking and airflow pulverization are common methods. Breaking is frequently used in the wine-making industry; wheat grain is crushed by mechanical forces.

Cooking is another important pretreatment; it has at least four aspects. First, the materials are sterilized in the cooking process. The steam has a strong penetrating ability, which gives the method higher efficiency than dry heat sterilization. Second, the plant cell components are destroyed in the cooking process, which facilitates the degradation of macromolecules. Third, the steam gelatinizes starch in the cooking process. The raw material starch granules swell by sucking water in the cooking process and increase the viscosity and volume to show a dissolved state, which is conducive to the subsequent hydrolysis of amylase and glucoamylase. Finally, high-temperature steam causes protein denaturation. The denatured protein is more likely to be degraded by protease, which can provide the easily absorbed nitrogen source for the microbes.

The cooking equipment used in traditional fermentation includes an atmospheric pressure digester, such as a steamer pot. In the cooking process, the steam penetrates into the raw materials from the bottom of the steamer pot. As the steam rises in the pot, the materials are heated by constant contact with steam. Because of batch operations, such equipment is labor intensive, has low efficiency, and is not suitable for large-scale production. So, such a batch cooking process has been replaced by continuous cooking. Using high-pressure steam can enhance the treatment effect on raw material; therefore, in addition to atmospheric pressure cooking, some pressure cooking equipment, such as rotary digesters and continuous digesters, is used in some current fermentation processes. The use of such high-pressure cooking equipment not only increases the pretreatment effects but also improves production efficiency because of the short treatment time at high pressure. High-pressure cooking has been used in rice wine and soy brewing and other products.

I established the new wine-brewing process that used steam explosion technology to pretreat cereal crops. The basic process is as follows: (1) The water content of various types of cereal grain is controlled at 10–25 %. (2) Using a segmented steam explosion pretreatment, low-pressure steam (0.5–1.2 MPa) is first aerated, followed by high-pressure steam treatment (0.8–1.6 MPa); the pressure is maintained for a certain time. (3) The pressure is relieved quickly, and the materials are collected. The materials are crushed according to the situation and brewed for wine production. Studies have shown that the following are advantages of using steam explosion technology to pretreat the grain: Processing time is shortened significantly, from hours to about 10 min; the starch is gelatinized quickly and is conducive to subsequent enzymatic hydrolysis; the costs of pretreatment are reduced, and the technique is easy to enlarge (Gutierrez-Rojas et al. 1995).

2.2.2.2 Pretreatment of Lignocellulosic Raw Materials

The chemical pretreatment of raw materials includes many means, such as acid treatment, alkali treatment, organic solvent treatment, wet oxidation, and ozone treatment. Chemical pretreatment degrades the components of lignocellulose using chemicals and separates the components. There are industrialized examples of acid treatment because of the early appearance of this mature technique. Acid pretreatment can destroy the hemicellulose and enhance the degradation of cellulose, which can also be directly used for the preparation of furfural (Modenbach and Nokes 1956). This method is the most common current experimental method. The problems of such methods are environmental contamination, equipment corrosion, and the high salt concentration because of neutralization.

Organic solvent treatment has many alternatives for solvent choice, which includes low-boiling alcohol, high-boiling alcohol, ionic liquid, and more. The main function of organic solvent pretreatment is to remove lignin by solvent extraction and enhance enzymatic hydrolysis. The biggest obstacle for scale-up of this technique is the high cost of solvent (Sun and Cheng 2002). Alkali pretreatment mainly uses sodium hydroxide, calcium hydroxide, and ammonia, and its major role is the relatively strong ability to remove lignin. The swelling cellulose materials after alkali treatment decrease the crystallinity and result in easier enzymatic hydrolysis. However, this approach is not promoted industrially because as more hemicellulose is lost, the degraded lignin is difficult to use and recover, and neutralization and washing of reagents are difficult. Oxidation is also an effective chemical pretreatment method, such as the wet oxidation process of water and oxygen under high-pressure conditions and ozone treatment. These methods show good selectivity and can produce relatively pure cellulose. However, costs are higher.

In addition to the chemical method, many physical methods are used for the pretreatment of straw, such as radiation treatment and supercritical processing. However, these methods cannot achieve the fractionation of lignocellulose and only destroy the internal structure of raw materials. Therefore, these methods can be regarded as auxiliary means. For example, the depolymerization of the cellulose chain is accomplished using gamma rays, electron beam, and microwave radiation to treat the lignocellulose raw materials. The supercritical method is also effective for pretreatment, and its medium can be CO_2, water, supercritical alcohol, isopropyl alcohol, or even some diverse mixture, such as acetate-water, acetate-supercritical CO_2, acetate-water-supercritical CO_2.

Steam explosion pretreatment was developed as a pretreatment method (Chen and Liu 2007). The technique was first used for the production of man-made fiberboard in 1928. The raw materials are heated to 180–235 °C by steam, and the pressure is maintained for a certain time. Then, the pressure is released rapidly, and the raw material structure is destroyed by mechanical forces. Since the 1970s, this technology has also been used in animal feed production and conversion of wood raw materials into related chemicals. The application of steam explosion has gradually expanded because of the further development of the specific technique.

The advantages of steam explosion pretreatment can be summarized as follows (Chen and Liu 2007):

1. There are fewer parameters for the reaction conditions in the pretreatment process, and they are easy to control.
2. The three components hemicellulose, lignin, and cellulose can be separated in different forms, including a water-soluble component, an alkali-soluble fraction, and an alkali-insoluble component.
3. After the steam treatment, the structure of lignin is preserved and can be used for the conversion of other chemical products.
4. Hemicellulose exists in the form of xylose and xylo-oligosaccharide, which facilitates further utilization.

5. There is a wide range of applications for both lignocellulosic materials such as straws and wood and nonlignocellulosic feedstock.

2.2.3 Material Parameters Affecting Solid-State Fermentation

The raw materials for solid-state fermentation are rich and vary widely. Even for the same kind of raw material, there are still great differences because of the different sources and pretreatment conditions. This has important influence on the effect of solid-state fermentation. Therefore, in addition to the consideration of types of raw materials, the characteristic parameters of the raw materials should be of concern in solid-state fermentation experiments. Many material parameter characteristics affect solid fermentation, mainly including particle size, porosity, and homogeneity.

The particle size of the raw material is a critical factor. It is related to the specific surface area and the density of the material. For aerobic solid-state fermentation, microbial growth generally starts from the particle surface and gradually penetrates into the interior of the particle. The large specific surface area of the smaller material particles is conducive to the growth of microorganisms and for obtaining nutrients. However, particles that are too small may also have an adverse effect, especially for aerobic microorganisms. Too small particles cause a material that is too dense, which makes oxygen become the growth-limiting factor.

The porosity of the raw materials mainly affects mass transfer. The pores inside the matrix can be regarded in two parts, the pores between the particles and the pores in the particle interior. Pores between particles mainly affect gas diffusion, which is particularly important for aerobic microorganisms; the effects of the pores inside particles on microorganisms are more complex. For example, they affect whether the enzyme produced by microorganisms or added from outside can penetrate into the interior of the particle and work and whether the microorganisms are able to enter into the interior of the particle and grow. Because of complexity, the impact of porosity in the solid-state fermentation process has not allowed formation of a mature theory. In production practices, especially the traditional fermentation of wine, workers use their experience to control fermentation.

The uniformity of the solid material is also an important issue in the consideration for solid-state fermentation. Solid-state fermentation cannot create an almost-homogeneous matrix environment like liquid fermentation. In contrast, the solid fermentation matrix generally has a high degree of heterogeneity. Such heterogeneity partly comes from the raw material. Solid-state fermentation raw materials are mostly agricultural products after simple processing. The differences among the particles are often relatively significant because of the distinct origins of plant tissue and the inconsistent processing means. Moreover, there are stirring difficulties in solid fermentation, and some fermentation processes are even without stirring, which also makes differences of material piling, water distribution, ventilation, and so on. These differences not only occur in the raw material before fermentation but also are present in the entire fermentation process.

2.3 Aseptic Techniques and Inoculation Techniques for Large-Scale Solid-State Fermentation

Fermentation contamination means that some other microorganisms besides the productive microbes have invaded the fermentation medium during the fermentation process. Traditional solid-state fermentation, such as for spices and wine manufacture, silage, or the composting process, is not a pure culture process but a process of natural microbial enrichment. Even if inoculated by koji, the starter microbes are not obtained from pure cultures. Therefore, most of the traditional fermentation processes do not need to consider the contamination problem or the use of aseptic techniques. However, solid-state fermentation technology in the modern sense is a thoroughbred training process or mixed culture with limited species of microbes, such as for biopesticide, enzyme, and specific metabolite production. As these processes are required to operate under conditions without contamination, some effective measures must be taken to eliminate the bacterial infection.

Various precautions have been developed to avoid solid-state fermentation contamination problems, for example, the cooking process before fermentation to provide a sterile environment for the pure culture; other measures such as closed bioreactors; inhibition of the growth of contaminating bacteria by reducing the moisture content and pH of the medium; the use of formaldehyde, antibiotics, and other fungicides; or increasing the inoculation amount to strengthen the advantage of the producing bacteria population. However, all these means cannot eliminate contamination problems completely, and their application in fermentation is also affected by the restrictions of the product. In fact, the solid fermentation process is contaminated in many ways, and avoiding contact with infectious bacteria is the key to avoid contamination. In other words, the complete elimination of contamination requires a successful aseptic technique in the inoculation and transportation process. On a smaller scale, especially in laboratory studies, there is the ability to mix the seed solution (sometimes the spore powder) with the sterilized medium thoroughly and then transferring it to the fermentation reactor under sterile conditions. However, in the industrial production process, because of the large space and long time for production operation in the process of moist solid materials delivery, mixing, and loading, contact with all kinds of bacteria and especially fungal spores in the air is unavoidable. It is necessary to design specialized devices for solid-state fermentation. This section provides discussion of this issue.

2.3.1 Large-Scale Aseptic Techniques for Solid-State Fermentation

Solid-state fermentation is a technology with a long history, and in most cases, traditional solid-state fermentation does not consider aseptic operation. However, an essential aseptic operation should be considered in modern solid-state fermentation with large-scale pure culture. The contamination caused by the solid-state fermentation processes and equipment is one of the bottlenecks of modernization of this traditional technique.

Scholars have made useful attempts to achieve aseptic operation for solid-state fermentation. The Chinese Planning and Design Institute of Agricultural Ministry designed a device for inoculation of solid-state fermentation that solved the uniformity problem of inoculated materials in an industrial-scale fermentation reactor by mixing (Han et al. 2008). Cheng et al. (2006) used high-temperature sterilization, material cooling, inoculation, and fermentation in the same device, but this device was only suitable for small-scale fermentation (Fig. 2.22).

In fermentation production, the key to avoid contamination is aseptic operation, which involves a series of processes, including seed preparation, medium sterilization, inoculation, and medium loading. Therefore, to achieve aseptic operation in solid-state fermentation, a set of solid-state fermentation aseptic methods and the corresponding operating devices are essential. Based on these considerations, Prof. Chen Hongzhang from the Institute of Process Engineering, Chinese Academy of Sciences, proposed a fermentation method and its dedicated devices for aseptic solid-state fermentation (Chen and Li 2002).

The aseptic operation technique for solid-state fermentation is accomplished by some associated devices, including solid media sterilization, mixed inoculation of liquid seed and sterilized media, and delivery and transfer of solid media and sterile culture. The set of devices consists of a liquid seeding tank, a tapered solid medium sterilization tank, sterile inoculation cylinder, drum screen inoculator, tray incubator, honeycomb incubator, conveyor incubator, 100-level clean room, 100-level clean operation platform, and the solid-state fermentation reactor under sealed pressure (Fig. 2.23).

The procedure is as follows:

1. Seeds (prepared from the liquid fermenter) are forced into the tapered sterilization tank (the solid medium has been sterilized at 120 °C and cooled to 30–40 °C) by aseptic air.
2. To achieve sufficient mixing of seeding liquid and solid medium, the tapered solid medium sterilization tank is whirled three to six times; in the 100-level operating platform, the under cover of the tapered solid medium sterilization tank and the upper cover of the sterile inoculation cylinder are opened quickly, and the mixed materials in the sterilization tank are dropped into the cylinder vertically.
3. In the 100-level clean room, the under cover of the sterile inoculation cylinder is opened to transfer the medium to the tray incubator, honeycomb incubator, and conveyor incubator by the drum screen inoculator; after that, the materials are delivered to the solid-state fermentation reactor by mechanical traction.

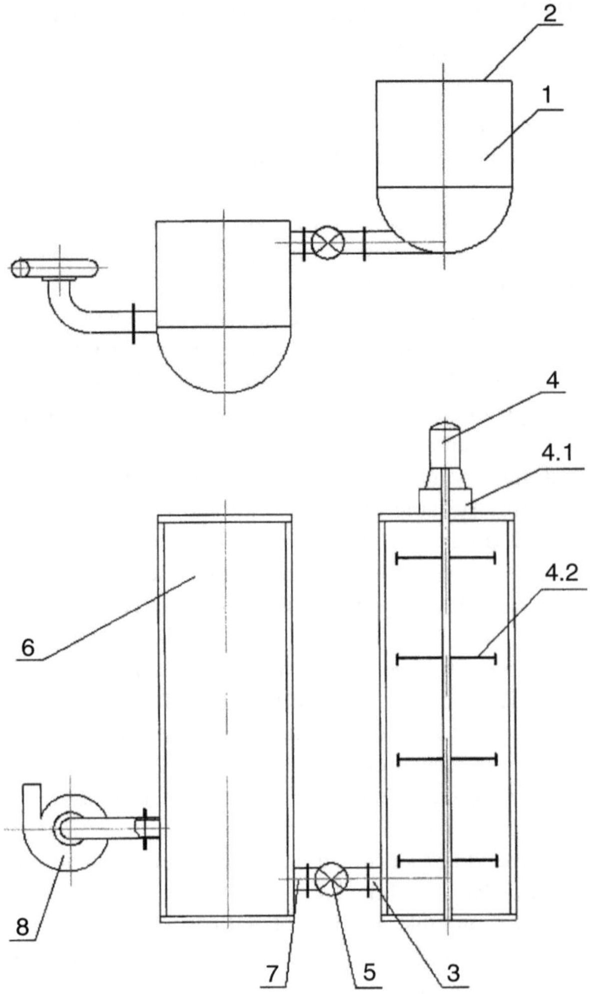

Fig. 2.22 Rapid inoculation device of fluid inoculum in horizontal solid-state fermentator (Chen and Liu 2007). *1* Container, *2* feed port, *3* discharge port, *4* mixing devices (*4.1* fixing, *4.2* movable parts), *5* valve, *6* storage, *7* feed port, *8* transfer pump

4. Based on the series of links from preparation, media sterilization, sterile inoculation, sterile loading, to sterile solid-state fermentation, the solid-state fermentation sterile operation device for large-scale production employs the conical solid medium sterilization tank for solid media sterilization and inoculation, which solves the problems that large-scale solid media are difficult to sterilize and inoculate without contamination. Meanwhile, the 100-level clean room is adopted in the material transfer process to avoid contamination; the sterile inoculation tank is placed vertically to reduce the adhesion of media to the wall, and mechanical traction to deliver incubators is used to reduce manual operation. In addition, the system uses a sealed pressure vessel for solid-state fermentation to avoid contamination in the fermentation process.

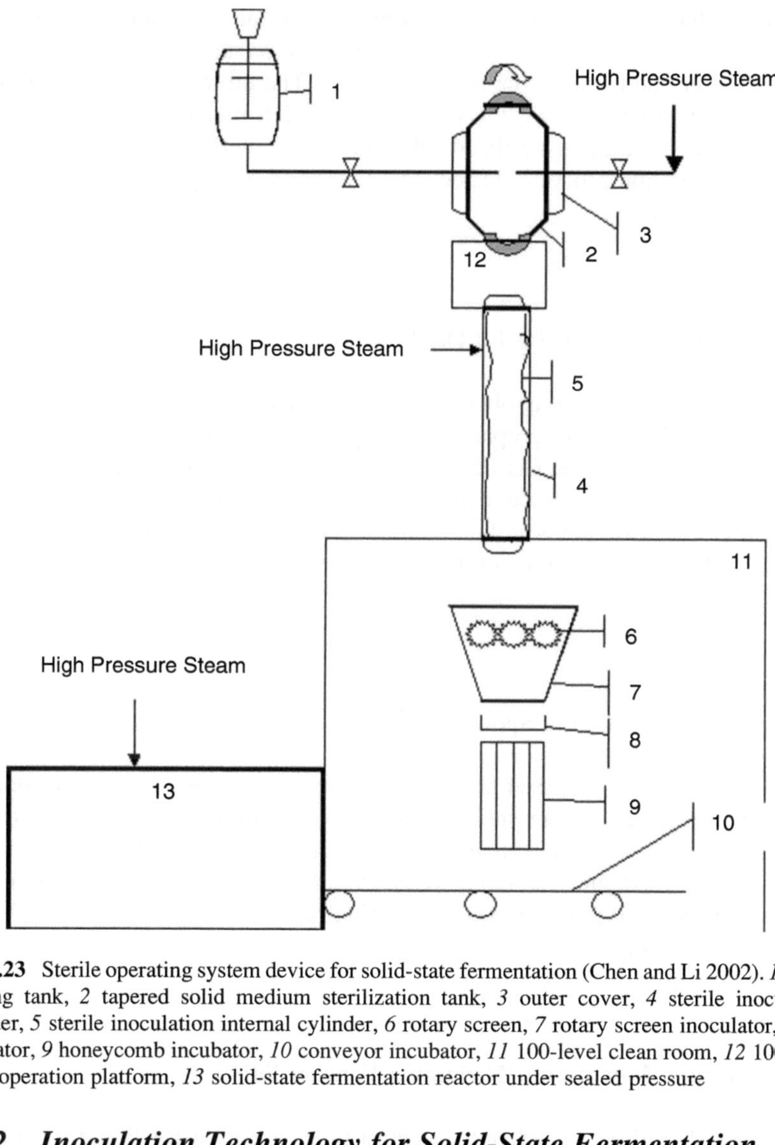

Fig. 2.23 Sterile operating system device for solid-state fermentation (Chen and Li 2002). *1* liquid seeding tank, *2* tapered solid medium sterilization tank, *3* outer cover, *4* sterile inoculation cylinder, *5* sterile inoculation internal cylinder, *6* rotary screen, *7* rotary screen inoculator, *8* tray incubator, *9* honeycomb incubator, *10* conveyor incubator, *11* 100-level clean room, *12* 100-level clean operation platform, *13* solid-state fermentation reactor under sealed pressure

2.3.2 Inoculation Technology for Solid-State Fermentation

2.3.2.1 Airflow Spores Inoculation Technology

According to the spread characteristics of fungal spores and the production practices of solid-state fermentation, I proposed the airflow spore inoculation technique. As the fungal spores are released by themselves or various external factors and then dispersed with the airflow under natural conditions, the air is the main medium for spread of the spores. In the solid-state fermentation process, the mobile phase is air. In the fermentation system, the ventilation process brings the necessary oxygen for microorganism respiration and takes away the heat, carbon dioxide, and other volatile metabolites. Therefore, the sterile ventilation process can also achieve airflow inoculation, which means transporting the dry spores into the large-scale sterilized solid-state fermentation reactor by the flow of air in the ventilation process. Besides the dry spores, the spore suspension can be inoculated by airflow.

Airflow Inoculation Processes for Dry Spores

A large number of spores is obtained by solid-state fermentation in strict sterile conditions (similar to the preparation of spores in penicillin production). After fermentation, the leavening is dried at a certain temperature, and then the dried spores in the dried inocula are dispersed into the solid medium by the dry air at a certain rate. The advantages of this technique are that the inocula can be prepared in quantity and at the same time a relatively high spore content can be gained in the inocula.

Atomization Inoculation of Spore Suspension

The spore suspension prepared from the obtained spores is atomized by an atomizer and then transferred to the solid medium by airflow delivery to complete the inoculation process. The advantages of this technique are that the inoculum does not need to be dried, and the moisture content of the medium will not experience a large loss because of the improved ventilation rate, but a special ultrasonic atomization device is needed.

2.3.2.2 Liquid Inoculation Through a Venturi Tube

Two issues should be considered if venturi tubes are used for liquid inoculation. First, the choice of the inoculum, the bacterial suspension, is optimal as a seed solution, and to reduce the viscosity and facilitate the extraction, the incubation time of the seed solution can be shortened. The next issue is the uniformity of inoculation. Artificial stirring of the solid medium should be avoided to achieve strict aseptic inoculation. Uniform inoculation can be accomplished by the flip effect of the high-speed gas stream from forced ventilation on the solid medium. However, the present method also has its limitations when the matrix layer is thick.

References

Chen HZ. Cellulose biological technology. Beijing: Chemical Industry Press; 2005.

Chen HZ, Li ZH. Asepsis fermentation method for solid state fermentation and its special device. Chinese Patent CN02126042.7; 2002.

Chen HZ, Liu LY. Principle and application of steam explosion technology. Beijing: Chemical Industry Press; 2007.

Chen HZ, Xu J. Principles and applications of modern solid state fermentation. Beijing: Chemical Industry Press; 2004.

Cheng YF, Cheng YF, Che XL. Hollow ball fermentator for sterilization and cooling inoculation. Chinese Patent CN200610150843.5; 2006.

Costerton JW, Cheng K, Geesey GG, Ladd TI, Nickel JC, Dasgupta M, et al. Bacterial biofilms in nature and disease. Annu Rev Microbiol. 1987;41:435–64.

Costerton JW, Lewandowski Z, Caldwell DE, Korber DR, Lappin-Scott HM. Microbial biofilms. Annu Rev Microbiol. 1995;49:711–45.

Desgranges C, Vergoignan C, Lereec A, Riba G, Durand A. Use of solid state fermentation to produce *Beauveria bassiana* for the biological control of European corn borer. Biotechnol Adv. 1993;11(3):577–87.

Dominguez M, Mejia A, Revah S, Barrios-Gonzalez J. Optimization of bagasse, nutrients and initial moisture ratios on the yield of penicillin in solid-state fermentation. World J Microbiol Biotechnol. 2001;17:751–6.

Donlan RM. Biofilms: microbial life on surfaces. Emerg Infect Dis. 2002;8(9):881.

Gelmi C, Pérez-Correa R, Agosin E. Modelling *Gibberella fujikuroi* growth and GA 3 production in solid-state fermentation. Process Biochem. 2002;37:1033–40.

Gervais P, Molin P. The role of water in solid-state fermentation. Biochem Eng J. 2003;13:85–101.

Gnansounou E, Dauriat A, Wyman C. Refining sweet sorghum to ethanol and sugar: economic trade-offs in the context of North China. Bioresour Technol. 2005;96(9):985–1002.

Gutierrez-Rojas M, Cordova J, Auria R, Revah S, Favela-Torres E. Citric acid and polyols production by *Aspergillus niger* at high glucose concentration in solid state fermentation on inert support. Biotechnol Lett. 1995;17:219–24.

Han J, Xiang X, Li X. Rapid preparation of fluid inocula and application for solid state fermentation. Chinese Patent CN200810111502.6; 2008.

Kolter R, Greenberg EP. Microbial sciences: the superficial life of microbes. Nature. 2006;441 (7091):300–2.

Kreth J, Zhang Y, Herzberg MC. Streptococcal antagonism in oral biofilms: *Streptococcus sanguinis* and *Streptococcus gordonii* interference with *Streptococcus mutans*. J Bacteriol. 2008;190:4632–40.

Li DJ. Studies on sustainable agro-ecology system of sweet sorghum. Sci Agric Sin. 2002;35:1021–4.

Li GX, Zhang FS. Solid waste composting and production of organic compound fertilizer. Beijing: Chemical Industry Press; 2000.

Loh W, Hubbard A. Encyclopedia of surface and colloid science. New York: Dekker; 2002.

Mitchell DA, von Meien OF, Krieger N, Dalsenter FDH. A review of recent developments in modeling of microbial growth kinetics and intraparticle phenomena in solid-state fermentation. Biochem Eng J. 2004;17(1):15–26.

Modenbach AA, Nokes SE. The use of high-solids loadings in biomass pretreatment-a review. Biotechnol Bioeng. 1956;109:1430–42.

Neufeld R, Zajic J, Gerson D. Cell surface measurements in hydrocarbon and carbohydrate fermentations. Appl Environ Microbiol. 1980;39(3):511–7.

Ooijkaas LP, Weber FJ, Buitelaar RM, Tramper J, Rinzema A. Defined media and inert supports: their potential as solid-state fermentation production systems. Trends Biotechnol. 2000;18:356–60.

Otsu N. A threshold selection method from gray-level histograms. Automatica. 1975;11:23–7.

Rahardjo YSP, Jolink F, Haemers S, Tramper J, Rinzema A. Significance of bed porosity, bran and specific surface area in solid-state cultivation of *Aspergillus oryzae*. Biomol Eng. 2005;22:133–9.

Rahardjo YSP, Tramper J, Rinzema A. Modeling conversion and transport phenomena in solid-state fermentation: a review and perspectives. Biotechnol Adv. 2006;24(2):161–79.

Shen P, Chen XD, Wei YB. Microbiology. Beijing: Higher Education Press; 2009.

Song JP, Chen HZ, Ma RY. Research on production of ethanol from sweet sorghum stalk by solid-state fermentation. Liquor Marking. 2007;34:81–3.

Sun Y, Cheng J. Hydrolysis of lignocellulosic materials for ethanol production: a review. Bioresour Technol. 2002;83:1–11.

Takahashi N. Acid-neutralizing activity during amino acid fermentation by *Porphyromonas gingivalis, Prevotella intermedia* and *Fusobacterium nucleatum*. Oral Microbiol Immunol. 2003;18:109–13.

Van Loosdrecht M, Lyklema J, Norde W, Zehnder A. Influence of interfaces on microbial activity. Microbiol Rev. 1990;54:75–87.

Wakelin SA, Anand RR, Reith F, Gregg AL, Noble RRP, Goldfarb KC, et al. Bacterial communities associated with a mineral weathering profile at a sulphidic mine tailings dump in arid Western Australia. FEMS Microbiol Ecol. 2012;79:298–311.

Weber FJ, Tramper J, Rinzema A. A simplified material and energy balance approach for process development and scale-up of *Coniothyrium minitans* conidia production by solid-state cultivation in a packed-bed reactor. Biotechnol Bioeng. 1999;65:447–58.

Xin LJ, Li MC. Ordinary mycology. Beijing: Higher Education Press; 1999.

Xin BC, Xu YL, Li YL, Liu TJ, Yang DQ. Communication and cooperation of different microorganisms within biofilms. Scientia Sincia Vitae. 2010;40:1002–13.

Yang SH. Chemistry of plant fiber. Beijing: China Light Industry Press; 2001.

Yu HY, Ding WX, Luo JF, Donnison A, Zhang JB. Long-term effect of compost and inorganic fertilizer on activities of carbon-cycle enzymes in aggregates of an intensively cultivated sandy loam. Soil Use Manag. 2012;28:347–60.

Zhu BY, Zhao ZG. Chemical basis of interface. Beijing: Chemical Industry Press; 1996.

Chapter 3
Principles of Solid-State Fermentation Engineering and Its Scale-Up

Abstract Industrial solid-state fermentation (SSF) is not widely used because of engineering difficulties and the lack of guided principles on the fermentation process and scale-up from the physics aspect and porous medium. From the nature of the biological processes, SSF can be featured as the continuous phase of the gas phase compared with the continuous phase of the liquid phase in submerged fermentation. It is important to recognize traditional SSF from the aspect of the gas-liquid-solid phase. In SSF, mass and heat transfer are crucial for understanding and applying this old technology. This chapter introduces the essence of SSF and its related influencing factors from the engineering aspect. It includes the essence of SSF, transfer principles, thermal physics phenomenon, and design and scale-up of bioreactors. The hope is to find novel means to solve the problems of mass and heat transfer in SSF and eventually achieve its industrialization.

Keywords Heat transfer • Mass transfer • Porous media • Fermentation process • Scale-up

3.1 The Essence of Solid-State Fermentation

Recognition of traditional solid-state fermentation (SSF) began from the solid matrix, which is not only carbon and energy sources for microbial growth and metabolism but also supplies a microbial growth microenvironment. It is difficult to reflect the scientific connotation in the definition of SSF. From the nature of the biological processes, SSF is featured as the continuous phase of the gas phase compared with the continuous phase of the liquid phase in submerged fermentation. This definition explains the characteristics of SSF and submerged fermentation essentially based on gas or liquid phase as the continuous phase.

H. Chen, *Modern Solid State Fermentation: Theory and Practice*,
DOI 10.1007/978-94-007-6043-1_3,

3.1.1 *The Relationship Between the Three-Phase Ratio in Fermentation and Microbial Physiology*

A porous matrix is the solid part in SSF; is always a porous structure with large surface area, which could reach 10^3–10^6 m^2/cm^3 (Chen and Xu 2004); and is the substrate for holding or transferring water, gas, solute, and heat. Broadly, a porous matrix not only means the solid part of the system but also includes the liquid and gas phases in a porous medium. Thus, the solid, liquid, and gas phases combined are the entire SSF system. Water and gas fill the pores in the matrix, gas occupies the macropores, and water occupies the micropores. The changing porosity ratios, especially ratios of macropores and micropores (i.e., constituent ratio of three phases in matrix affected by its characteristics and structure), are another important factor in SSF besides surface area and chemical factors.

3.1.1.1 Interpretation for Properties of a Porous Matrix

Analysis of Solid-Liquid-Gas Phases in a Porous Matrix

Solid Phase

To achieve industrialization of SSF, one of the problems solved first is adequate cognizance of the solid carrier. Two types of SSF systems can be distinguished depending on the nature of the solid phase used. The first, also the most commonly used (and most often described), system involves cultivation on a natural material; this system is referred to as *cultivation on natural substrates*. The second system, which is not as frequently used, involves cultivation on an inert support impregnated with a liquid medium. As a nutrient carrier, its chemical and biological characteristics on SSF mainly need to be considered; the influencing factors contain carbon and nitrogen sources, pH, microelements, and so on. As a support carrier, the role of some physical factors in SSF need to be considered, including material, structural parameters such as specific surface area, porosity, distribution of pores, water-holding ability, ventilation ability, and matrix size.

Because nutrients in the matrix can be regulated by adding other nutrient sources, the effects of chemical properties and biological characteristics of the matrix are less important in SSF than its physical characteristics. These physical properties are all related to the porosity of the matrix; thus, study from the aspect of the porous characteristics of the matrix is an important breakthrough for SSF industrialization.

Larger total porosity means more water and air pores in the matrix. Generally, pores with equivalent diameter less than 0.002 mm are called inactive pores; these pores are affected by the bounded water film between particles, which cannot afford to capillary effect. Pores with an equivalent diameter of 0.2–0.002 mm are called capillary porosity pores; these pores can maintain water by capillary forces. Ventilation pores are pores larger than 0.02 mm.

The total porosity of the matrix can only reflect the sum of water and air space accommodated by the matrix; it cannot reflect their respective water and air space. Large pores in the matrix refer to the space occupied by air, also known as ventilation pores; small pores reflect space occupied by water in the matrix, called water-holding pores. The ratio of aeration porosity to water-holding porosity R is an important parameter.

The value of R can reflect matrix water and gas conditions. A larger R means less water-holding capacity and a bigger ventilation characteristic, indicating the matrix is loose, with lack of water retention and excess ventilation, and must increase water supply in the entire SSF. On the other hand, if R is small (has a small air volume and a large water-holding capacity, i.e., inadequate ventilation and excessive water retention), it could easily lead to water storage within the matrix.

Liquid Phase

In the liquid phase, water contained in the matrix, called matrix water, can be divided into four kinds: absorption water, wilting water, capillary water, and gravity water. Among those, the first three occupy the water-holding pore, and gravity water fills the aeration pore.

Dried matrix could absorb water vapor molecules in the air and attach water on the surface; this water is called absorption water. It relates to the relative humidity of air; when the relative humidity is close to saturation, the absorption water in the matrix reaches its maximum. Absorption water cannot transfer pressure, so it is unusable to microorganisms. The maximum absorption water content correlates with matrix characteristics and quantity, temperature, humidity, organic matter content, and other factors. The maximum absorption water content is about 1.25–2.00 times that of wilting water. Wilting water is the least-effective water for microorganisms and forms the smallest continuous liquid film adsorbed on the outer of substrate.

Water that remains in the matrix capillary pore and relies on capillary forces is called capillary water. It is not dominated by gravity; it is the main source of water required for microbes and is the transporter of solvent and nutrients during fermentation.

Gravity water is not maintained by the matrix but flows downward by gravity. It exists in large pores (aeration pores) temporarily and relates to nutrient leaching in the matrix. Excessive water often causes insufficient air and waterlogging; it is harmful to growth of microorganisms (excess water).

Gas Phase

The gas phase matrix content, affected by aeration porosity, is associated with fermentation ventilation.

The matrix gas composition is similar to but not the same as the air above the substrate.

1. Matrix gas contains less O_2 but more CO_2 than the air above.
2. There is higher water vapor content in matrix air.

3. Matrix air contains a small amount of reducing gas.
4. There are unstable compositions in matrix air.
5. Matrix air can be divided into free-state gas, adsorbed gas, and dissolved gas by its physical properties.

Constituent Ratio of Three Matrix Phases

The constituent ratio of the three matrix phases is always expressed by a weight or volume proportion; because of the low density of gas, a volume ratio is generally used. The three-phase composition can be calculated according to the discussion that follows.

The total porosity of the matrix *f* is the pore volume percentage of the dry unit volume of substrate, including the gas and water volume in the matrix, calculated as follows:

$$f = \frac{V_t - V_s}{V_t} = \frac{\frac{m_s}{\rho_b} - \frac{m_s}{\rho_s}}{\frac{m_s}{\rho_b}} = 1 - \frac{\rho_b}{\rho_s} \tag{3.1}$$

V_t = total volume of matrix;
V_s = volume of matrix solid phase;
m_s = mass of matrix solid phase;
ρ_s = density of matrix solid phase;
ρ_b = bulk density of matrix solid phase;
f_a = aerate porosity, which refers to the relative air content of the matrix and is an important indicator characterized by aeration.

For a nonswelling matrix,

$$f_a = f - \theta \tag{3.2}$$

where θ is the volumetric water content:

$$\theta = \theta_m \frac{\rho_b}{\rho_w} \tag{3.3}$$

θ_m = water content calculated by mass ratio;
ρ_w = density of water at room temperature.

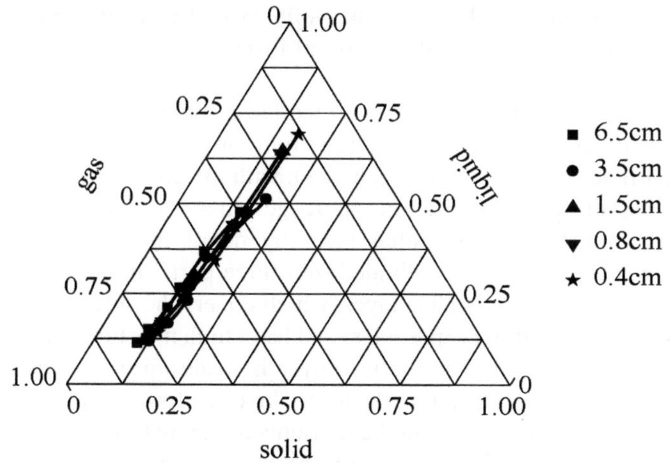

Fig. 3.1 Three-phase proportion of solid matrixes with different particle size and water content

Table 3.1 Variation coefficient (VC) of three-phase volume

	Solid phase	Liquid phase	Gas phase
Mean level (%)	14.638	31.172	56.40
Standard deviation	2.292	17.532	18.371
Variation coefficient	0.157	0.562	0.326

The volume ratios of liquid, gas, and solid in an SSF system are θ, f_a, and $\frac{\rho_b}{\rho_s}$, respectively (Shao et al. 2006; Mitchell et al. 2006).

Preliminary studies of the three-phase composition of different lengths and different water contents of steam-exploded straw (Duan and Chen 2012) are shown in Fig. 3.1. Using different matrixes in SSF, there was a smaller range of the solid phase (0.104–0.204) but a larger range of the liquid phase (0.112–0.739) and the gas phase (0.144–0.784).

Further investigation of the effect of straw size and moisture content variation on the three-phase composition (i.e., the three-phase variability) was quantified by a variation coefficient (VC), calculated as follows:

$$VC = \frac{\sigma}{\bar{x}} \quad (3.4)$$

where σ is the standard deviation of the phase ratio, and $\bar{x}$ is the average value for the phase content. VC analysis of each phase (Table 3.1) showed that liquid phase variability played a dominant role in the structural changes, followed by the gas phase. Solid phase volume change was very small for the same weight but different particle lengths.

3.1.1.2 The Relationship Between the Solid-State Fermentation Three-Phase Ratio and Microbial Physiology

In general, bacteria and filamentous fungi tend to grow in SSF. Bacteria generally grow on the surface of a solid substrate; the fungi mycelia generally grow into the surface of particles (aerobic microorganisms), with the growth process exemplified by the process of spore inoculation, spore germination, hyphae extension, and formation of branches and mycelial layer by dense mycelia. For filamentous fungi, the mycelial layer can be divided into three parts: biofilm, aerial mycelia, and substrate mycelia. The hyphae layer, which covers the surface of the particles (or a shallow surface within the particles) and has a high moisture content, is always considered a biofilm layer. The biofilm layer is made up of microbial cells and water; most microbial mycelia enrich in this layer to form a close hyphae layer; outside oxygen enters this area; and the nutrients of the material inside the particles also penetrate into this region. Therefore, the region is often rich in nutrients and oxygen; it is the main place for mass transfer and exchange. Few hyphae penetrate into the interior of the substrate; it is difficult for aerobic microorganisms to grow because of hypoxia inside the particles, and generally oxygen cannot be measured at a depth of 100 μm. There is a region of the aerial mycelia that could insert into the gas phase of the particles. However, even if the mycelia fill throughout the gap space, they would not completely occupy the gap tightly because the mycelia are loose, occupying up to about 34 %.

Microcosmically, the microbial cell concentration varies in the different parts of the substrate; the substrate and the product concentration gradient exist in the SSF. Mass transfer limitations often are the main factor limiting microbial growth and product formation. Microorganisms in SSF are substantially in a stationary state; without stirring, there is almost no material convection because of the lower water content. Macromolecules (such as polysaccharides, proteins) cannot be dissolved in water, and the nutrients, products, microorganisms, and enzymes cannot transfer easily; this causes mass transfer difficulties. The mycelial growth process is actually a tendency to seek nutrients because the nutrients are exhausted in the vicinity. With penetration into the matrix, mycelia extend into the substrate and secrete enzymes, and macromolecular substance near the enzymes is decomposed into small molecules. The small molecules are then dissolved in water and used by microbes.

3.1.2 Characteristics and Roles of the Matrix Gas Phase in Solid-State Fermentation

The essential difference between SSF and submerge fermentation (SMF) is the continuous phase of the gas phase in the system, which explains the importance of the gas phase in SSF research. The following discussion analyzes the characteristics and roles of the SSF matrix gas phase.

The SSF gas phase mainly includes ventilated air in the reactor, the carbon dioxide produced during the fermentation process, the volatile gases, and water vapor. In the SSF bioreactor, gas is mainly divided into three parts: gas in the headspace on the upper part of the matrix layer; gas in the surface of the material particles, including the surface of the biofilm; and gas in the pores of the internal particles. The transfer of the gas phase is as follows: At the top of the matrix layer, the aerial hyphae consume oxygen and release carbon dioxide. Oxygen and carbon dioxide pass through the liquid membrane at the surface of the particles. Oxygen and carbon dioxide diffuse in the material particles. Mycelia, which are immersed in a liquid environment, absorb oxygen and release carbon dioxide. Finally, oxygen passes to the microbial cells through a series of processes.

Aeration occurs for heat and mass transfer of the gas phase between the interior and the exterior of the bioreactor. There are two ways this happens; one is forced ventilation, which forces pressurized air through the main body of the material layer, and the main means for heat and mass transfer is convection. Another is nonforced ventilation; that is, air is in a natural flow state, and heat and mass transfer are primarily achieved through diffusion. For example, in tray fermentation, gas contacts the solid material through natural diffusion; however, in a horizontal drum reactor, air enters from one end of the reactor, and when the drum is moving (in a variety of ways), a portion of the gas contacts part of the material to accomplish the exchange of gas.

When the air enters a layer of material, the distribution of air depends on the size of the particles and bulk density of the material. If the particles are uniform in size and stacked, the pressure drop in each horizontal plane and the airflow rate are the same. As the distribution of gas through the material layer becomes more uniform, the possibility increases for providing essential oxygen for microbial fermentation, strengthening heat and mass transfer. The gas flow resistance varies in different parts of the material layer; most of the airflow is always a priority through the path of least resistance. With the enhanced gas flow, the material layer is easy to crack, resulting in a material layer gap, which often leads to an air short circuit, reduced air utilization, inadequate oxygen in most areas, and metabolic heat. The phenomenon also causes mycelial agglomeration and substrate shrinkage because of dehydration.

3.1.3 The Roles of the Evapotranspiration Process in Solid-State Fermentation

3.1.3.1 Moisture-Related Parameters

In the solid-state cultivation process, there is almost no free water. However, the water content is still one of its main factors: As the solvent, the nutrients must be dissolved in water before they are used by microbes; as the medium of high thermal entropy, water can play a role in regulating the fermented material temperature.

In solid-state cultivation, the main factors causing changes in the moisture content of the internal fermentation system are the following: intake of water by cell growth; water produced or consumed in metabolic processes; water used by hydrolysis of starch or lignocellulose; and water brought into the medium by humid air.

The moisture content of SSF generally refers to the water content of wet or dry material.

The water content of the wet material (kg/kg) is calculated for the wet material as follows:

$$W = \frac{\text{water weight}}{\text{dry material weight} + \text{water weight}} \tag{3.5}$$

The water content of the dry material (kg/kg) is calculated for the dry material as follows:

$$W = \frac{\text{water weight}}{\text{dry material weight}} \tag{3.6}$$

In some research processes, the water activity parameter is generally used rather than the moisture content for better SSF process control. Water activity is more significant than the moisture content because it reflects material water affinity and indicates the amount of available water in SSF. The growth of microorganisms on the solid matrix depends on the water activity; the driving force for evaporation of water from solid material is the deviation between the water activity of the solid material and saturated water activity. The water activity α_w is defined as

$$\alpha_w = \frac{f}{f_0} \tag{3.7}$$

In the formula,

f = fugacity of solvent (fugacity means the trend of the solvent escaping from the solution);
f_0 = fugacity of pure solvent.

For pure water, $\alpha_w = 1$; for completely anhydrous solvent, $\alpha_w = 0$.

The water activity is often close to 1 in material that is more than 0.5 kg water/kg dry matrix. There are some relationships between water activity and moisture content, but they are not in proportion. They are also related to temperature and the nature of the materials. With temperature as an example, for the same material with the same water content, the higher the temperature is, the greater the water activity will be. This can be explained from the definition of water activity. As the temperature increases, the water increasingly escapes. For different types of materials with the same moisture content, the water activity is not necessarily the same; for example, the water activity of material not inoculated and of fermented substrate will vary greatly. Different solute concentrations also result in different water activities; for example, a high glucose concentration will lead to a serious decline in water activity.

During fermentation, the water activity of the medium is dynamic. The solid matrix dehydration and soluble solute accumulation on the solid matrix (such as glucose and amino acids and other low molecular weight hydrolyzates) will reduce the solid matrix water activity. Most water activity research focuses on the preservative effect in food microbiology. So, further research is needed on the effect of water activity on microbial growth and the form of the biological macromolecules (such as enzymes) present, particularly in the active center of the enzyme molecules and mode of action; the relationship between water activity and moisture content in the medium; and the impact of culture conditions (such as temperature, humidity, pressure, amount of ventilation, different media ingredients, etc.).

3.1.3.2 Water Evaporation and Evaporation Heat Removal

The Rate of Water Evaporation and Calculation of the Heat Transfer Rate

In the SSF process, microbiological growth produces a lot of heat. The maximum temperature gradient even could reach 3 °C/cm in disk fermentation. In SSF, evaporative cooling is the most important measure to decrease temperature and remove heat. However, evaporative cooling leads to loss of moisture; to ensure the normal growth of the microorganism, supplemental water needs to be added in a continuous mix of materials. That means monitoring and controlling the solid substrate moisture content are important for the evapotranspiration process. Although the total moisture content of the culture can be measured, the online detection and control of water are difficult.

In nonforced ventilation SSF, evaporative cooling is the main means of media cooling; the amount of water evaporation relates to the bioheat in the fermentation. When calculating moisture evaporation, the material layer is often considered a two-phase system.

The evaporation speed of water R_e is calculated according to the changes in the water activity of the solid material. The evaporation speed of water is proportional to the difference between the actual water activity α_w and water activity α_w* under the conditions for which the material reaches equilibrium with the gas phase, proportional to the contact area A of the solid and vapor phase and to the mass transfer coefficient k_w of water vapor,

$$R_e = k_w A(a_{ws} - a_{ws}^*) \quad (3.8)$$

In the formula,

R_e = evaporation speed of water, kg water/h;
A = contact area of solid and vapor phase, m^2;

k_w = water vapor mass transfer coefficient, kg/(m^{2}h);
a_{ws} = actual solid substrate water activity;
a^*_{ws} = water activity under the conditions at which the material reaches equilibrium with the gas phase.

Because A is difficult to determine, researchers often regarded k_wA as the overall transfer coefficient of the product of the mass transfer coefficient and the contact area.

In forced ventilation SSF, temperature reduction depends mainly on two kinds of mechanisms: the change in sensible heat of air intake and exhaust to achieve the cooling (i.e., when the temperature of the inlet air is less than the temperature of the material, the exhausted gas takes heat away) and the latent heat of evaporation (also known as the latent heat during the phase change), which is the main mechanism of reducing substrate temperature in SSF. The relative humidity of the ventilation air is less, the moisture content of the material is higher, the relative humidity of the particle gap gas phase is almost 100 %, the water vapor in the material pores comes into the air together with the air to form the gas phase of the fermentation substrate, and the water in the material is continuously vaporized and migrates with the migration of the air. Evaporation of water requires a large amount of heat to be used as the latent heat of vaporization or heat of evaporation. The water evaporation speed depends on a variety of factors, such as temperature, humidity, surface area of the liquid, airflow on the surface of the liquid, and so on. Elevating temperatures will increase the evaporation rate; increasing the evaporation surface area of the liquid can also increase the evaporation rate. SSF medium is almost particulate. The specific surface area is generally large, and coupled with forced ventilation, the evaporation speed is greater; therefore, the consumed heat of vaporization is large, and the cooling rate is correspondingly increased.

The material layer is considered a two-phase system: the rate of evaporation and heat transfer of the water as it evaporates from the solid phase into the gas phase Q_v:

$$Q_v = \lambda k_w A(a_w - a^*_w) \tag{3.9}$$

In the formula,

Q_v = the rate of evaporation-heat transfer, J/h;
λ = entropy of the evaporation of water (i.e., evaporation heat), J/kg water.
Other symbols are the same as the formula in Eqs. 3.7 and 3.8.

Moisture Loss and Heat Removal in Forced Ventilation Solid-State Fermentation

The material layer is regarded as uniform; material layer moisture loss could occur when forced airflow goes through the material layer because of humidity differences between import and export air. The speed of the total evaporative loss of water could be calculated as

$$R_{con} = G_{air}A_a(H_{out} - H_{in}) \tag{3.10}$$

In the formula,

G_{air} = quality of air getting through materials per unit time and per unit cross-sectional area, kg dry air/(m^2 h);
A_a = cross-sectional areas of the material layer, m^2;
H = air humidity, kg water/kg dry air.

If we regard the particles and air in the pores of particles as two phases, the humidity difference dH/dz exists in different sites of the material layer. Thus, the formula mentioned could be modified as

$$R_{con} = G_{air}A_a\frac{dH}{dz}\Delta z \tag{3.11}$$

When the materials are mixed completely and in forced ventilation, the heat of evaporation is calculated as follows:

$$Q_v = \lambda G_a A_a(H_{out} - H_{in}) \tag{3.12}$$

In the formula,

H_{out} = air humidity outlet, kg water/kg dry air;
H_{in} = air humidity inlet, kg water/kg dry air;
A_a = cross-sectional areas of the bioreactor, m^2;
G_a = fluxes of dry airflows through the material layer, kg/(m^2s).

3.2 Solid-State Fermentation Transfer Principle

The porous medium (mainly the pore structure and physical and chemical properties of the porous medium), the fluid (mainly the components of fluid and its physical and chemical characteristics), and the flow status (mainly the environment and conditions of flow and the interaction between the fluid and solid) determine fluid flow regulation through porous media. The SSF system is a three-phase system consisting of porous media and can be seen as a typical porous media system; thus, it can be analyzed by a corresponding theory of porous media. Study of the SSF transfer process can refer to the porous media process, which involves a multidisciplinary theory: flow porous medium theory, capillary theory, diffusion theory, fluid mechanics, heat and mass transfer, thermodynamics, and more.

3.2.1 Introduction to Heat and Mass Transfer in Solid-State Fermentation

Bioheat is generated by the growth and metabolism of microorganisms and results in the rise of material temperature; to reduce the material temperature and supply enough oxygen, ventilation with air of a certain temperature and humidity is needed. The air enters the layer of material (for a packed bed reactor) or the headspace layer (for a drum reactor). The airflow entering the material layer is a mixture of air and water vapor. Ventilation could decrease the material temperature: The air carries the water vapor and its heat of vaporization is directed away from the material layer, enters the headspace layer, and then discharges through the exhaust pipe. Generally, the temperature and moisture content of air discharged from the reactor are higher than for the air entering the reactor. The parameters most closely related to heat exchange are the gas flow rate, temperature, and moisture content of air in and out of the reactor.

3.2.1.1 Heat Transfer Process in Porous Media

Analysis of the heat transfer process in porous media shows that the process includes the heat conduction process of solid skeleton contact with the fluid in the particle gaps; convective heat transfer of fluid in the gap (it could be forced convection, natural convection, or mixed convection of both and comprises the liquid boiling, evaporation, and condensation of steam); and the radiation heat transfer between solid skeleton and the gases. Many experimental studies and theoretical analysis results showed that, for a particle not more than 4–6 mm in diameter and with $G_rP_r < 10^3$, the contribution of convective heat transfer between the fluid is negligible, and the radiation heat transfer contribution is obvious, only when the great temperature difference between solid particles is large, and the pores are vacuum or occupied by gas. Thus, for a SSF system, evaporative cooling, convection are more important ways of heat exchange (GutierrezRojas et al. 1996).

3.2.1.2 Mass Transfer Processes in Porous Media

The mass transfer processes in porous media include the following two aspects (Shao et al. 2006):

1. Molecular diffusion. This is caused by the random motion of the fluid molecules or solid microscopic particles. It corresponds with the heat conduction mechanism in heat transfer.
2. Convective mass transfer. This is caused by the macroscopic motion of fluid; it corresponds with convective heat transfer. Briefly, it includes both the mass

transfer between the fluid and the solid skeleton wall and the convective mass transfer between two immiscible fluids (including gas-liquid phase). Single-phase fluid convective mass transfer is divided into laminar and turbulent flow according to different fluid states. The gas-liquid two-phase flow (i.e., nonsaturated flow in porous media) has many different forms of convective mass transfer. Obviously, the macroscopic motion of the fluid in the gap is caused by capillary force, pressure, gravity, and so on.

It must be pointed out that there is mutual influence and a coupling effect among the transfer process of momentum, energy, and mass in porous media.

Some scholars (Martynenko and Pavlyukevich 1998) recently summarized their research work in porous media and suggested that studies of heat and mass transfer in porous media should focus on the following aspects:

1. Combine the macro and micro aspects of research; use theoretical analysis, experimental research, and numerical simulation to establish and improve the micro and macro models of porous media.
2. Develop measurement principles and methods, especially measurement technologies for heat and moisture transfer characteristics in porous media; enrich and improve the basic database of porous media; explore measurement methods for permeability, porosity, capillary force, surface tension, and contact angle tests.
3. Strengthen basic research concerning heat and mass transfer in porous media at the background of engineering applications, which also become the main research directions of heat and mass transfer in porous media.

3.2.2 Theoretical Basis of Heat, Moisture, and Solute Transfer Processes

The SSF system is one of the most typical porous media systems. It contains a solid skeleton, an aqueous solution, and a gas phase and involves moisture and heat transfer.

Heat and mass transfer properties of media have a great deal to do with media moisture content and energy types of water in the SSF system. When studying matrix heat and mass transfer, the substrate is divided into three categories based on the amount of moisture in the substrate: the water content is 100 % (i.e., the pore space is filled with the liquid water [water saturated]); moisture in liquid and gaseous forms in nonsaturated matrix; and only moisture vapor in the pore space of the dry saturated matrix (this state is clearly not suitable for the growth of microorganisms and is excluded in the subsequent discussion of this topic). In most cases, the substrate is an unsaturated multiphase system that contains solid particles, water, steam, and air.

In this section, the transfer process of heat, moisture, and solute and its influencing factors are discussed.

3.2.2.1 Heat and Water Transfer Mechanism in Solid-State Fermentation

Water Vapor Transfer Mechanism

Water Transfer Mechanism

In the SSF process, water presents in the form of water vapor in the gas phase; liquid free water (water film or droplets on the surface of the material particles, capillary water, water in the particle gaps); and bound water of materials.

The water vapor in the gas phase includes water vapor in the air phase of the headspace layer and gaps between particles.

Most of the moisture in the material is present in the form of bound water. The solid matrix is commonly porous; these pores can be regarded as capillary tubes. The different curvatures of the liquid level in and out of the pores lead to different vapor pressures. For a concave liquid surface, the equilibrium vapor pressure is lower than the normal vapor pressure of the liquid. Therefore, water condensation can occur in the capillary (capillary condensation) when the vapor pressure of the system is lower than the normal saturated vapor pressure. This explains the normal growth capability of microorganisms in a solid matrix with a low moisture content; there is still some residual moisture in the capillary, so the water activity is still relatively high.

Free water can be used by microbes; however, bound water cannot. The combined water can be divided into (1) chemically combined water (i.e., crystallization water of the compound as well as moisture linked by hydrogen of certain compounds); (2) physically bound water (i.e., the moisture absorbed on the outer layer of material particles); and (3) solution moisture, which comprises the water solvent of the liquid phase, the solution in the biological cells, and solution discharged after cell rupture or that permeates extracellularly.

The bound water is closely combined with the dissolved macromolecules to form a hydrate. The bound water includes all of the water molecules near the hydrated layer as well as water molecules close to the outside of the hydration layer. The hygroscopic water is monomolecular layer water covering a nonwater component, and it is combined closely to some specific hydrophilic group of the nonaqueous components. The film water is the hydrogen-bonding water combined with the solute; its water activity is greatly reduced. The combined water is closely combined with these objects; therefore, there is a great difference between its nature and the nature of free water. Bound water does not have the nature of a solvent; there is difficulty in crystallization, evaporation, boiling, and freezing even at −40 °C, and it and is not readily squeezed out under atmospheric pressure. The binding force between the water and the object is very strong; the water integrated with the macromolecules relies on the ionic bonding and hydrophilic polar group. This combination reduces the vapor pressure and chemical potential of water.

There is also another type of bound water combined with biological macromolecules; it is the composition of the nonwater component. The content of this type of water is low; it is deep inside the folded macromolecules and does not substantially participate in chemical reactions.

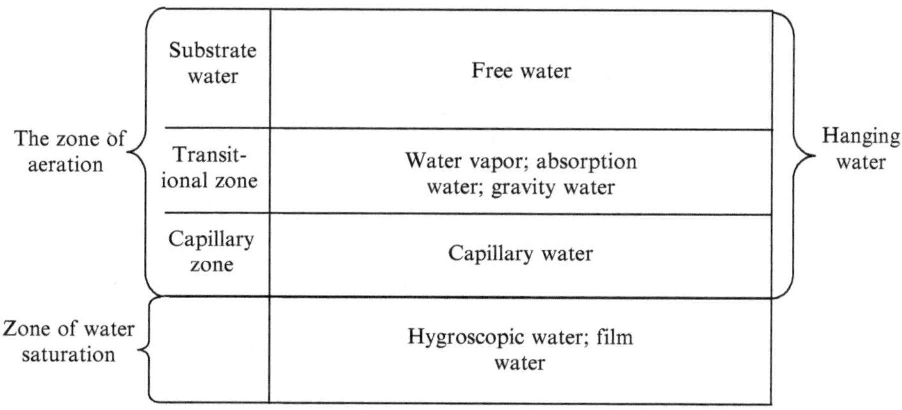

Fig. 3.2 Vertical distribution maps for matrix water in solid-state fermentation

From the perspective of water used by microorganisms, the moisture includes intracellular and extracellular water (i.e., water outside all the mycelia, including the moisture in the material).

Figure 3.2 shows a diagram of the vertical distribution of the substrate water. From the previous analysis, the film water, capillary water, and gravity water behave according to migration characteristics. The moisture movement is driven not only by capillary force but also by a variety of forces, such as the Darcy resistance of solid particles to liquid, the interaction between the gas and the liquid, the inertial force of liquid film movement, the liquid gravity, and so on. Figure 3.2 also provides the status and characteristics of the fermentation substrate and the water in the SSF system.

There is a variety of ways of moisture transfer in solid materials, including molecular diffusion and convection. In the pretreatment of raw materials (such as soaking, cooking), materials have absorbed enough water. However, in the fermentation process, because of water consumed by microbial growth and metabolism and water evaporation, the partial material moisture content decreases and forms amoisture concentration gradient in different parts of the material, from the macroscopic point of view. From the microscopic point of view, the moisture transfer exists in internal and external particles as well as inside and outside microbial cells.

For moisture transfer inside the material particles, the moisture contents of the surface and the interior of the material particles are different; the diffusion of water occurs automatically in the material particles because of the presence of a moisture concentration gradient. Limited by the oxygen supply, aerobic microorganisms are mainly grown on the outer surface of the material particles. Evaporation of moisture and water intake of new cells also occurs on the surface of the particles, so the moisture content of the surface of the particles is relatively lower; the water could reach the surface of the material from internal particles by diffusion. In addition, microbial hyphae always utilize nearby sugar; this will cause a solute concentration gradient in different parts of the material particles (Nagel et al. 2002). The moisture concentration gradient caused by the solute concentration gradient further generates water diffusion.

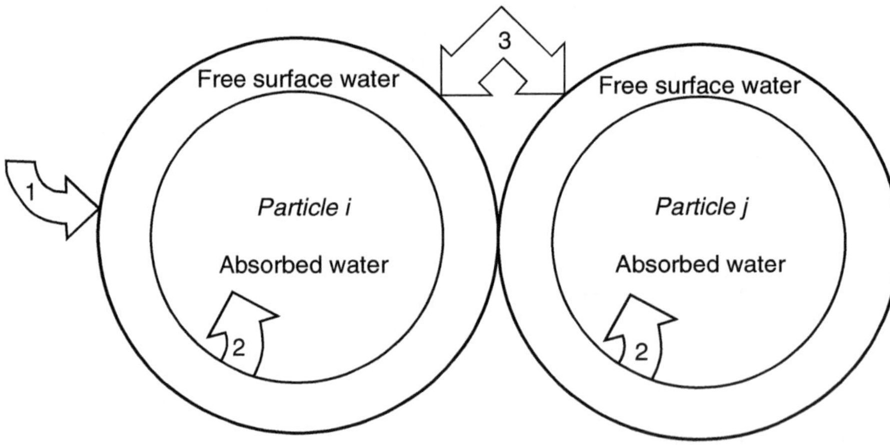

Fig. 3.3 Scheme of the moisture distribution model for two particles (*i* and *j*) (Schutyser et al. 2003)

The water flowing between the material particles is mainly free water. Assuming that the moisture absorbed on particles would not be transmitted to the other adjacent particles, the transfer of moisture between particles is limited to the free water on the surface of the particles. To model the water distribution of the sprayed water throughout the mixed substrate bed, Schutyser et al. (2003) distinguished three different processes: (1) external addition of water to the particles; (2) absorption of water by individual particles; and (3) transfer of water between neighboring particles (Fig. 3.3).

Schutyser et al. (2003) considered the transfer of water between particles can only take place via the fraction of water that is freely present on the surface of particles, which for convenience is called the free surface water volume. The precise location of the water on the surface is not considered, but it may be envisaged as a film on the surface of the particles. Experimental data indicated that the absorption process could be divided into two cases (a and b). If the amount of total water was below a critical value, the water present at the surface was absorbed instantaneously into the particle (case a). If this critical volume was exceeded, absorption took place at a constant rate (case b).

Particle *i* is located at the surface of the bed. Two varying water fractions are distinguished: the absorbed water and the free surface water. In Fig. 3.3, the different water transport processes are depicted as open arrows: (1) the addition of water; (2) the absorption of water by the grains; and (3) and the transfer of water between two grains.

Vapor Diffusion Mechanism

Matrix vapor diffusion has a close relationship with the internal structure of the matrix. The shape, size, and gas density of the internal pores determine the diffusion mechanism. When denser gas gets through the matrix pores, the vapor molecules have fewer opportunities for a collision with the pore wall; this diffusion still follows Fick's law, and it is Fick-type molecular diffusion. When the pore diameter is small, collision occurs mainly between the gas molecules and the pore wall surfaces; when the less-dense gas gets through the pore, collision between the molecules is relegated to a secondary position. The diffusion resistance is mainly caused by collision between the gas molecule and the pore wall; this diffusion does not follow Fick's law and is called Knudsen diffusion. When the diameter of the pores equals the mean free path of the gas molecules, the collision between the molecules and the collision between the molecules and the pore wall are both important; that is, both Fick diffusion and Knudsen diffusion exist, with this diffusion called transition zone diffusion.

The structure of a matrix particle determines the form of the vapor diffusion, typically Fick diffusion, Knudsen diffusion, or both. Because of the large capillary channel in general research, Fick diffusion predominates.

Substrate Gas Phase Movement Mechanism

In the nonsaturated matrix, gas is mainly composed of air and water vapor; air and water vapor fill the pore space. The migration of the gas in the matrix may occur in four ways: Knudsen diffusion, viscous flow, continuous diffusion, and surface diffusion.

Heat Transfer Mechanism

Generally, for studying heat migration of internal particulates, three common forms are mainly considered:

1. Heat conduction. When there is a temperature gradient in the substrate, heat is transferred by heat conduction in a solid matrix, liquid water, vapor, and air.
2. Convection. Convective heat transfer is caused by flow of liquid water, vapor, and air within the matrix pore channels.
3. Phase change. Liquid vaporization and vapor condensation may occur on the liquid surface within the matrix because of the temperature gradient, which produces heat by phase change. Meanwhile, because of the humidity difference between the substrate surface and gas on the top of the matrix, the evaporation phenomenon will occur on material surfaces, subsequently causing phase change heat.

Theoretical Model of Heat and Moisture Transfer

Researchers have made considerable progress in theoretical studies of water movement in unsaturated or saturated porous matrix under isothermal conditions. The early studies showed that under nonisothermal conditions, the distribution and temperature changes in the matrix had an impact on the physical and chemical properties of the matrix water and subsequently affected matrix potential, solute potential, and hydrodynamic parameters. These parameters affected the fermentation temperature in turn. Therefore, in the study of water and heat transfer, people should highlight the mutual coupling relationships between them.

In a brief introduction to the theoretical model of water and heat transfer in porous media under nonisothermal conditions, the Philip and de Vries model (Philip and De Vries 1957; De Vries 1958) provided a water and heat-coupling migration model as follows:

$$\frac{\partial \theta}{\partial \tau} = \nabla(D_T \nabla T) + \nabla(D_\theta \nabla \theta) - \frac{\partial K}{\partial z} \tag{3.13}$$

$$C_{\mathrm{h}} \frac{\partial T}{\partial \tau} = \nabla(\lambda \nabla T) - L\nabla(D_{\theta \mathrm{v}} \nabla \theta) \tag{3.14}$$

In the formula,

θ = material water content;

τ = time;

D_T = matrix water diffusivity under the temperature gradient;

D_θ = matrix water diffusivity under the moisture gradient;

T = temperature;

K = hydraulic conductivity;

C_{h} = heat capacity;

λ = thermal conductivity;

L = latent heat of vaporization;

z = the perpendicular distance.

The equations take into account the liquid-gas two-phase flow under the water content and temperature gradients at the same time. They describe the two coupled nonlinear partial differential equations to explain the relationships between heat and water transfer. The wide application of the theory of Philip and de Vries promotes the coupled study of water and heat transfer in the matrix.

Cassel et al. (1969) applied this theory to predict porous medium moisture movement under a temperature gradient; they proved that the theory of Philip and de Vries could predict soil water movement. Especially in substrates with lower water content, the calculated and measured values of water flux fit well.

3.2.2.2 Heat and Moisture Transfer Process in Moisture Stratification

Often, SSF substrate belongs to a typical water-unsaturated porous medium; however, in some cases water is seriously lost and utilized along with strong ventilation

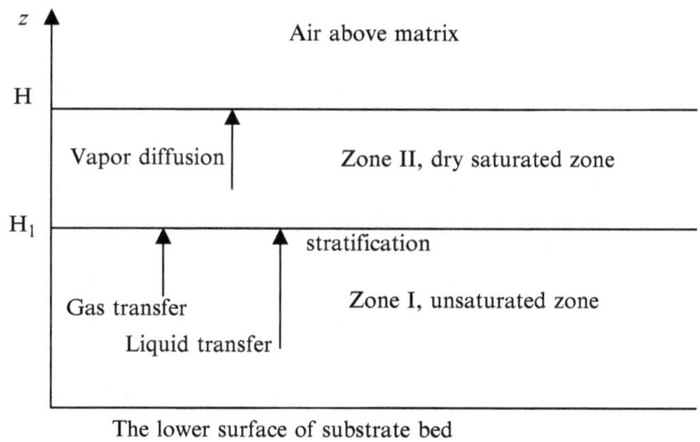

Fig. 3.4 Schematic diagram of moisture stratification

in the fermentation process. The surface water will be depleted; thus, the internal pore water will be used. When the water inside the pores also is dissipated completely, the substrate will form a dry saturated layer (moisture content about 10 %). When the dry saturated layer appears, the evaporation has not taken place in the substrate surface, but internally in the matrix. The moisture evaporation intensity decreases, and the loss of matrix moisture is controlled by diffusion from vapor to air; this is related to dry saturated layer thickness. Therefore, the study of the actual system must also consider the impact of moisture stratification on its internal heat and moisture migration. Figure 3.4 is a schematic diagram of substrate moisture stratification.

3.2.2.3 Effect of Temperature on Heat and Moisture Transfer in Solid Matrix

The changes in temperature exhibit variations in heat loss and water evaporation macroscopically. Essentially, the temperature changes affect the physical parameters of the matrix. Gardner (1959) put forward that there was a positive correlation between the temperature and the matrix water potential according to capillary theory. Zhang and Bai (1990) considered that water suction of a porous matrix decreased with increasing temperature and thus improved water potential and water energy. Philip and de Vries (1957) first quantitatively studied the effect of temperature on soil water potential; they pointed out that soil water suction of the wetting front decreased 1 cm when the temperature was increased by 0.002 °C.

Preliminary studies showed that with the increase of temperature, the moisture transfer accelerated (Constantz and Murphy 1991).

The temperature changes can also cause differences in the density of the water vapor and subsequently cause water vapor movement.

Table 3.2 Diffusion coefficient of NaCl in water ($\times 10^{-9}$ m^2/s) under different temperatures (°C)

Temperature (°C)	5	15	25	35
Diffusion coefficient ($\times 10^{-9}$ m^2/s)	0.919	1.241	1.612	2.031

In addition, temperature variations will affect some physicochemical properties of water, such as viscosity, density, surface tension, and so on, which affects water potential and water movement.

Philip and de Vries (1957) attributed the effect of temperature on the water potential to its effect on the surface tension of water. Haridasan and Jensen's (1972) experiments showed that the impact of temperature on the unsaturated hydraulic conductivity of a matrix at a given water content can be completely attributed to its influence on water viscosity.

3.2.2.4 Impact of Thermal Effects on Matrix Solute Transport

Temperature greatly affects moisture movement. For water-soluble organic salts and organic nutrients, the solute diffusion coefficient is also associated with temperature. Table 3.2 gives the diffusion coefficient of NaCl in water under typical temperatures.

From the table, the diffusion coefficient of NaCl in water at 35 °C is nearly twice that at 5 °C, indicating that temperature greatly influences the diffusion of salt. The relationship between the diffusion coefficient of NaCl in water and the temperature can be fit as follows:

$$D_{\mathrm{w}}(t) = 0.77669 + 0.0272\ t + 0.00025\ t^2 \tag{3.15}$$

Fermentation substrate is a porous medium; diffusion of salt in water is different from that in the matrix solution. The impact of substrate water content and the matrix skeleton tortuosity factor need to be considered.

3.2.3 Physical Parameters Affect the Solid Matrix Transfer Process

In the control equation for the heat and mass transfer process of the nonsaturated porous medium, some physical parameters depend not only on the characteristics of the solid porous skeleton but also on the flow of fluid in the interstices of the porous medium and the percentage ratio of each phase of the fluid to the void volume. These parameters are discussed next.

3.2.3.1 Physical Parameters That Affect Solid-State Fermentation Heat Transfer

The accumulation of heat is a typical temperature effect on SSF, and these effects are spontaneous with the growth of cells. As the fermentation process proceeds, metabolism heat accumulates in the material, and this heat is difficult to diffuse in the material in a timely manner because of the poor thermal conductivity of the matrix. Meanwhile, the matrix shrinks, and the porosity decreases, further hindering the convective cooling of gas.

The thermal characteristics of the matrix are the essential cause that determines matrix thermal conductivity, including the matrix heat capacity, thermal conductivity, and so on (Schutyser et al. 2001). (1) The heat capacity of the matrix could be divided into quality thermal capacity and volumetric heat capacity; the former refers to heat required for warming of 1 °C per unit mass of matrix, and the latter refers to heat required for warming of 1 °C per unit volume of matrix. Owing to the three phases, the heat capacity of SSF substrate is the heat capacity of all components. Because the volumetric heat capacity of air is too small (0.003 cal/cm^3/°C), the heat capacity of the solid phase is less than that of the liquid phase (1 cal/cm^3/°C); the matrix heat capacity is mainly dependent on matrix moisture. (2) Matrix thermal conductivity mainly depends on the nature and state of the composition itself (i.e., the three-phase composition and porosity, the arrangement of the solid particles, and the contact surface area of solid and liquid phases). Thermal conductivity of matrixes with the same humidity is related to their cohesive strength and porosity (Mitchell et al. 2006).

Heat conduction as the main heat transfer path for a static SSF reactor or compost fermentation makes the packed thickness another important factor for heat transfer (Shao et al. 2006); with the increase in height, the heat transfer path increases, the temperature gradient forms, and the temperature is more difficult to regulate. For packed bed reactors and other forced ventilation reactors, the heat transfer rate from the surface of the solid medium into the gas phase generally increases with an increase in forced ventilation speed; thus, physical indicators that affect matrix permeability (such as the air vent rate, oxidation reduction potential, permeability, and other indicators) have an impact on convection. Forced ventilation also affects moisture content, and evaporative cooling can be adjusted by adjusting the speed of the ventilation airflow and moisture humidity of the air, so the physical indicators mentioned also indirectly affect evaporative cooling (Jou and Lo 2011).

In addition to the physical properties of the material itself, the changes in physical and chemical properties of the matrix caused by microbial growth and metabolism during the fermentation process affect the transmission properties of the matrix; therefore, the chemical composition of the solid phase affects the transfer process by affecting its physical properties. For example, in a laboratory-scale drum reactor, the maximum temperature does not exceed 35 °C when gel is used as an SSF solid phase, but under the same conditions, the maximum temperature of the substrate is up to 45–50 °C when wheat bran is used as the substrate.

The reason for this phenomenon is the different carbon sources in the two substrates. The starch content of the brain exceeds 15 %; because of its easily utilization, the particles are reduced, resulting in deterioration of ventilation (Costa et al. 1998; Ashley et al. 1999).

In summary, matrix properties that directly affect heat transfer include thermal properties with an impact on the heat conduction of the matrix bed (heat capacity, thermal conductivity, and bed density) and matrix porosity properties with an impact on matrix convective cooling and evaporative cooling (such as the air vent rate, the oxidation reduction potential, permeability, specific surface area) and water-holding capacity. The described two properties together determine the indirect indicators of the matrix (e.g., thermal conduction coefficient, the chemical composition of the matrix bed, and its applicability to microorganisms).

3.2.3.2 Physical Parameters That Affect Solid-State Fermentation Gas Transfer

In fermentation of aerobic filamentous fungi, the mycelia could grow both on the substrate surface and internal to the matrix particles; oxygen transfer often is one of the limiting factors on microbial growth and product formation. There are two forms of the oxygen transfer process from reactor operation at the macroscopic level. The first is diffusion transfer, which is represented in static tray fermentation. In this transfer process, the air only enters the headspace above the layer following diffusion to provide oxygen to the microorganisms. Intense convection occurs in forced ventilation operating conditions. The transfer process is mainly affected by factors such as material thickness, surface humidity, material layer bulk density, and particle size (Stuart et al. 1999).

For certain material properties, the dissolved oxygen level is related to the particle radius size, and there exists a critical radius. The dissolved oxygen in the particles below the critical radius is approximately zero. The diffusion distance of oxygen within the particles is only 0.5 mm; mycelia nearby cannot grow within the particle. Also, because of the emissions of carbon dioxide during microbial growth at the same time, the actual critical radius value is larger than the theoretical value. On the other hand, related to the water film thickness of the particle surface, the water film is the major limiting factor in oxygen diffusion to the biofilm, and the water film thickness is determined by the water-holding capacity and the water content of the matrix (Mitchell et al. 2006). In addition, the matrix porosity greatly affects gas diffusion and convection. The size of the pore is determined by particle size, shape, and matrix water content itself, and these factors are further decided by matrix density, bulk density, and other macroscopic properties. Sometimes, to increase porosity, rice husk and other loose material are specifically added to make ventilation conducive. The matrix water content is also concerned with oxygen transfer in the matrix; excess free water obstructs the flow of air (Corona et al. 2005; Oostra et al. 2001). In addition, the physical properties of the matrix have an impact on matrix water vapor flux, including hygroscopicity. A matrix with good hygroscopicity is higher in vapor flux than a weakly hygroscopic matrix.

Generally, the following measures are adopted to improve mass transfer: choice of particulate, porous, or fibrous material as a substrate; reduced thickness of the matrix layer; increased gaps between the substrate; stirring or using a drum reactor to prevent agglomeration of matrix, which results in decreased porosity.

3.2.3.3 Physical Parameters That Affect Solid-State Fermentation Moisture Transfer

Whether microorganisms could grow on the substrate depends on the water activity of the substrate (Oostra et al. 2001; Ikasari and Mitchell 1998; Thibault et al. 2000). The water activity is defined as the ratio of the fugacity of the solvent to the fugacity of the pure solvent. It is approximately equal to the vapor pressure of the food in the sealed container P to water vapor pressure of pure water at the same temperature P_0 (Ashley et al. 1999; Nagel et al. 2001). Water activity expresses the amount of unbound water concerned with microbiology; the water activity value is more important than matrix moisture in maintaining the physiological activity of microorganisms.

There is no free-flow water in the SSF substrate, and moisture transfer in the fermentation process is mainly the evaporation of moisture from the substrate surface and the delivery of steam between gaps in the solid phase. Previous studies showed that, when there was less evaporation, the water content and water activity were not significantly changed (Weber et al. 2002). Migration of steam is similar to migration of air in the matrix; therefore, it can be considered that the main parameters that affect gas transfer, such as matrix porosity, pore morphology, and the like, also affect water vapor migration in the matrix. However, there is less related research. For moisture evaporation from the solid phase surface, the specific surface area is the main factor affecting the rate of evaporation. In addition, the temperature gradient is the essential reason for evaporation (Ashley et al. 1999); thus, the thermal conductivity of the matrix and its water retention properties are also closely related.

3.2.3.4 Physical Parameters That Affect Solid-State Fermentation Bioavailability

Microorganisms grow on the substrate and produce metabolites; the process is affected by physical factors of the matrix itself (particle size, shape, porosity, fiber content, viscosity, etc.) (Weber et al. 2002). The particle size and humidity of the substrate in SSF are critical to microbial growth and its activity. Generally, substrates with smaller particles can provide a larger surface area for microbiological attack and significantly improve the SSF reaction rate, which is considered to be an ideal choice (Mitchell et al. 2003; Favela-Torres et al. 1998). Once mycelia cover the entire surface of the particles, even if the nutrients in the particle interior are not being used, the biomass also no longer increases. Thus, the specific surface area of the particles determines the maximum value of the biomass. The main parameters that have an impact on the specific surface area of the particles are the particle size and shape.

However, in many cases, particles that are too small are likely to cause substrate agglomeration; interparticle porosity is also decreased, resulting in increased resistance that adversely affects heat and mass transfer, impedes microbial respiration or ventilation, and eventually leads to undesirable growth of microorganisms. On the other hand, large particles are conducive for improving mass and heat transfer efficiency because of the presence of large gaps, and they can provide better breathing and aeration conditions, although they provide a smaller surface area for microbial attack. With mycelial growth, the size of the gap will be reduced during the reaction, and the effective diffusion coefficient of oxygen and carbon dioxide will decrease. Previous research suggested that the changes in substrate bed pressure drop and the amount of mycelial growth prediction are closely related. With the expansion of mycelia, the gap between the particles decreases, and the substrate bed pressure drop increases. Therefore, in SSF, the choice of an optimum particle size is necessary (Sangsurasak et al. 1996).

Among the mathematical models for the fermentor, the growth kinetic equation does not contain the factors that influence particle size; rather, they describe the growth curve empirically. Therefore, the optimum size of the particles must be obtained from the perspective of dynamics.

3.2.4 Changes in Nutritional Carrier Transfer Properties During Cell Growth

The physical properties of the nutritional carrier are mainly affected by cell growth, thus altering the transfer properties and eventually affecting the fermentation temperature, oxygen content, and water activity. The changes in the transfer properties of the matrix in a biochemical reaction are the nature of SSF, which is different from submerged fermentation and chemical reaction. In the process of cell growth, the substrate bed pressure drop presents a gradually increasing trend, and the nutritional matrix pressure change is greater than for the inert carrier. It has been observed and is believed that the volume and particle sizes of nutritional supports will change during fermentation. However, because of the difficulty in sampling caused by mycelial adhesion, it is difficult to study the changes in the matrix alone. Therefore, the former studies that examined the effect of cell growth on fermentation mainly focused on the changes in matrix substrate concentration, moisture content, oxygen content, particle length, and heat. These studies quantified the environmental parameters affected by microbial metabolism, but they ignored the impact of cell growth on the transfer properties of the matrix and failed to study the overall effect on substrate bed heat transfer.

I measured the transfer features of nutritional carrier matrix with different water contents and particle sizes in the fermentation process (including air permeability,

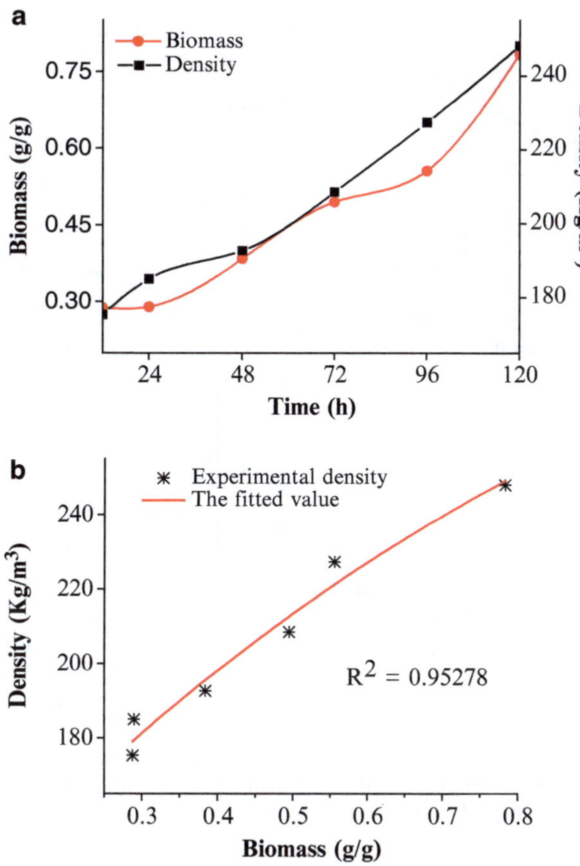

Fig. 3.5 Density variation of SEWS-bran substrates against fungal growth during *SSF* (solid-state fermentation). (**a**) Experimental density variation of SEWS (Steam explosion wheat straw) bran substrates in SSF; (**b**) fitting of the density variation of substrates to original data

thermal conductivity, heat capacity, and the transfer coefficient of oxygen) and determined the internal oxygen distribution in fermentation. The established model supplied a new basis for understanding the SSF process, could be used to improve the existing SSF mathematical model in quantification, and provided new ideas for substrate selection and process optimization.

3.2.4.1 Changes in Solid Matrix Density During Cell Growth

As shown in Fig. 3.5a, the density of the steam-exploded wheat straw was positively correlated with cell growth in the fermentation process, and it met the following power function relationship through analysis:

$$\rho = a_1 e^{(-X/a_2)} + a_3 \tag{3.16}$$

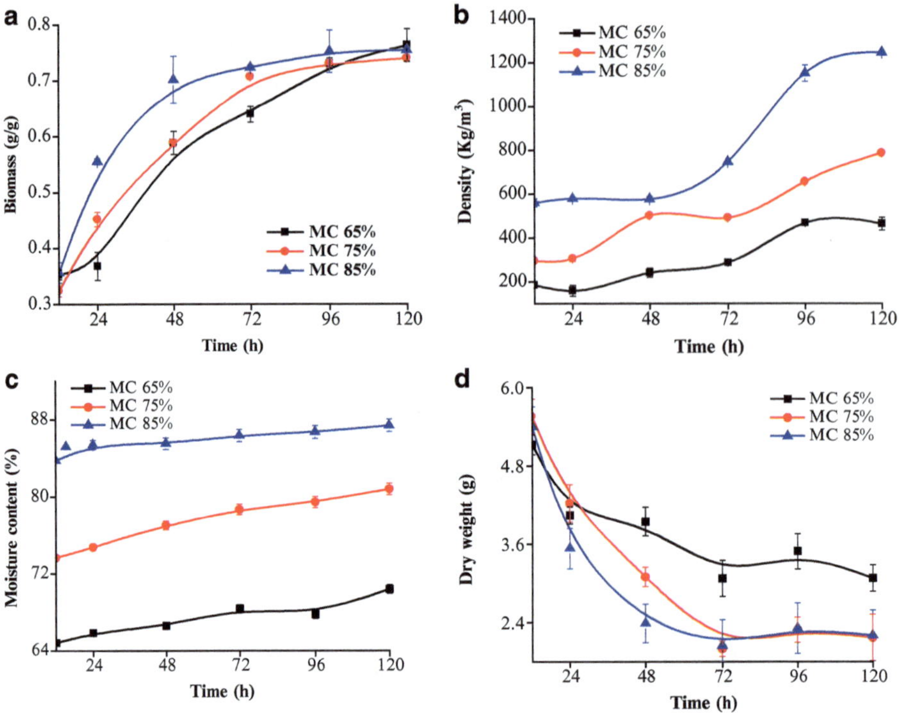

Fig. 3.6 Effect of moisture content (*MC*) on the density variation of SERS-bran substrates against fungal growth. (**a**) Fungal growth in SERS (Steam explosion wheat straw) bran substrates; (**b**) density variation of SERS-bran substrates; (**c**) moisture variation of substrates; (**d**) dry weight of substrates

In this formula,

ρ = the density of the substrate, kg/m^3;
X = cell biomass, g/g (dry matrix);
a_1, a_2, and a_3 = model parameters, determined by the matrix.

Figure 3.5b displays the fitting results, and it shows that the model can characterize the relationship between the matrix density and cell growth well. The corresponding model determination coefficient R^2 is approximately 0.9528.

Figure 3.6 shows the effect of microbial growth on the density of steam-exploded straw 1.5 cm in fiber length with different moisture contents and with a moisture content of 75 % (w/w) with fiber lengths of 4, 1.5, 0.4 cm, respectively. The maximum specific growth rate of the cell μ_M increased with the water content gradually and decreased with the increase in the matrix fiber sizes; the corresponding linear change rate of matrix density k also showed the same trend, positively correlated with cell growth. As shown in Table 3.3, the determination coefficient R^2 of the model showed that the power function model could characterize the quantitative relationship between the microbial biomass and matrix density

Table 3.3 Results of model fitting for substrate density

Sample	μ_M	k	a_1	a_2	a_3	R^2
A_2B_1	0.0324	3.095	0.783	−1.060	137.869	0.928
A_2B_2	0.04612	4.507	0.954	−1.222	295.136	0.7403
A_2B_3	0.09044	6.927	2.24E-12	−0.227	559.408	0.988
A_2B_2	0.05709	5.660	3.74E-08	−0.325	353.194	0.961
A_3B_2	0.03575	4.481	0.101	−0.875	278.679	0.993

well; it had a good reference value. In in-depth discussion of the reasons for the matrix density changes, it can be found that the dry weight of substrate itself continuously decreased in the fermentation process; while the biomass rapidly rose, the corresponding moisture content of the matrix also continuously increased, which indicated good metabolism of the microorganism in the entire fermentation process. Growth of mycelia enabled increased weight of the fermentation product (the mixture of matrix and mycelia), accompanied by reduction of the matrix volume; eventually, the overall density of the matrix increased correspondingly.

A_1, A_2, and A_3 respectively represent particle lengths of 0.4, 1, and 3.5 cm; B_1, B_2, and B_3 respectively represent the initial moisture contents of 65, 75, and 85 %. The combination of A_iB_i represents the substrates with a certain particle length and moisture content. Here, k is the density variation rate, and μ_M is the maximum specific fungal growth rate.

Further analysis of the impact of the water content and fiber length on the change in matrix density was the same as previous reports; the growth of mycelia in matrix with high water activity was superior to matrix with low water activity (Fig. 3.6). In the early stage of fermentation, it was possible to provide a good supply of oxygen and nutrients to minute fibers. Thus, in the early stage, with the matrix fiber length decreased, the cell growth rate sequentially increased (Fig. 3.7). Because of the porosity of the minute fibers, the substrate was filled by cell growth in the late stage of fermentation (72 h later); there was insufficient space and nutrient supply for further microbiological growth. Thus, cell growth in long fibers was superior to that of minute fibers. Because of large pores in the 4-cm matrix, the result was inferior nutritional availability compared to a 1.5-cm matrix; therefore, cell growth on a 1.5-cm matrix is preferable in late fermentation.

Therefore, matrix density and cell growth are positively correlated. The reasons for differences in cell growth are precisely the main reason for matrix density variation.

3.2.4.2 Gas Transfer Performance Changes of Nutritional Matrix in Fermentation

Permeability Changes of Nutritional Substrate

The permeability variation of steam-exploded wheat straw at a height of 8 cm and steam-exploded rice straw at a height of 5–6 cm in the fermentation process was investigated. As shown in Fig. 3.8, with microbial growth, the permeability of the

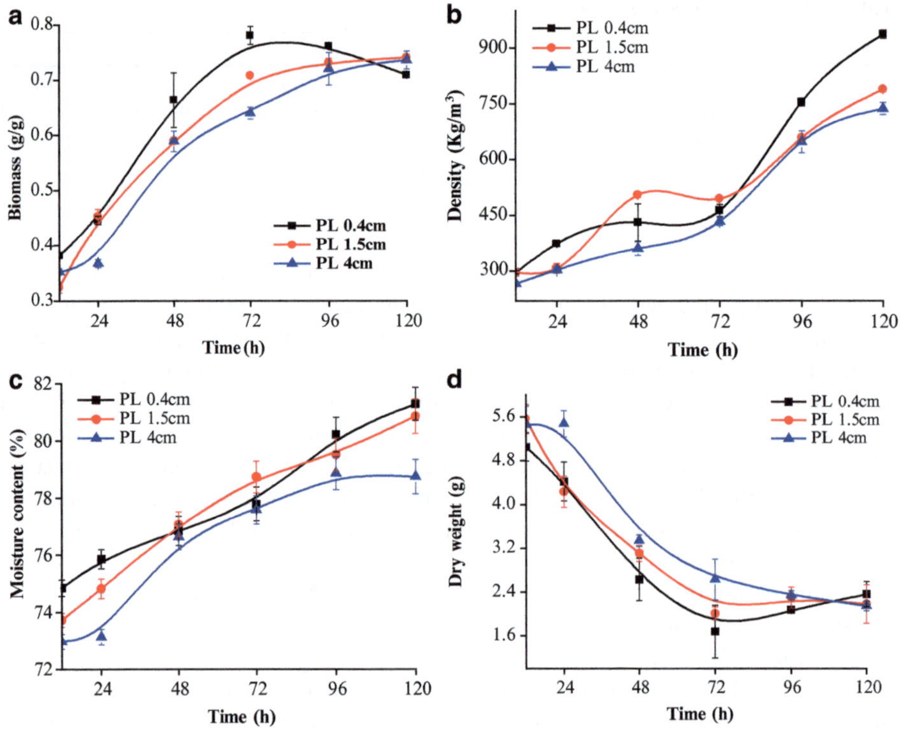

Fig. 3.7 Effects of particle length (*PL*) on the density variation of SEWS (Steam explosion wheat straw) bran substrates against fungal growth. (**a**) Fungal growth in SERS-bran substrates; (**b**) density variation of SEWS-bran substrates; (**c**) moisture variation of substrates; (**d**) dry weight of substrates

solid matrix first decreased and then increased. The trend was consistent with the trend of the fractal dimension in the matrix. It can be inferred that the reasons for matrix permeability change were closely related to matrix structure. At cell growth initiation, the matrix pore morphology has strong permeability; with the continued growth of microbial cells, the internal pores are filled by mycelia, and this results in reduced porosity and weakened permeability. This occurs in the logarithmic growth phase and the early stationary phase; when cell growth goes into the stationary phase, further use of the matrix damages the entire structure. The increased quantity of matrix debris makes its porosity increase; thus, permeability begins to increase. Late in the stationary phase, the used nutritional matrix further degrades into smaller fragments, so that porosity increases further, as does permeability. However, because of the impact of compaction, the permeability growth rate is less than the increased fractal dimension rate.

To further confirm the reasonableness of the inference, as well as the effect of cell growth on matrix permeability under different matrix properties, I further investigated growth of *Penicillium decumbens* in matrixes with different water contents and lengths.

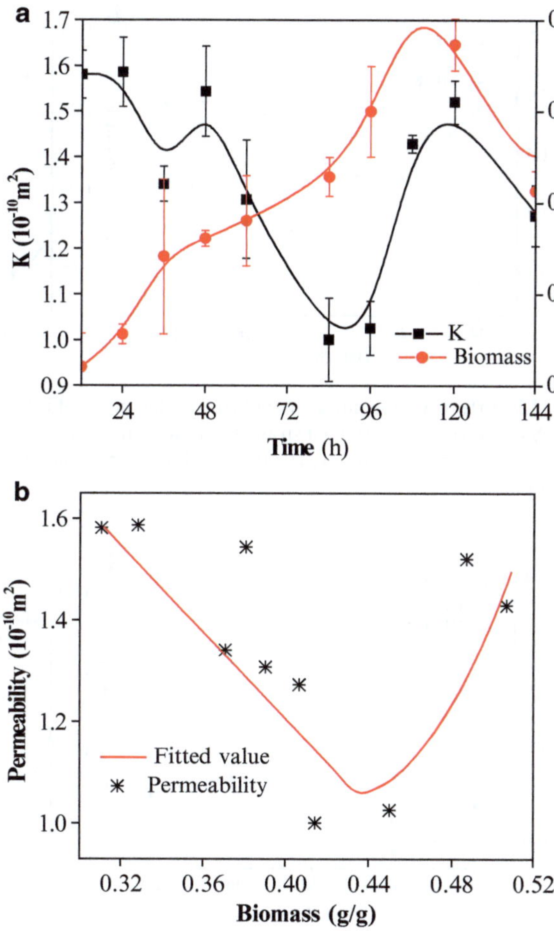

Fig. 3.8 Permeability variation of SEWS (Steam explosion wheat straw) bran substrates along with fungal growth. (**a**) Experimental permeability variation of SEWS (Steam explosion wheat straw) bran substrates in *SSF* (solid-state fermentation); (**b**) fitting of the permeability variation of substrates to original data

The results are shown in Figs. 3.9 and 3.10; the effect of cell growth on matrix permeability under different conditions also met the previous rules. So, we can infer that the matrix permeability changes were closely related to the structural change of matrix morphology; meanwhile, and the specific growth rate is conformed to the kinetic equations:

$$\frac{dX}{dt} = \mu_M X\left(1 - \frac{X}{X_M}\right) \tag{3.17}$$

$$\left\{\frac{dK}{dt} = \frac{dX}{dt}\frac{k_1\alpha}{\mu_M} = \frac{dX}{dt}\frac{\omega}{\mu_M}, K > K_m\right. \tag{3.18}$$

$$\left\{\frac{dK}{dt} = \left[\frac{dX}{dt}\left(\frac{1}{X_{g'}} - \frac{1}{X}\right)\right]\frac{k_2\beta}{\mu_M} = \left[\frac{dX}{dt}\left(\frac{1}{X_{g'}} - \frac{1}{X}\right)\right]\frac{\zeta}{\mu_M}, K_m < K < K_M\right. \tag{3.19}$$

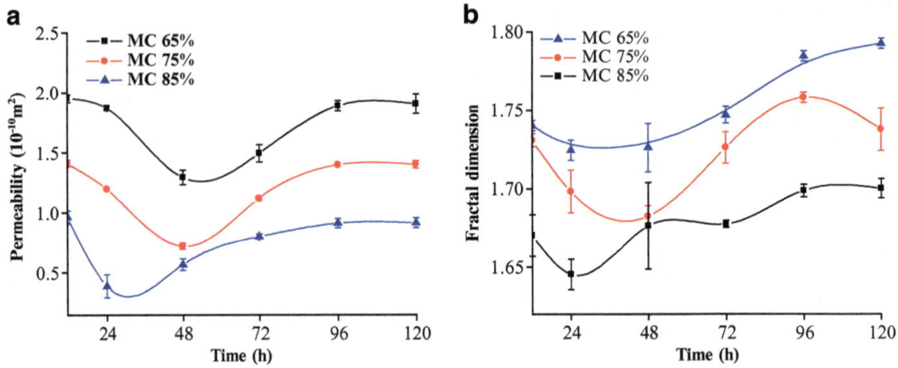

Fig. 3.9 Effect of moisture content (*MC*) on the permeability and fractal dimension variations of SERS (Steam explosion wheat straw) bran. (**a**) Permeability variations of SRWS (Steam explosion wheat straw) bran substrates; (**b**) fractal dimension variations of SERS (Steam explosion wheat straw) bran

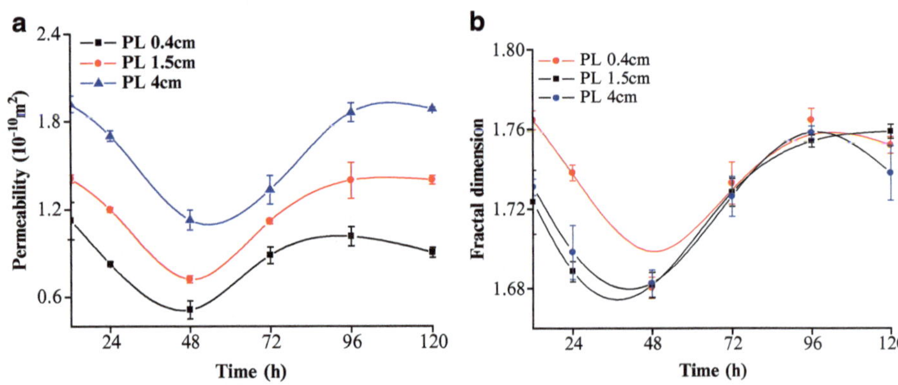

Fig. 3.10 Effect of particle length (*PL*) on the permeability and fractal dimension variations of SERS-bran. (**a**) Permeability variations of SRWS (Steam explosion wheat straw) bran substrates; (**b**) fractal dimension variations of SERS-bran

In the formulas,

K_m = the air permeability minimum value in the fermentation process;
K_M = the air permeability maximum value in the fermentation process;
μ_M = the maximum specific growth rate of the microorganisms;
$X_{g'}$ = the corresponding biomass at the air permeability turning point in the fermentation process;
α, β = the specific growth rate and decreased rate of air permeability respectively;
k_1, k_2 = equivalent coefficients associated with biomass change;
ω, ζ = the equivalent growth rate and decreased rate of air permeability associated with mycelia respectively.

The formulas express the matrix permeability change with cell growth directly.

Table 3.4 Results of model fitting for substrate permeability

Sample	ω (×(−10^{-3}) /h)	ζ (×10^{-1}/h)	μ_M (×10^{-2}/h)	$X_{g'}$(g/g)	RE of K (×10^{-3})	RE of biomass (×10^{-2})
A_2B_1	7.9	7.9	3.24	0.589	1.55	3.6292
A_2B_2	10.3	9.9	4.612	0.590	0.4544	3.2566
A_2B_3	26.8	8.1	9.044	0.556	0.6512	2.7616
A_2B_2	9.7	13.5	5.709	0.664	1.6811	5.2203
A_3B_2	10.6	12.1	3.575	0.589	1.1957	1.8694

RE relative error

Based on the model, the fit of the data using the least-error method and the obtained parameters are shown in Table 3.4. This table shows that the model can express the gas permeability changes with the cell growth well; therefore, the changes in matrix morphology represented by the fractal dimension behave with a similar trend as the changes in matrix permeability. This is in agreement with a similar kinetics model. But, from the turning points of the biomass curve, it can be found that the time when the permeability reaches a minimum value slightly lagged that of the fractal dimension. It can be inferred that, when the microbial cells occupy most of the matrix space, the stack effect is not obvious, and the corresponding permeability could not reach the minimum value. Further microbiological use of the matrix slightly compacts the matrix and thus causes the lowest air permeability. After that, further use of the matrix increases matrix permeability. This inference can also be confirmed by the change in interior oxygen distribution of the matrix in the fermentation process.

Oxygen Concentration Changes in Nutritional Substrate

During fermentation, the matrix oxygen concentration decreases because oxygen is utilized by cell growth. Because the fermentation substrate is a porous medium, oxygen concentration is low in the cell growth area and high in the pore area without cell growth. Oxygen is not distributed evenly in the matrix in different fermentation stages. The distribution and fluctuation of oxygen concentration with matrix height are also coincident with changes in the fractal dimension of the matrix (morphology) and permeability.

Figure 3.11 shows matrix oxygen distribution during *Penicillium decumbens* fermentation. Take the fermentation substrate of steam-exploded straw as an example: The substrate had an 85 % moisture content and was 1.5 cm in length. At the early fermentation stage (12 h), the oxygen in the matrix was evenly distributed, with continuous distribution as the matrix depth increased. At 24 h, the growth of microorganisms increased the matrix agglomeration; the dense microorganisms made the oxygen concentration lower in the matrix, and the porosity was reduced. These changes coincided with the changes in matrix fractal dimension and air permeability. At 48 h, the further growth of cells caused the agglomerated matrix to start disintegrating into small pieces; its porosity increased.

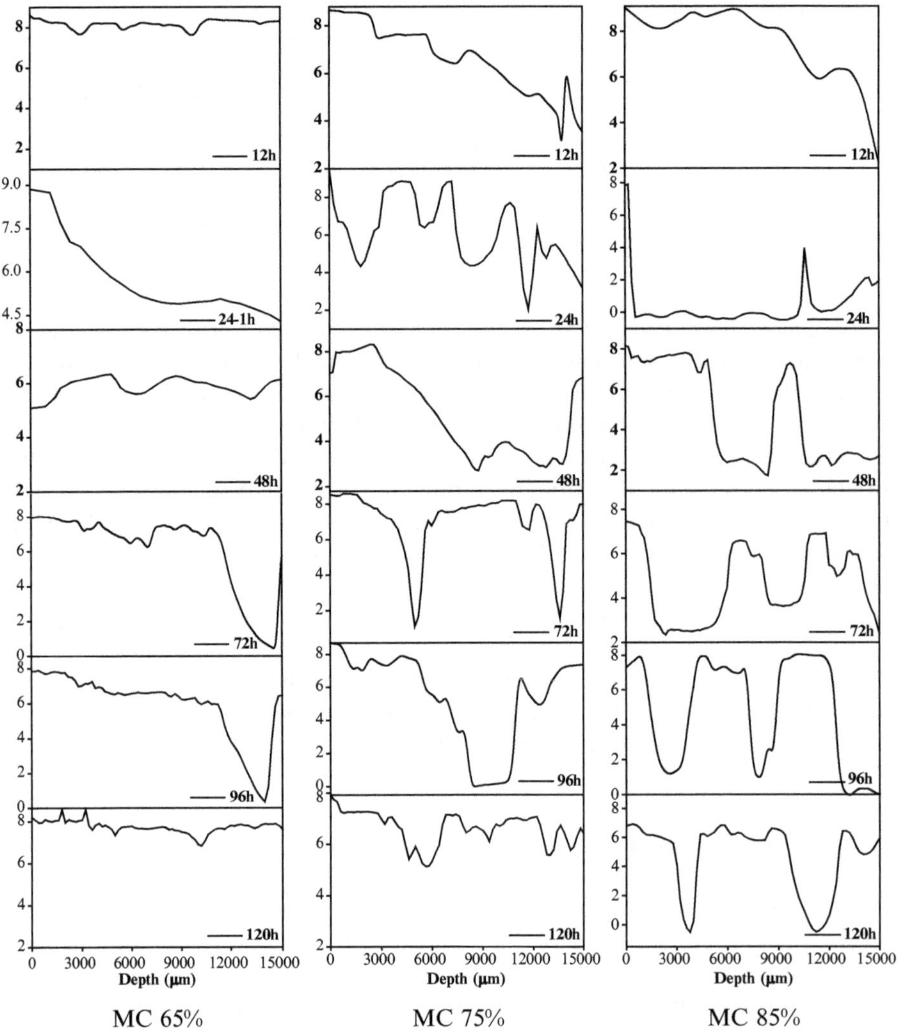

Fig. 3.11 Oxygen profile in SERS-bran substrates with different moisture contents (*MC*) during *SSF* (solid-state fermentation)

Meanwhile, the fractal dimension of the matrix-bacterial junction also started to increase, and the corresponding matrix air permeability also gradually increased. During 72–120 h, the matrix further degraded into smaller clumps, accompanied by cell growth; the corresponding pores (region of high oxygen concentration), the matrix fractal dimension, and the gas permeability also increased. Therefore, the oxygen concentration change in the region was able to reflect detailed changes in matrix morphology in fermentation. Compared to the detailed changes of water content of the matrix morphology, with the increase of water content the agglomerate strength of the matrix increased, and the corresponding decomposition of clumps was more obvious.

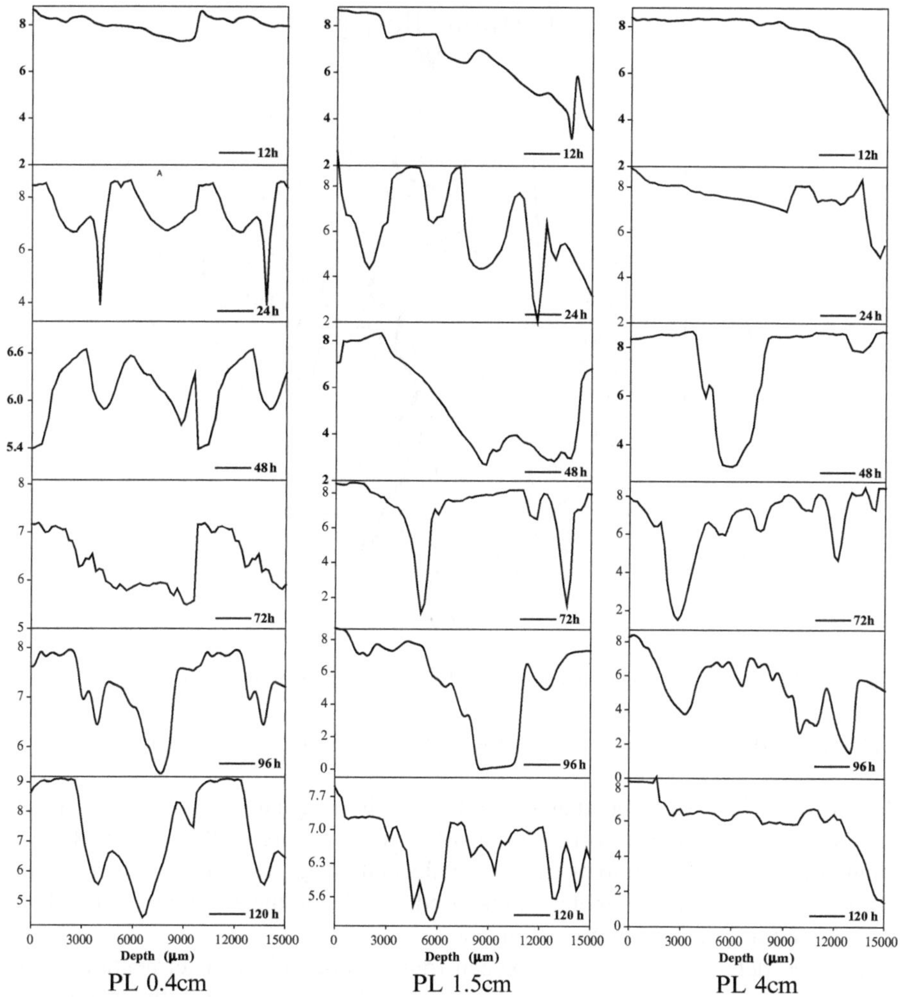

Fig. 3.12 Oxygen profile in SERS-bran substrates with different particle lengths (*PL*) during *SSF* (solid-state fermentation)

The change in oxygen concentration with the cell growth of different lengths of steam-exploded straw is shown in Fig. 3.12. From the figure, it can be found that the oxygen was more uniformly distributed in the 4-cm matrix than smaller particles at 24 h, and the agglomeration was not obvious in late fermentation. Pores were large throughout the fermentation process; because of the longer fiber length of the 4-cm matrix and the larger distance between pores, the adhesion strength between particles was weaker. The shorter 0.4-cm distance between pores and 1-cm matrix made it more prone to agglomeration; the corresponding decomposition was also more evident. From the oxygen consumption, the cell growth in the 0.4-cm matrix was more exuberant than that in longer-fiber matrix. This was consistent with the results of previous experiments.

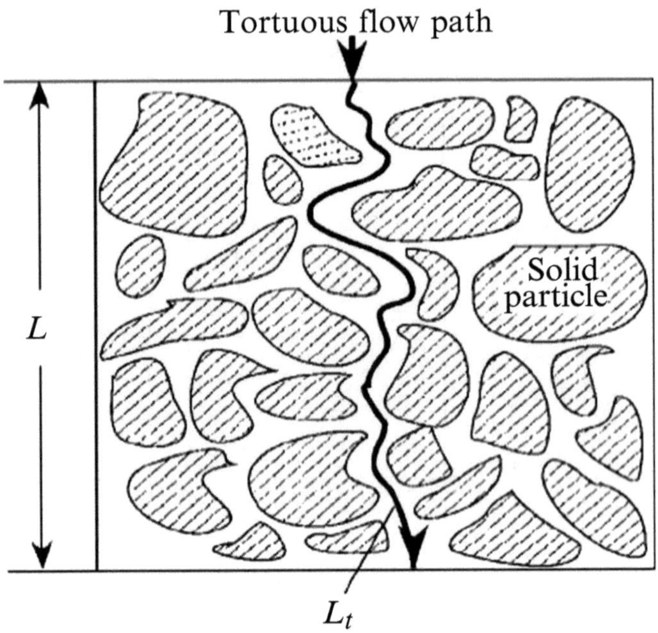

Fig. 3.13 Schematic drawing of pores in the solid substrate (Liu et al. 2006)

Calculated Matrix Oxygen Diffusion Rate Based on Fractal Dimension

As shown in Fig. 3.13, the characterization of the internal structure of the matrix can be quantified by tortuosity. Tortuosity τ is the square of the ratio of the true length of a pore to the straight length; it reflects the degree of bend of the pore in a porous medium. This parameter can be used to indicate the effect of channel bend degree on mass transfer in porous media.

Routine determination of the oxygen transfer coefficient in fermentation substrate can only measure the transfer characteristics macroscopically; it does not reflect the impact of the porosity on the transfer coefficient. Taking into account the effect of irregular pore structure in fermentation medium on oxygen transfer, the fractal dimension of the matrix is used in this study to calculate the oxygen diffusion rate in the matrix. According to previous studies, the tortuosity degree is expressed by the fractal dimension as follows:

$$\tau = \left(\frac{L_0}{\lambda}\right)^{D_B - 1} \tag{3.20}$$

In this formula,

L_0 = the linear distance of the channel; in this chapter, it is specified as the length of the matrix height (i.e., 0.015 m);

λ = the maximum channel diameter, which can be calculated by the distance between the matrix particulates;
D_B = the fractal dimension of the medium.

The corresponding oxygen diffusion coefficient D is expressed as

$$D = D_0/\tau^2 \tag{3.21}$$

In Eq. 3.21, D_0 is the macro oxygen diffusion rate; the mean free path of air under the standard condition is 0.069 μm and is much smaller than the pore size of the matrix. In this case, the air diffusion is Fick diffusion, which is expressed as follows:

$$D = \frac{2k_B T^{1.5}}{3\pi l^{1.5} d^2 p m^{0.5}} \tag{3.22}$$

In this formula,

k_B = the Boltzmann constant, 1.3806×10^{-23} J/K;
m = the molar mass of the gas;
T = temperature, K;
l = free path of molecular motion, m;
p = pressure, Pa;
d = molecular diameter, m.

The merger of the three formulas is

$$D = \frac{2k_B T^{1.5} \lambda^{2D_B - 2}}{3\pi^{1.5} d^2 p m^{0.5} L_0{}^{2D_B - 2}} \tag{3.23}$$

The oxygen diffusion rate could be calculated by the porous structure of the matrix and the fractal dimension (Figs. 3.14 and 3.15).

The calculation result is consistent with the inferred result; the oxygen diffusion rate changes caused by the water content and the matrix fiber length are similar to the changes of the permeability. But, note that the oxygen diffusion rate is not related to the intrinsic characteristics of the matrix; it is closely related to the internal temperature of the matrix, the oxygen concentration, and its solubility in the matrix.

3.2.4.3 Changes in Nutritional Matrix Heat Transfer in the Fermentation Process

Changes in Thermal Conductivity of Steam-Exploded Straw During Cell Growth

As shown in Fig. 3.16, the thermal conductivity of steam-exploded wheat straw and biomass yield were positively correlated. The thermal conductivity was mainly

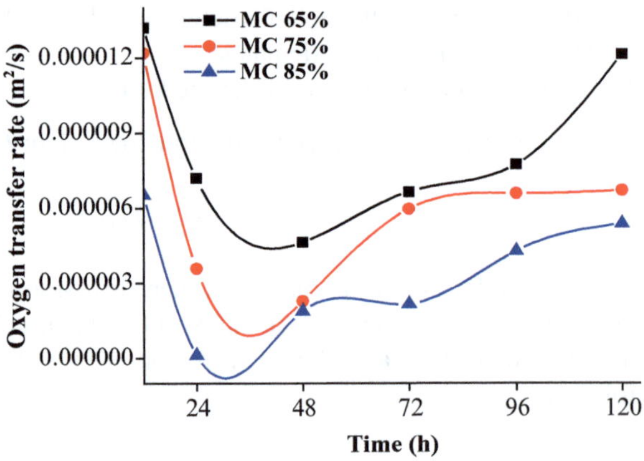

Fig. 3.14 Oxygen transfer coefficient moisture of SERS-bran substrates with different moisture contents (*MC*) in *SSF* (solid-state fermentation)

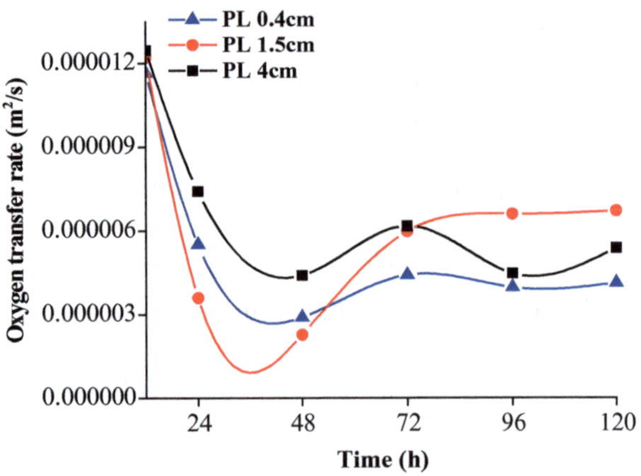

Fig. 3.15 Oxygen transfer coefficient moisture of SERS-bran substrates with different particle lengths (*PL*) in *SSF* (solid-state fermentation)

affected by the liquid content of the three phases of the matrix, and the water content increased with cell growth increases; therefore, it is possible to infer that the thermal conductivity of the matrix and cell growth were related. The relationship between the two parameters fit the following model:

$$\mathrm{TC} = a_1 e^{(-X/a_2)} + a_3 \tag{3.24}$$

The fitting results showed that the model could be used to characterize the relations between the thermal conductivity of steam-exploded wheat straw and cell growth. By further study of the influence of the water content and matrix fiber length on thermal conductivity, it can be found that the power function model can well characterize the changes in matrix thermal conductivity with cell growth at different conditions.

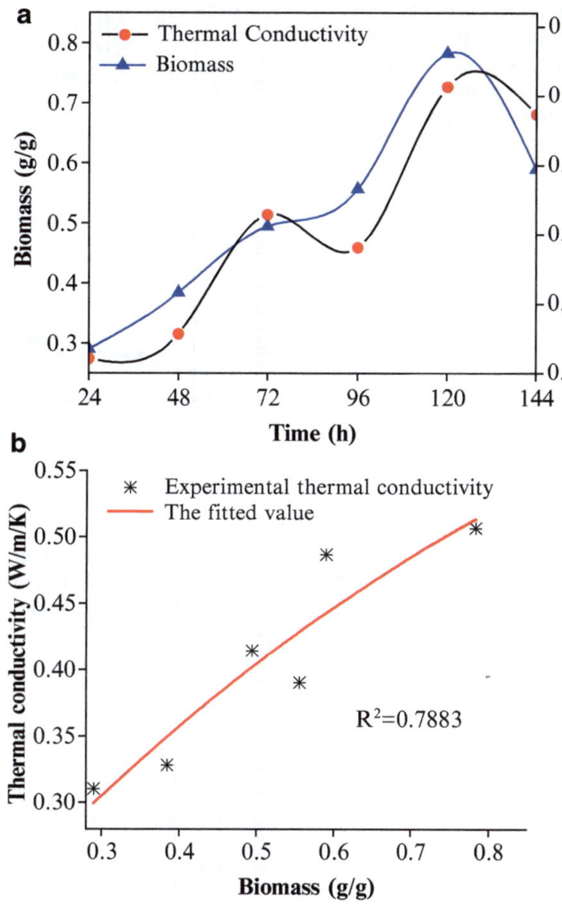

Fig. 3.16 Thermal conductivity variation of SEWS-bran substrates against fungal growth during SSF. (**a**) Experimental thermal conductivity variation of SEWS-bran substrates; (**b**) fitting of the thermal conductivity of substrates to original data

As shown in Fig. 3.17, the changes in thermal conductivity of the matrix showed a significant difference with the changes in water content; it had a linear relationship with water content. However, in the final analysis, the change in water content was the result of cell growth and metabolism. Therefore, from this sense, cell growth was the root cause of the thermal conductivity changes in the fermentation process.

As shown in Fig. 3.18, the thermal conductivity of the matrix was reduced with increased fiber length. The reason is that another factor affecting the thermal conductivity was the linear distance of thermal conductivity in porous media (Table 3.5). The thermal conductivity of solid steam-exploded straw itself is greater than the thermal conductivity of the air, so the thermal performance of a matrix with smaller pore size is stronger than for the larger pore in long fibers. However, according to results of previous studies, the effect of fiber length on the thermal conductivity of substrate is weaker than that of water content. Therefore, the statistical analysis results also showed that there was no significant difference in thermal conductivity of different fiber lengths.

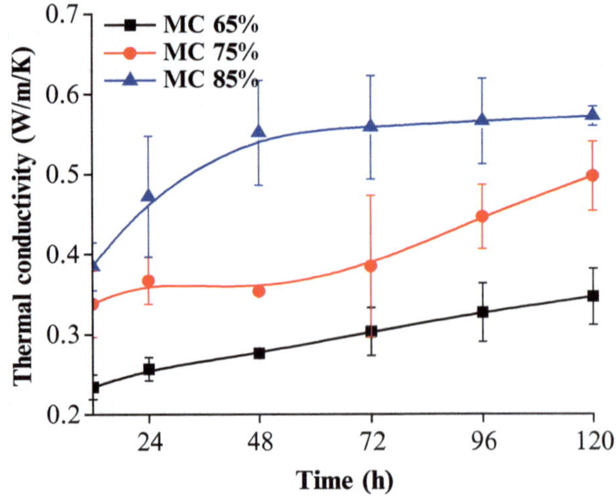

Fig. 3.17 Thermal conductivity profile in SERS-bran substrates with different moisture contents (*MC*) during *SSF* (solid-state fermentation)

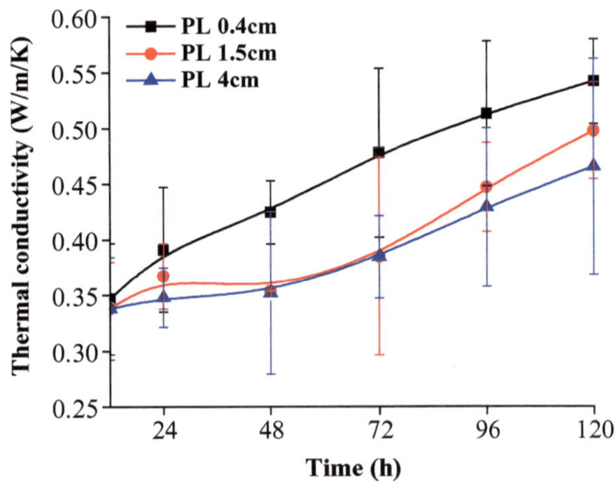

Fig. 3.18 Thermal conductivity profile in SERS-bran substrates with different particle lengths (*PL*) during *SSF* (solid-state fermentation)

Table 3.5 Results of model fitting for substrate thermal conductivity

Sample	a_1	a_2	a_3	R^2	RE of K ($\times 10^{-4}$)
A_2B_1	0.012	−0.250	0.179	0.997	0.150
A_2B_2	2.61E-15	−0.023	0.353	0.963	1.398
A_2B_3	1.471	−3.63	−1.238	0.996	0.247
A_2B_2	0.057	−0.434	0.218	0.653	29.8
A_3B_2	7.42E-06	−0.076	0.342	0.962	0.983

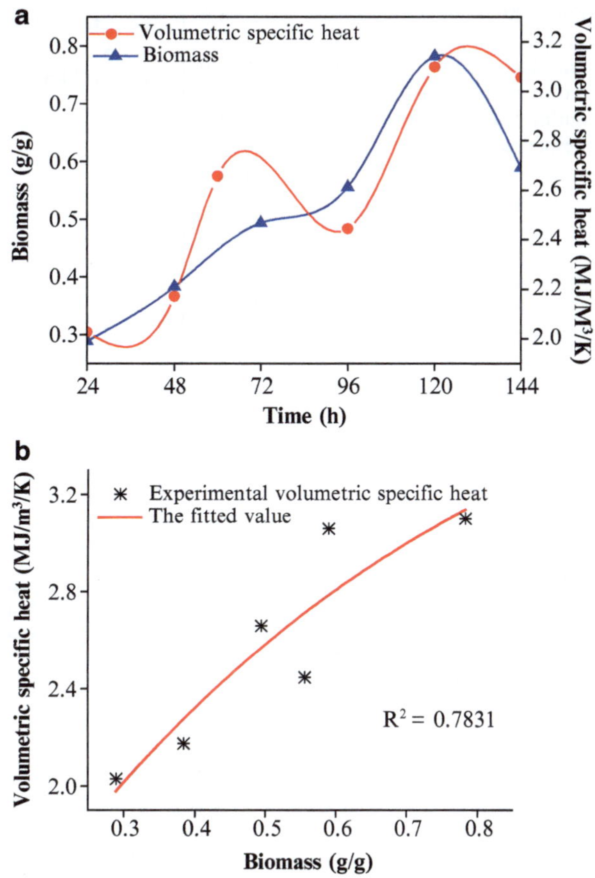

Fig. 3.19 Volumetric specific heat variation of SEWS-bran substrates with fungal growth during *SSF* (solid-state fermentation). (**a**) Experimental volumetric specific heat variation of SEWS-bran substrates; (**b**) fitting of the volumetric specific heat of substrates to original data

Changes in Steam-Exploded Straw Heat Capacity with Cell Growth

As Fig. 3.19 shows, the volume specific heat capacity of steam-exploded straw matrix and cell growth were in direct proportional relationship. Previous studies also showed that the volume specific heat capacity of the matrix was significantly affected by water content; therefore, it can be considered that the increased moisture content by the metabolism of cell growth was the main reason for the increases in specific heat capacity. Further analysis showed that the volume specific heat capacity of the matrix was similar to changes in the cell growth; therefore, it also fit the power function model:

$$\mathrm{VSH} = a_1 e^{(-X/a_2)} + a_3 \tag{3.25}$$

As shown in Figs. 3.20 and 3.21, the effect of water content on the heat capacity was similar as its effect on thermal conductivity. However, the effect of different fiber lengths was different. Although the effect of fiber length was similar, the

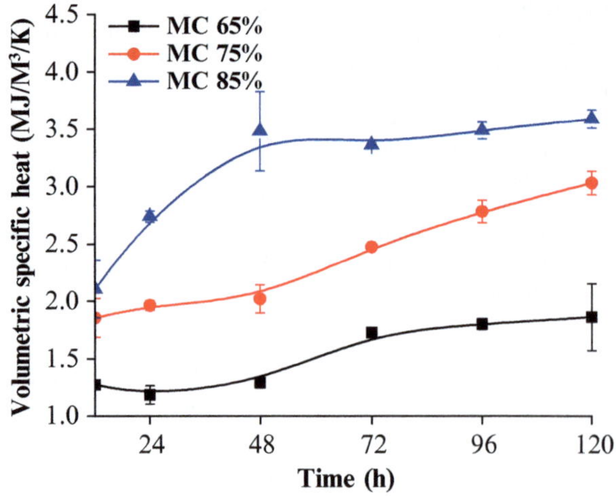

Fig. 3.20 Volumetric specific heat profile in SERS-bran substrates with different moisture contents (*MC*) during *SSF* (solid-state fermentation)

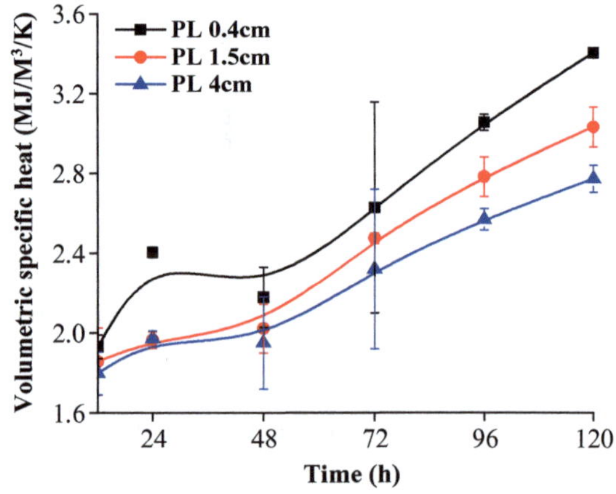

Fig. 3.21 Volumetric specific heat profile in SERS-bran substrates with different particle lengths (*PL*) during *SSF* (solid-state fermentation)

difference between thermal conductivity of the 1.5- and 4-cm matrixes was lower than its difference in heat capacity. The reason was that the specific heat capacity was in closer relationship with the volume of the matrix (or porosity); relatively speaking, the effect of matrix fiber length on specific heat capacity was stronger than that on thermal conductivity. Table 3.6 shows that the constructed power function model could characterize the changes in specific heat capacity with microbial growth.

The study showed that the effect of cell growth on the heat capacity and thermal conductivity was not caused by the cell itself or the chemical composition of the matrix but was closely related with the water content of the microbial cells themselves as well as metabolic water. In the fermentation process, the effect of fiber length on thermal conductivity and specific heat capacity had the same trend,

Table 3.6 Results of model fitting for substrate volumetric specific heat

Sample	a_1 ($\times10^{-1}$)	a_2 ($\times10^{-1}$)	a_3	R^2	RE of K ($\times10^{-2}$)
A_2B_1	0.186	−1.681	1.026	0.817	1.696
A_2B_2	4.35E-06	−0.504	1.926	0.986	0.323
A_2B_3	43.557	−16.608	−3.304	0.970	1.022
A_2B_2	8.582	−6.505	0.364	0.657	12.004
A_3B_2	0.0073	−1.034	1.848	0.900	1.498

but the impact of the latter was even stronger. Interaction between the cell and matrix changed thermal conductivity overall, which is a primary concern for understanding the SSF heat transfer process. The constructed power function model of thermal properties and cell growth fit the heat transfer process of the matrix well.

3.3 Thermal Physics Phenomenon in Solid Substrate Covered by Organisms

As we know, microbial growth depends on many factors. From the thermal physics point of view, the two most important physical parameters are moisture and heat. In addition, the microbial culture conditions have some effect on microbial growth.

3.3.1 Microorganism-Matrix System Transfer Problems

Microbial mycelia generally grow on the surface of material particles (aerobic microorganisms); their growth is influenced by inoculation, spore germination, mycelial extending and branching, and mycelial layer formation. For filamentous fungi, the mycelial layer can be divided into three parts: biofilm, aerial mycelia, and substrate mycelia (Fig. 3.22). The hyphae layer, which covers the surface of the particles (or a shallow surface within the particles) and has a high moisture content, is always considered a biofilm layer. The biofilm layer is made up of microbial cells and water. Most microbial mycelia are enriched in this layer to form a closed hyphae layer; outside oxygen enters this area, and the nutrients of the material inside the particles also penetrate this region. Therefore, the region is often rich in nutrients and oxygen; it is the main place for mass transfer and exchange.

Few hyphae penetrate into the substrate interior, but growth of aerobic microorganisms is difficult because of hypoxia inside the particles; generally, oxygen cannot be measured at a depth of 100 μm. Part of the aerial mycelia will insert into the gas phase of the particles. However, even if the mycelia fill throughout the gap space, they would not completely tightly occupy the gap because the mycelia are loose; they occupy up to about 34 %.

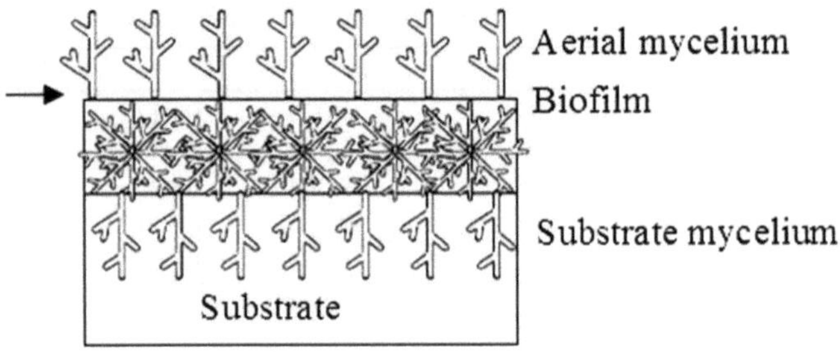

Fig. 3.22 Distribution of filamentous fungi in solid-state fermentation substrate

3.3.1.1 Substrate Temperature Distribution

Factors that affect substrate temperature distribution are generally divided into two categories: Bioheat makes the temperature rise, and various factors exist in matrix heat dissipation. The bioheat produced in the fermentation process is the main reason for the elevated temperature of the substrate and the temperature gradient. Evaporation of water is the main means of heat dissipation in fermentation, and the effects of natural convection are generally not large. In forced ventilation conditions, forced convection is expected to become the most important means of cooling.

The operation mode has important effects on matrix cooling. Different means of heat transfer in the matrix layer are in different operating modes. For static SSF, thermal radiation and natural convection cooling are the main ways. For forced ventilation fermentation, forced convection and heat conduction play an important role. As airflow speed, direction, and mode have important implications for convection cooling intensity, the cooling capacity under different operating modes is different.

3.3.1.2 Matrix Moisture Migration

Substrate water is a basic requirement for the growth of microorganisms. Microorganisms could not absorb nutrients without a water solvent. Changes in the moisture content also affect substrate aeration and temperature conditions; water will be enriched and consumed around the cell and then evaporated to the air above the mycelia to process the heat and mass exchange. The bacterial cell contains water, known as the intracellular water; moisture outside bacteria is called extracellular water, including water in the water film and materials.

3.3.1.3 Matrix Air Conditions

The matrix air is an important factor for the growth of microorganisms; gas composition and content directly affect the generation of microbial mycelia, spores, and secondary metabolites. The air permeability of the matrix (absolute permeability) and its corresponding oxygen diffusion coefficient are important indicators. It should be noted that the absolute permeability and the moisture transfer capacity of the matrix are directly related according to the transfer theory of porous media (Membrillo et al. 2011), so that the parameter can simultaneously reflect the transfer efficiency of air and moisture to the matrix.

It can be inferred from studies that the reasons for changes in matrix permeability are closely related to the effect of the cell on the matrix structure. At the beginning of cell growth, the matrix pore morphology is complete and behaves with strong permeability. With the continued growth of microorganisms, the internal pores of the matrix are filled with the mycelia, resulting in reduced porosity and permeability, which occurs in the logarithmic growth phase and early stationary phase. When cells grow into the stable phase, further use of the matrix will damage the whole structure of the substrate, and the porosity increases are caused by increased matrix debris; thus, the permeability of the matrix starts increasing. In the late stable phase, the use of the nutritional matrix by cells will generate further degraded smaller fragments, so the pore quantity is further increased, and permeability increases.

3.3.2 Effect of Microbial Growth on the Matrix Heat Transfer Process

Metabolic heat is generated in the process of cell growth. The heat transfer process includes the thermal conductivity of contact between the solid skeleton and the fluid in the particle gap; there is convective heat transfer of fluid in the gap and radiation heat transfer between solid skeleton and the gases (Liu 2004). The effect of heat exchange is closely related to the composition of the matrix, fiber length, temperature, and moisture content.

For solid substrate, the change in internal energy of a system is mainly caused by changes in sensible heat and latent heat of materials, as shown in Table 3.7. As the heat source for the substrate for microorganisms, the substrate temperature is determined by the heat production capacity in different periods of the microorganism. Thermal conductivity and thermal diffusivity of the matrix itself are the intrinsic factors that affect the temperature. The thermal conductivity of a matrix is affected by the porosity, moisture content, and chemical composition of the matrix itself; thus, the thermal conductivity in the fermentation process varies because of the utilization of the material layer by microbial growth.

3.3.2.1 Effect of Microbial Growth on Substrate Thermal Conductivity

The effect of microbial growth and matrix water content on substrate thermal conductivity was studied. As shown in Fig. 3.23, the thermal conductivity of

Table 3.7 Heat transfer in solid substrates

	Inflow	Outflow	Heat transfer
Accumulation of sensible heat	Sensible heat of dry air inlet	Sensible heat of dry air outlet	Metabolic heat
	Sensible heat and latent heat of water vapor inlet	Sensible heat of water vapor outlet	
	Sensible heat of water supply	Heat loss of convection and conduction	
	Radiation of environment	Heat loss of radiation	
		Heat loss of evaporation	

steam-exploded wheat straw was positively correlated with biomass yield. The relationship between the two was in line with the following model:

$$\mathrm{TC} = a_1 e^{(-X/a_2)} + a_3 \tag{3.26}$$

The fitting results showed that the model could be used to characterize the thermal conductivity of steam-exploded wheat straw substrate against cell growth. Further study of the influence of the water content on thermal conductivity showed that the power function model can well characterize the thermal conductivity of the matrix changes with cell growth.

As shown in Fig. 3.24, the thermal conductivity of the matrix showed a significant difference with water content changes. It was in a linear relationship with water content.

3.3.2.2 Effect of Microbial Growth on Substrate Specific Heat

As shown in Fig. 3.25, the volume specific heat of steam-exploded straw and cell growth was directly proportionally related. Further analysis found a similar result with the changes in thermal conductivity with bacterial growth; therefore, it was also in line with the power function model:

$$\mathrm{VSH} = a_1 e^{(-X/a_2)} + a_3 \tag{3.27}$$

These studies showed that the cell growth affected the heat capacity and thermal conductivity of the matrix. Previous research had shown that the growth of *Penicillium decumbens* was affected by the combined effect of the specific heat capacity and thermal conductivity (Fig. 3.26). Therefore, the interaction between the microorganism and the matrix made the thermal conductivity change in the fermentation process, which is a primary concern for understanding the SSF heat transfer process. The constructed power function model of thermal properties against the cell growth could clarify the heat transfer process in SSF well.

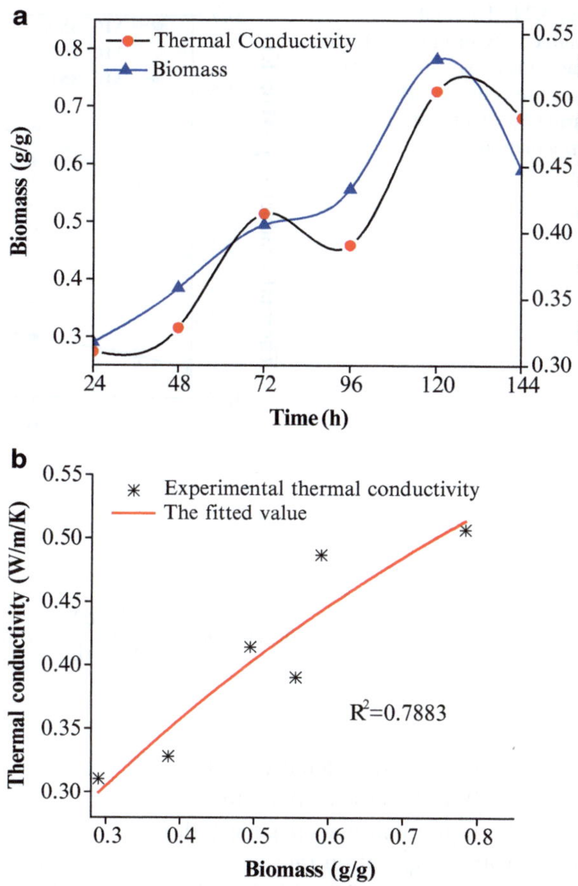

Fig. 3.23 Thermal conductivity variation of SEWS-bran substrates against fungal growth during *SSF* (solid-state fermentation). (**a**) Experimental thermal conductivity variation of SEWS-bran substrates; (**b**) fitting of the thermal conductivity of substrates to original data

3.3.3 Effect of Microbial Growth on Matrix Moisture and Oxygen Transfer

3.3.3.1 Effect of Microbial Growth on Matrix Moisture Transfer

Because of the small amount of free water in the solid matrix, moisture transfer mainly includes the migration of water vapor in the matrix layer and partially unsaturated flow of water caused by the water supply. The moisture generated or consumed by microbial growth and metabolism has an important influence on the moisture balance of the whole matrix layer (Liu et al. 2006).

In fermentation systems that require a water supply, moisture transfer in the fermentation substrate mainly includes seepage of unsaturated water (Liu et al. 2006; Gervais and Molin 2003), the migration of water vapor in the matrix porosity, and moisture changes caused by microbial growth.

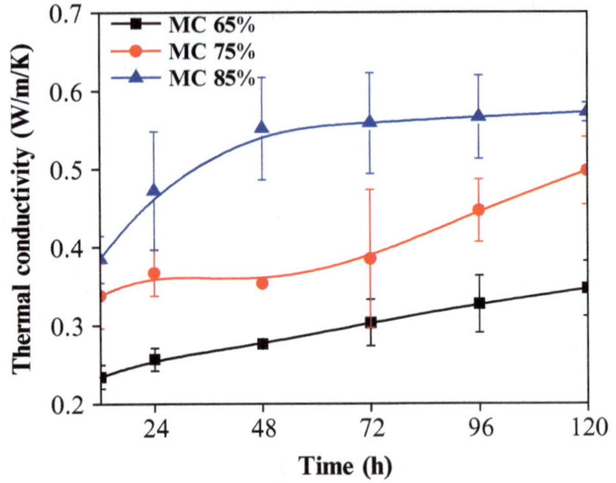

Fig. 3.24 Thermal conductivity profile in SERS-bran substrates with different moisture contents (*MC*) during *SSF* (solid-state fermentation)

$$\frac{d(\rho_1 V_1)}{dt} = -(\rho_1 V_1)\nabla.v_1 - m + \left(\frac{1}{Yx_1}\frac{dX}{dt} + m_1 X\right); V_1 = \phi_w; \tag{3.28}$$

In the formula,

ρ_1 = the liquid phase density, g/cm^3;
V_1 = the liquid phase volume, m^3;
ν_1 = the liquid phase seepage speed, m/s;
m = water evaporation rate, g/s;
Yx_1 = consumption of liquid phase by cell growth, g/g (biomass/water);
m_1 = production of water by cell metabolism, g/g;

ϕ_w = water content (v/v).

Analysis of the moisture movement mechanism of unsaturated porous media showed that the water potential gradient was the inherent water movement power. In a state of equilibrium, the water potential is equal everywhere; the water is in a relatively static state, and there is no continuous flow of water (Zambra et al. 2011). Further analysis of the moisture movement of unsaturated substrate found that the main driving force was the gravitational potential and matrix potential, and the unsaturated hydraulic conductivity was low. Thus, the moisture transfer of unsaturated fermentation substrate was as follows:

$$\frac{dv_1}{dt} = -\frac{gD_1}{K_1}\nabla V_1 + +\mu\nabla.\nabla v_g - g - \frac{\mu_1}{K_1}(v_1 - v_g)V_1 \tag{3.29}$$

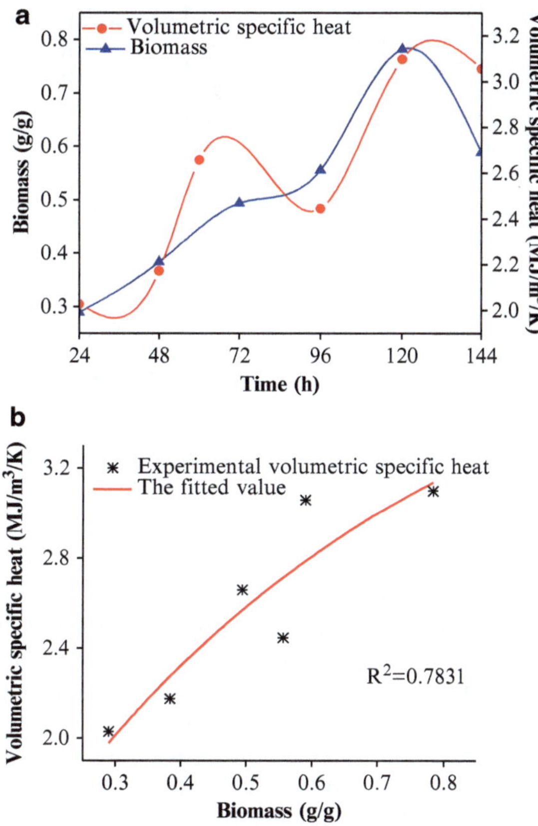

Fig. 3.25 Volumetric specific heat variation of SEWS-bran substrates with fungal growth during *SSF* (solid-state fermentation). (**a**) Experimental volumetric specific heat variation of SEWS-bran substrates; (**b**) fitting of the volumetric specific heat of substrates to original data

In the formula,

D_1 = unsaturated hydraulic conductivity, m/s;
K_1 = substrate permeability, m^2;
g = gravitational potential, bar;
ν_g = water vapor transfer speed, m/s;
μ_1 = dynamic viscosity of water, Pa/s.

For fermentation systems with environmental humidity more than 90 %, there is no need for continuous replenishment of water. Therefore, water input and output can be ignored and only moisture content changes in situ considered. Thus, the water balance includes only the evaporation of moisture and effect of microbial growth and metabolism in a fermentation process without replenishment:

$$\frac{\mathrm{d}(\rho_1 V_1)}{\mathrm{d}t} = -m + \left(\frac{1}{\mathrm{Y}x_1}\frac{\mathrm{d}X}{\mathrm{d}t} + m_1 X\right) \tag{3.30}$$

Despite the small proportion of water used by microbial growth and metabolism, because of its proximity to the microbes, the effect of microbial growth on water

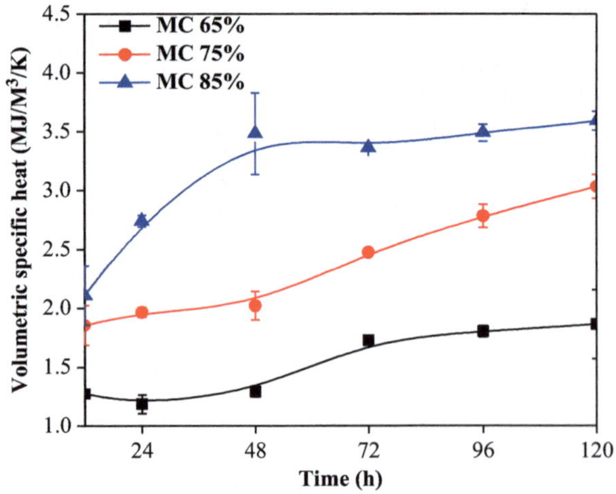

Fig. 3.26 Volumetric specific heat profile in SERS-bran substrates with different moisture contents (*MC*) during *SSF* (solid-state fermentation)

variation cannot be ignored. Nagel et al. (2001) established a mathematical model to calculate the total moisture content of material and the intracellular and extracellular moisture content. This was accomplished via the online determination of O_2, CO_2, air inlet moisture, and air outlet dew point and via the element balance and the amount of carbon dioxide generated to calculate the loss of the matrix dry weight and the water generated or consumed in microbial metabolism and the enzymatic process. Calculate the water used for the synthesis of new cells via the consumption of oxygen. The following gives the reaction formula derived from growth of *Aspergillus oryzae* on glucose and wheat protein; the formula cell amount was obtained based on *Aspergillus niger*.

$$\begin{aligned} &Y_{S/O_2}CH_2O + O_2 + Y_{N/O_2}CN_{1.98}O_{0.63}N_{0.26} \\ &\quad \rightarrow Y_{X/O_2}CH_{1.72}O_{0.52}N_{0.17} + Y_{CO_2/O_2}O_2 + Y_{W/O_2}H_2O \end{aligned} \tag{3.31}$$

In the formula,

Y_{S/O_2} = yield coefficient for substrate S on oxygen, mol/mol O_2;
Y_{N/O_2} = yield coefficient for protein N on oxygen, mol/mol O_2;
Y_{X/O_2} = yield coefficient for biomass X on oxygen, mol/mol O_2;
Y_{CO_2/O_2} = yield coefficient for carbon dioxide on oxygen, mol/mol O_2;
Y_{W/O_2} = yield coefficient for water W on oxygen, mol/mol O_2.

Nagel et al. (2001) constructed the moisture equilibrium equation (3.32) by microbial metabolism, enzymatic hydrolysis, and so on. The left side of the equation is the formation rate of the water; the right side of the equation is the moisture taken away from the reactor by air (i.e., the evaporation of water), water used for new cell growth, moisture produced by microbial cultivation, and moisture needed by starch hydrolysis:

$$\frac{\mathrm{d}W\mathrm{wh}}{\mathrm{d}t} = F_{\mathrm{air}} \cdot (C_{\mathrm{Win}} - C_{\mathrm{Wout}}) + X_{\mathrm{W,X}} \cdot Y_{\mathrm{X/O_2}} \cdot r_{\mathrm{O_2}} \cdot M_{\mathrm{WX}} - Y_{\mathrm{W/O_2}} \cdot r_{\mathrm{O_2}} \cdot M_{\mathrm{WW}} + Y_{\mathrm{hyd}} \cdot Y_{\mathrm{S/O_2}} \cdot r_{\mathrm{O_2}} \tag{3.32}$$

In the formula,

F_{air} = volumetric gas flow at 273 K and 1.013×10^5 Pa, $\mathrm{m^3/s}$;
C_{Win} = water concentration in air inlet recalculated at 273 K and 1.013×10^5 Pa, kg water/$\mathrm{m^3}$;
C_{Wout} = water concentration in air outlet recalculated at 273 K and 1.013×10^5 Pa, kg water/$\mathrm{m^3}$;
$r_{\mathrm{O_2}}$ = oxygen production rate, mol/s;
M_{wx} = molecular weight of biomass, kg/mol;
$X_{\mathrm{w,x}}$ = water content of biomass, kg/kg;
M_{ww} = molecular weight of water, kg/mol;
$Y_{\mathrm{X/O_2}}$ = yield coefficient for biomass X on oxygen, mol/mol O_2;
$Y_{\mathrm{W/O_2}}$ = yield coefficient for water W on oxygen, mol/mol O_2;
Y_{hyd} = water needed to hydrolyze starch, kg/mol;
$Y_{\mathrm{S/O_2}}$ = yield coefficient for substrate S on oxygen, mol/mol O_2.

In the SSF process, the water content of fungal hyphae is up to 3 kg/kg biomass dry weight. According to facts and the model, it is estimated that the water required for new cells is about 45 % of water evaporation.

Another study (Liu 2004) reported the effect of microbial growth and metabolism on moisture; according to the results of the chemical composition balances, the generation rate is

$$r_{\mathrm{w}} = r_{\mathrm{CO_2}} \bullet M_{\mathrm{H_2O}} \bullet \frac{\beta}{\alpha} \tag{3.33}$$

$r_{\mathrm{CO_2}}$ = carbon dioxide production rate, mol/s

$M_{\mathrm{H_2O}}$ = molecular weight of water, kg/mol

Take wheat bran as an example; the chemical formula is $\mathrm{CH_{1.75}O_{0.68}N_{0.05}}$; the chemical balance is

$$\gamma \mathrm{O_2} + \delta \mathrm{CH_{1.75}O_{0.68}N_{0.05}} \rightarrow \alpha \mathrm{CO_2} + \beta \mathrm{H_2O} + \text{remains} \tag{3.34}$$

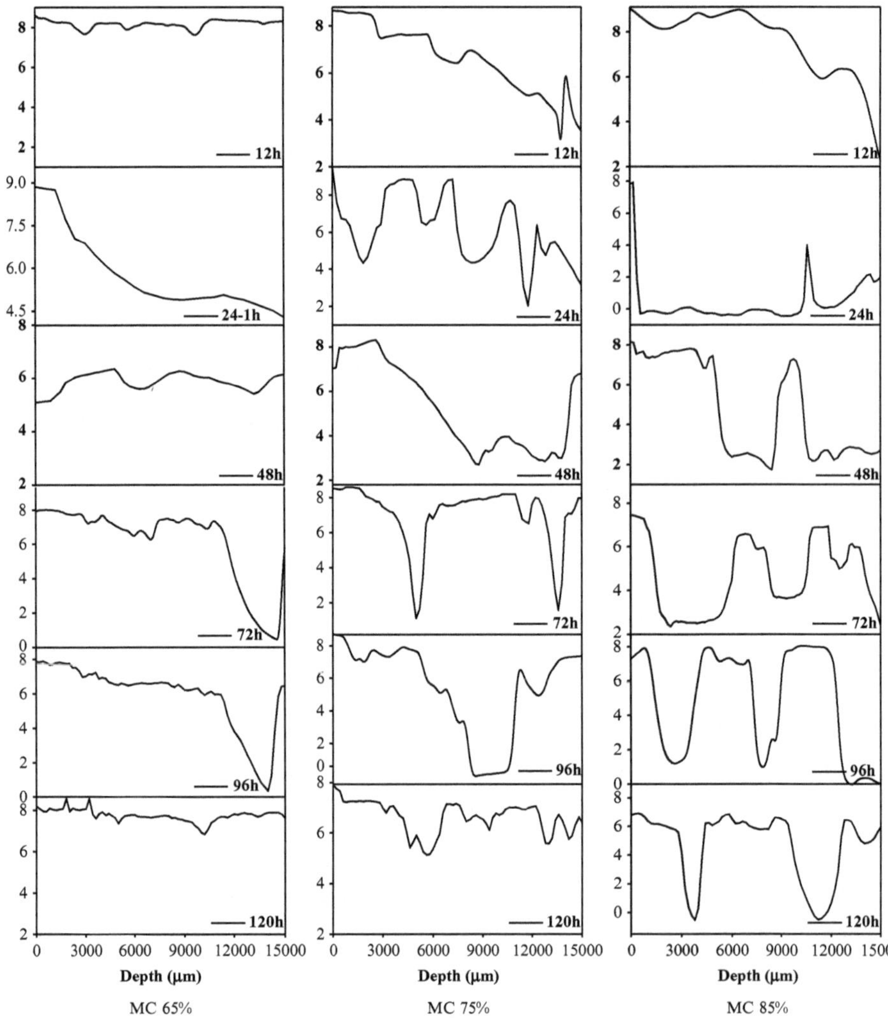

Fig. 3.27 Oxygen profile in SERS-bran substrates with different moisture contents (*MC*) during *SSF* (solid-state fermentation)

According to this formula, the generation rate of water can be obtained from the carbon dioxide generation rate.

3.3.3.2 Effect of Microbial Growth on Matrix Oxygen Transfer

Figure 3.27 shows the oxygen distribution in the matrix during *Penicillium decumbens* fermentation. During fermentation, the oxygen concentration in the matrix decreased because of oxygen utilization by cell growth. Because the fermentation substrate was a porous medium, oxygen concentration was low in the cell growth area and high in the pore area without cell growth. It could be found that the oxygen was not distributed evenly in the matrix at different fermentation stages.

Take the fermentation substrate of steam-exploded straw as an example. The substrate has 85 % moisture content and is 1.5 cm in length. In the early fermentation period (12 h), the oxygen in the matrix is evenly distributed, with continuous distribution with the increase of matrix depth. At 24 h, the growth of microorganisms increases matrix agglomeration; the dense microorganisms make the oxygen concentration lower in the matrix and reduce porosity. These changes coincide with the changes in matrix fractal dimension and air permeability. At 48 h, the further growth of cells causes the agglomerated matrix to start disintegrating into small pieces; its porosity increases. Meanwhile, the fractal dimension of the matrix-bacterial junction also starts to increase, and the corresponding matrix air permeability also gradually increases. During 72–120 h, the matrix degrades into smaller clumps, further accompanied by cell growth. The corresponding pores (region of high oxygen concentration) also increase, and the matrix fractal dimension and gas permeability also increase. Therefore, the oxygen concentration change in the region is able to reflect detailed changes in matrix morphology during fermentation. Compared to the details changes in different water content of matrix morphology, it can be found that with the increase of water content, the agglomerate strength of the matrix increases, and the corresponding decomposition of clumps is more obvious.

The particles in SSF may be regarded as a certain thickness of layers; the aerial mycelia are located in the layer of wet cells, and the gap between the aerial mycelia is filled with air. This was confirmed experimentally: The cells grew in the different layers, and the oxygen concentration remained substantially stable at different aerial hyphae layer distances. Oostra et al. (2001) discovered very small oxygen gradients in the aerial upper layer, which was to be expected because the air-filled pores gave negligible resistance to oxygen transfer. In the wet part of the fungal mat, however, oxygen transfer was obviously hampered. Figure 3.27 shows that there was a smooth oxygen gradient in the aerial hyphae layer, but steep oxygen profiles developed in the wet layer of the fungal mat as growth proceeded. After approximately 36 h, oxygen was depleted at a depth of about 60 mm in both media. It was further concluded that oxygen transfer from the gaseous bulk phase to the microorganisms mainly occurred via the oxygen dissolved in the liquid film because the contribution of aerial hyphae in overall oxygen uptake was negligible. The thickness of the liquid layer and the available gas-liquid interface were identified as key parameters in optimizing oxygen transfer to the microorganisms in SSF (Oostra et al. 2001).

The zero position in Fig. 3.27 indicates the interface between the aerial and wet layer of the fungal mat. The profiles have been shifted horizontally to let their zero positions coincide and make direct comparison possible. The oxygen concentrations in the gas phase are presented as equilibrium values in the liquid phase at 30 °C.

A theoretical analysis showed that four unknown parameters contributed to this progression: the thickness of the wet fungal layer L, its packing density r_x, the specific respiratory activity q_o of the mycelia in the wet aerobic layer, and the diffusion coefficient D_e of oxygen in this layer. The effective diffusion coefficient in a fungal pellet was related to the diffusion coefficient through the bulk medium, the porosity of the biolayer (void fraction in film), and the tortuosity factor.

The rate of oxygen transfer to the fungus was limited because of poor oxygen transport through the wet fungal layer. Aerobic growth took place in a thin layer (about 60 mm) near the gas-liquid interface, while in deeper regions, anaerobic growth and anaerobic conversions took place. With these results in mind, the authors (Oostra et al. 2001) discussed some possibilities to optimize aerobic growth in an SSF bioreactor. One way to increase aerobic growth is by increasing the oxygen partial pressure in the gas phase because higher interfacial oxygen concentrations will increase the oxygen penetration depth. However, increasing the oxygen concentration in the gas phase is costly, and high oxygen concentrations can lead to growth inhibition. A more practical way to increase aerobic growth is by increasing the oxygen transfer rate from the gas phase to the fungus via enlargement of the interface area between the gas phase and the moist fungal layer. It is one of the best ways to reduce the diameter of the material particles to increase the specific surface area. Increases in interparticle and intraparticle porosity are also effective measures to strengthen the oxygen supply. A small internal aperture is usually conducive to water filling; a larger aperture is conducive to gas filling.

Generally, the smaller the particle sizes of the solid matrix, the greater the surface area microorganisms can use. However, too small a particle size will often lead to cohesion agglomeration of the matrix, thereby affecting the ventilation and oxygen transfer effects and respiratory metabolism of the microorganisms, even resulting in delayed growth or death of cells. In contrast, a larger particle size would improve ventilation, oxygen transfer, and metabolic efficiency because of the increase in gap size between the particle but limit the growth area for microbes. Pandy (1991) studied the effect of particle size on enzyme activity in the SSF process; the experimental results showed that the particle size distribution of the solid matrix had a significant impact on the growth of microorganisms and enzyme production. When the particle size distribution of wheat bran was 425–500 and 500–600 μm, glucoamylase produced the highest enzyme activity by *Aspergillus niger*. The results for particle size less than 180 μm and greater than 1.4 mm were equivalent; the glucoamylase enzyme activities were low.

Traditional fermentation is often mixed fermentation using a porous substrate such as rice husks or wheat bran with raw food materials or pretreated raw material to increase porosity within the particles to expand the area for gas exchange.

Because oxygen must be dissolved in the water film before use, the moisture content of the solid matrix, the water film thickness, and the area of the water film are key factors in determining the oxygen concentration. Water fills into the pore between particles, and biomass sucks up water from the solid matrix. Therefore, the thickness of the layer of wet cells depends on the water content of the solid matrix. Nagel et al. (2001) found that there was a linear correspondence relationship between the water content in the bacterial layer and the water content of solid particles. Therefore, adjusting the flow rate and humidity of the air inlet, supplying water in a timely fashion, and thereby controlling the moisture content of the particles are key for controlling the thickness of the wet cell layer and ensuring SSF aerobic conditions.

3.3.4 *Effect of Microbial Growth on the Solid Matrix*

The essential difference between the nutritional matrix and inert matrix is the decomposed components and destroyed structure that caused by microorganisms. These changes can be characterized by the matrix density macroscopically. From the angle of the porous medium, the movement of the solid phase in the nutritional support mainly is by mechanical dispersion and molecular diffusion between the liquid film and the solid phase of the organic macromolecule. The enzyme migration in the liquid film has been studied previously, but for the matrix layer, the movement of the macromolecular substances is accompanied by the transfer of the liquid flow. However, the migration of the macromolecules in the matrix layer pores is very weak because of little free water.

From the angle of the operation mode, another movement form of the nutritional support can be found: the movement of overall solid phase in situ by the pressure-driving force (Mitchell et al. 1991), with the movement speed and direction closely related to the gas flow velocity and direction, but rarely diffusion.

Thus, the solid phase movement of nutritional support matrix on the one hand is because of in situ migration by gas flow; on the other hand, it is caused by the changes in its volume, density, porosity, and structure.

$$\frac{\mathrm{d}(\rho_s V_s)}{\mathrm{d}t} = -(\rho_s V_s)\nabla . v_s + \left(\frac{1}{Y_s}\frac{\mathrm{d}X}{\mathrm{d}t} + m_s X\right) \tag{3.35}$$

$$\frac{\mathrm{d}(\rho_s V_s)}{\mathrm{d}t} = r_s \tag{3.36}$$

In the formula,

ρ_s = the density of the solid phase, g/cm^3;
V_s = the solid phase volume, m^3;
v_s = the solid phase transfer speed, m/s;
Y_s = consumption of solid phase by cell growth, g/g (biomass/dry weight of substrate);
m_s = consumption of solid phase by cell metabolism, g/g;

r_s = substrate weight loss rate.

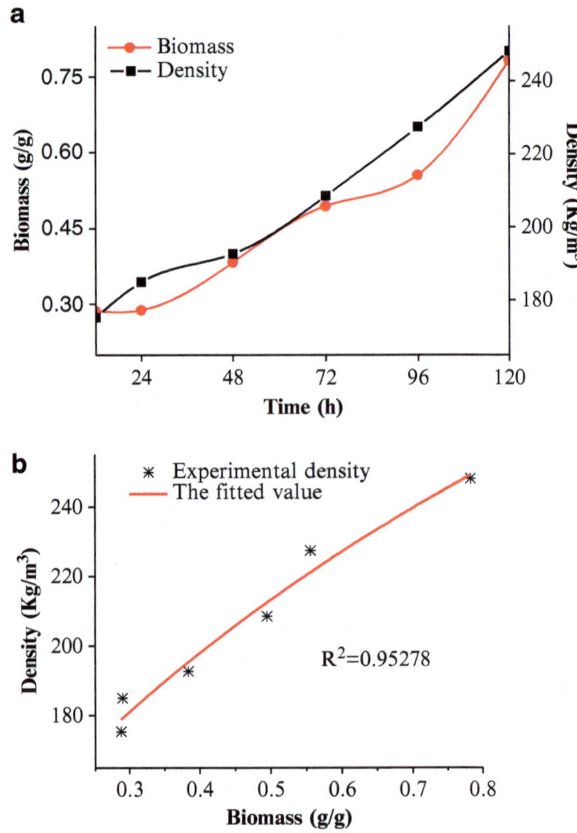

Fig. 3.28 Density variation of SEWS-bran substrates against fungal growth during *SSF* (solid-state fermentation). (**a**) Experimental density variation of SEWS-bran substrates in SSF; (**b**) fitting of the density variation of substrates to original data

As shown in Fig. 3.28a, the density of steam-exploded straw was positively correlated with cell growth in the fermentation process. The relationship was in line with the power function as follows:

$$\rho = a_1 e^{(-X/a_2)} + a_3 \tag{3.37}$$

In the formula,

ρ = the density of the solid phase, kg/m^3;
X = biomass, g/g (biomass/dry weight of substrate);
a_1, a_2, a_3 = the parameters of the model, determined by the substrate.

Figure 3.28b displays the fitting result and shows that the model could characterize the relationship between the matrix density and cell growth. The determination coefficient of the corresponding model R^2 was approximately 0.9528.

3.4 Design and Scale-Up of Solid-State Fermentation Bioreactors

Bioreactors provide space for suitable environments and conditions for the growth of microorganisms on the solid material and metabolite production. The reactor should meet the following basic requirements: accommodate substrates (closed or semiclosed room); prevent the contamination of microorganisms from outside as much as possible; prevent the fermentation microorganisms from getting out to the environment; maintain proper temperature and humidity for fermentation; provide enough oxygen for aerobic microorganisms; provide an anaerobic environment for anaerobic microorganisms; provide easy mixing and movement materials; facilitate the extraction of the fermentation products as much as possible; and distribute the material as evenly as possible.

3.4.1 Solid-State Fermentation Bioreactors

Since penicillin was discovered by Fleming and was successfully put into industrial production through cooperation between microbiologists and chemical engineers in 1945, the submerged fermentation technique has spawned a modern fermentation industry. SSF has not fulfilled the requirements of the modern fermentation industry and has thus been ignored because it has no engineering means to solve such problems as transportation, agitation, oxygen supply, and control of temperature, humidity, and pH (Chen et al. 2003; Hölker et al. 2004). The key point is that there has not been a good solid-state fermentor that meets the requirements of the modern fermentation industry.

A bioreactor is the heart of a fermentation process; in contrast to SmF systems, SSF bioreactor systems have yet to reach a high degree of development, mainly because of the problems associated with solid beds, like poor mixing, heat-transferring characteristics, and material handling (Raghavarao et al. 2003). Especially, heat removal is typically the major concern in the case of SSF (Chen and Xu 2004; Mitchell et al. 2006).

To date, many types of reactors for SSF (including laboratory, pilot, and industrial scales of production) have been developed. Lonsane et al. (1992) summarized them into nine types: (1) drum type, (2) wooden box type, (3) capped plate type, (4) vertical cultivation box type, (5) inclined culturing box type, (6) tray type, (7) belt conveyor type, (8) cylinder type, and (9) mixed type. They can be summarized in two categories (static and dynamic fermentation) according to the state of the culture medium. The static state means a stationary culture medium, which causes difficulties in mass transfer, heat transfer, oxygen supply, and control of temperature, humidity, and pH. The dynamic state means that the culture medium is in intermittent and continuous movement, which significantly improves mass transfer, heat transfer, and oxygen supply, but the mechanical parts used are unfavorable for aseptic operation, energy consumption on material agitating is high, mycelia are likely to be damaged, and engineering scaling up is difficult. Regardless of the bioreactor types, the following issues must be considered: (1) inoculation; (2) sterilization; (3) characteristics of the fermentation substrate; (4) air supply; (5) measurement of the parameters and control; (6) sampling and analysis; and (7) simplicity of the structure so it is easy to operate.

3.4.1.1 Static Bioreactor

Currently, the static reactor for SSF is common in laboratory studies, especially the cylindrical reactor. The bioreactors reported in the literature are mostly one or more static cylindrical reactors placed parallel in a thermostatic chamber and ventilated with saturated air. The advantages of this kind of bioreactor are as follows:

1. The system is simple, inexpensive, and easy to operate.
2. It overcomes the shortcomings of much basic research in a shake flask for SSF, operates parallel experiments for a multitude of conditions, and keeps the culture conditions (e.g., temperature, humidity) uniform.
3. The system is easy to sterilize.

Its disadvantages are as follows:

1. It cannot accurately control the humidity of the gases and materials, only supply saturated wet air.
2. It cannot be sampled and analyzed.
3. It is difficult to eliminate the effect of bed diameter enlargement in the amplification process.

Regardless of volume and the height diameter ratio, a static reactor has the same basic form, but the gas supply, insulation, and temperature control system are vastly different. A sound system should behave according to the following characteristics: (1) measure and control the humidity and temperature of the inlet air; (2) measure the exhaust gas composition and feedback regulation of humidity and temperature of the intake air; (3) provide gas supply circulation in a relatively large-scale application; and (4) have sound gas filtration equipment.

3.4.1.2 Dynamic Bioreactor

With respects to the "static" in static closed SSF, the dynamic closed SSF process means that the substrate media is in continuous or intermittent agitation. In large-scale production, because of microbial metabolism heat caused by the static SSF process and difficulty in metabolite transfer, there are many restrictions in its wide utilization. Therefore, it is always a dynamic process in practice. An efficient cooling system promotes metabolic heat transfer; in addition, stirring promotes mass transfer, so that the fermentation environment is more uniform and it is easier to realize control of important process parameters.

3.4.2 Factors that Influence Bioreactor Design

The crucial problem for industrial SSF is the scale-up of an SSF bioreactor. When choosing reactors, the performance of the various reactors must be taken into account. The selected reactor should meet the fermentation requirements; at the same time, the cost of investment and running the reactor must also be considered.

Generally, large-scale SSF could greatly reduce investment and operating costs compared with submerged fermentation. But, this also depends on reasonable design, selection, and operation of the SSF bioreactors.

Analysis of the performance of an SSF reactor should focus on the following aspects:

When choosing bioreactors, the *impact of investment and running costs* on the total cost should be considered. End devices (such as drum bioreactors) would be considered for high value-added products if necessary. Generally, the ordinary ventilation fermenting cellar can meet the requirement for most low value-added products.

Ventilation used in the bioreactor: This can be divided into natural convection ventilation and forced ventilation. Most SSF is aerobic. The primary operating variables for ventilation are pressure volume flow; air supply pressure drop; air velocity in the entrance or internally in the reactor; and air temperature and humidity. Intermittent ventilation and agitation can basically meet the technological needs for many SSF products, but air volume and stirring speed parameters need to be determined experimentally.

Agitation mode: SSF can be divided into three types: completely stationary fermentation, intermittent stirred fermentation, and continuous stirred fermentation. If microbes can withstand continuous stirring, continuously stirred fermentation is undoubtedly the most desirable solution. Agitated operation has a positive impact on oxygen supply and ventilation, heat removal, and carbon dioxide. However, the negative impact of stirring is obvious; for example, agitation causes fracture of mycelia, which affects the growth of microorganisms and even affects the synthesis of metabolites. Stirring also causes sticky material agglomerate and internal hypoxia. Single-cell microorganisms are not sensitive to a shearing force; most filamentous fungi are sensitive. Therefore, caution must be exercised in selecting a reactor with a mechanical stirring device. Mechanical stirring is more common, but different types of mechanical agitation generate material flow in various directions, mainly including radial flow and axial flow. The major operating variables of a stirring operation are stirring intensity (such as speed) and stirring duration.

Control of the air temperature and humidity: Air is generally heated or cooled through the heat exchanger outside the reactor. Sometimes, the temperature and humidity of the air can be adjusted at the same time.

Method of material temperature control: The most difficult technical problem in SSF is how to remove metabolic heat effectively. The method and efficiency of heat removal vary with bioreactor structure variety. The work capacity of a reactor also has a large impact on temperature. The cooling methods include a cooling jacket installed in the reactor and cooling media added in the stirring shaft. The most common method is forced ventilation; air can take away the water evaporated with evaporation heat.

Material moisture control method: Saturated moist air would replenish water in the material; a more common method is spraying water into the material directly. The former method can distribute added water more evenly to the material; the latter method needs stirring to achieve uniform distribution.

Energy consumption: The energy consumption of the reactor is mainly used for sterilization of raw material, ventilation, stirring, and cooling. In addition, feed and discharge of material require energy consumption.

Prevention and control of contamination: From the point of view of ensuring fermentation safety, pure fermentation is necessary. However, polluting microorganisms in a controllable range are also acceptable, even indispensable, in some fermentation flavor products. Most SSF, because of the low water activity of selective media and materials, is not conducive to bacterial growth; thus, there is no need to pursue complete sterility. In addition, different preventive measures can be taken for different products, focusing on raw materials, bioreactors, and ventilation systems.

Equipment productivity: The amount of product produced per unit time and volume of the fermentation vessel is an important basis for equipment selection. Fermentation time is determined according to the characteristics of the microorganisms and products, so the key question is the loading coefficient. For the commonly used SSF reactor, the highest is for the cellar; the second highest is the packed bed and fluidized bed. The general equipment, such as the drum, disk, and pool, have a relatively lower loading coefficient, generally only 30–40 %.

Since the twentieth century, with the expanding applications of SSF technology, traditional fermentation equipment cannot meet needs. Many new SSF bioreactors were invented, especially those suitable for large-scale fermentation. Bioreactors with a high degree of automation and mechanization receive special attention. The requirements for a modern bioreactor are an entire compact production line; multifunctional fermentor; completion of most of the process within the reactor; sealed fermentor and high pollution prevention ability; good insulation effect of temperature and moisture; energy conservation; mechanization; automatic detection and control of process parameters; simple operation; low labor intensity; high production efficiency; and small investment and low operating costs.

3.4.3 The Principle of Scale-Up of a Solid-State Fermentation Bioreactor

3.4.3.1 Scale-Up

Research and development of biological products usually need to go through three stages: laboratory stage, pilot stage, and industrialized-scale stage. Even though the biological reaction of the various stages in the bioreactor is the same, the mass and heat transfer are often not the same. Estimating the state of the biological reaction in bioreactors of different scales, especially in an amplification process, and maintaining cell growth and the biological reaction rate similar to that of small bioreactors are the main objectives of the scale-up of SSF bioreactors (Jia 2003).

The scale-up of bioreactors includes a series of steps that depend on whether the amplification process can provide sufficient data for subsequent production.

According to the basic steps of scaling up for a submerged fermentation process, the steps of the amplification process for SSF can be summarized as follows:

1. *Flask scale*: The work capacity is 50–1,000 g; mainly, this is for study of media composition and optimum culture conditions. It is lower cost, and the desired data is obtained in a short time.
2. *Laboratory scale*: The operation capacity is 5–20 kg; mainly, the study at this scale is of the inoculation amount, media sterilization method, aeration, stirring, downstream processes, and quantitative study of various fermentation parameters, such as O_2 transfer rate, CO_2 generation rate, biomass accumulation speed, and product synthesis. In addition, study can be made of the effect of pH and aeration rate on fermentation, the fermentation mode (continuous or intermittent), the right control strategies for the equipment, and the technical and economic feasibility of the process.
3. *Pilot scale*: The working capacity is 50–5,000 kg; mainly, this scale verifies the reliability of the data obtained in the laboratory fermentor; selects the optimal means of inoculation, the media sterilization method, and downstream process methods; and produces a certain amount of product for physical chemical analysis.
4. *Industrialized scale*: The working capacity is 25–1,000 tons; mainly, the study optimizes the production process and eventually obtains good economic benefit.

The main problem for the scale-up of SSF is that an increase in the fermentation scale affects the fermentation yield. Corresponding factors that affect the scale-up process are discussed next.

Heat Transfer

A large amount of metabolic heat is produced in the fermentation process. In addition, the kinetic energy of the agitation and aeration in the fermentation process will eventually be converted into heat. The heat generated by the fermentation process is about 80–3,200 kcal/kg dry material. The maximum temperature difference in the medium can be achieved above 30 °C. The temperature can be controlled by the thermostatic chamber, water bath, and other methods in an experimental bioreactor. With an increase in reactor volume, the specific surface area of the reactor decreases and results in the relative reduction of the cooling surface; the heat is difficult to control effectively (Chinn and Nokes 2003). In SSF, the heat transfer is achieved by conduction and convection; the cooling effect is poor because of the static material and the low coefficient of heat transfer of air and solid substrate. Therefore, the heat transfer problem is the bottleneck of an ordinary SSF reactor. Evaporative cooling is used on many occasions; its cooling efficiency is much higher than conduction and convection. It was reported that 80 % of the heat is removed by evaporative cooling, thus causing media dryness. Therefore, in the SSF process, an effective combination for controlling temperature and humidity is essential (Barstow et al. 1988).

Mass Transfer

The amount of water in the SSF process is important; too low or too high a water content will affect the fermentation yield. The moisture content affects the physicochemical properties of the fermentation substrate significantly; these physicochemical properties in turn affect the fermentation yield. Therefore, the water content and water activity of the media must be strictly controlled. Different moisture control strategies have been reported, such as using a high initial media water content and a high air water content (96–98 %) to maintain the humidity of the media. Evaporative cooling is a preferred way to lose heat, but it will cause uneven water distribution in the media (Gervais and Molin 2003).

At the laboratory flask scale, ventilation by simply shaking the flask is sufficient. However, in a large-scale SSF process, forced ventilation is needed. Ventilation not only can provide O_2 but also can remove CO_2 and other volatile metabolites and promote heat transfer. The transfer of oxygen from the gas to the liquid phase is limited by its diffusion speed to the media rather than the transfer speed from the liquid phase to the cell. In a static reactor, the aerobic mycelia deep in the particles cannot grow and cannot get enough O_2 because of the diffusion difficulty of oxygen between the particles. In large-scale SSF, ventilation is needed, whereas other problems will arise once operating conditions are adopted that meet the O_2 transfer requirements. Generally, higher efficiency can be achieved with a combination of forced ventilation and stirring. To prevent channeling, an air distributor or other methods, such as changing the direction of airflow or the air pressure inside the reactor periodically, are needed (Liu and Yang 2006).

Agitation is one of the important SSF operation parameters, and it plays an important role in guaranteeing consistent temperature and humidity between the medium and gaseous environment in SSF. Stirring can increase the gas-liquid interface area of the media, promote mass and heat transfer, and distribute the nutrients evenly in the medium. Although stirring is beneficial and even essential in some fermentation processes, opposite reports exist; for example, stirring made *Rhizopus oligosporus* grow poorly (Sabu et al. 2002) and reduced the ethanol production of *Saccharomyces cerevisiae* (Roukas 1994). Agitation will have a negative impact on the porosity of certain materials and affect the normal growth or cut mycelia of filamentous fungi. The scale-up of a reactor with stirring does not create new problems because the stirring speed is generally slower and usually intermittent rather than continuous; this can avoid injuring by shear force on the mycelia or the adhesion of mycelia on solid particles.

Aseptic Operation

Except for autoclaving, many of the steps are not strictly sterile in SSF operations. Therefore, if strict precautions are absent, contamination often occurs in large-scale SSF. Usually, measures to enlarge the inoculation amount avoid contamination. An appropriate type of bioreactor is needed in large-scale fermentation.

3.4.3.2 Approaches for Scale-Up

The scale-up of SSF bioreactors is a bottleneck in the application of SSF. It is really difficult to design an efficiently operating large-scale SSF bioreactor because of the complex system and obvious gradient of mass and temperature. There is no common theory for scale-up. Scale-up of SSF bioreactors not only enlarges the size of the apparatus but also ensures that, in larger equipment, the microbial reaction conditions are substantially identical with the reaction conditions at the small experimental scale. In the bioreactor reaction system, there are three important different types of process: a thermodynamic process, a microscopic kinetic process, and the transfer process. The transfer process (mainly mass transfer and heat transfer) is the most important and is a core issue for scale-up.

Various quantitative approaches have been proposed for scale-up of SSF bioreactors, including the use of mathematical models and various "simplified approaches" that have some similarity with the "rule-of-thumb" approaches to scaling-up SmF bioreactors. Given the complexity of SSF systems, models will be more powerful tools and should be preferred if possible, especially because various fast-solving models are available in the literature and can be adapted to new systems without requiring an onerous amount of work.

Method of Trial and Error

The trial-and-error method is often used in scale-up of SSF bioreactors. Direct extrapolation to an industrial scale according to reports from the column reactor in the laboratory scale will encounter many difficulties. Elimination of a large amount of metabolic heat produced during fermentation is a difficult problem. In the trial-and-error method, the various problems encountered in the experimental stages must constantly be adjusted until the experiment is successful.

Geometric Similarity Method

The scale-up of a gas-solid fluidized bed is a classic example of enlarged bioreactors using the geometrically similar method. Because of good mixing of substrates, a scale-up method similar to that for SmF is adopted. Based on experimental studies of an increase in column diameter, the material layer height, temperature distribution, and effect of metabolic heat on fermentation yield in the laboratory-scale, a reactor was successfully enlarged to 3,300 L using the geometric similarity method (Chen et al. 2008). Although this method is simple, it cannot guarantee that the results obtained for a large-scale case are similar to those for a small experiment.

Quantitative Analysis Method

Presently, there is still no rational scale-up method to achieve reactor amplification, but much recent experimental work is approaching this goal. There are two amplification methods that are basically derived from the mass and energy balance, but the complexity of the mathematical tools required by these two methods is different. Relatively complex methods, such as mathematical models, require the solution of partial differential equations that describe the state of the reactor. There are relatively simple methods, such as a dimensionless method that is a simplified mathematical model proposed by Saucedo-Castaneda et al. (1992) and calculated by the algebraic equations.

There are reports of the scale-up of bioreactors by mathematical model methods: The SSF process is simulated at the laboratory scale, and then the amplification of the reactor is guided. The reactor types include drum, ordinary packed bed, and Zymotis packed bed fermentors.

For an ordinary packed bed reactor, if the temperature in any position of the reactor should not exceed the set value, then the height of the material in the packed bed will be limited. The height depends on the heat production dynamics of strains used and superficial gas velocity in the packed bed. If height is restricted in the amplification process, it only increases the width of the material bed to increase the volume of the reactor. For ordinary microorganisms, substrate, and operating conditions, the critical height of the material layer is 1 m. With an increase in height of the material layer, the ventilation flow rate will significantly increase and even reach a fluidized state (Mitchell et al. 1999).

For a Zymotis-type reactor, because of the added cooling plate in the internal reactor, the height restrictions can be cancelled, thereby making it possible to design reactors several meters high. Further study is needed regarding whether there is an excessive pressure drop (Mitchell and von Meien 2000) . For a drum reactor, high ventilation (2 vvm) and low air humidity (15 % relative humidity) could achieve better temperature control under a working capacity of a few hundred kilograms of substrate. The fermentation process needs to add a lot of moisture, which can be realized through intermittent spraying of water (Mitchell et al. 2000).

For the dimensionless method, the key is the description of heat generation and heat removal from the energy balance equation and given a dimensionless number by a certain percentage. A dimensionless number can be used to build the operating block diagram and for displaying the value of the operation parameters. It is necessary to control the bed temperature in different bioreactor volumes (Saucedo-Castaneda et al. 1992).

In the past several years, great progress has been made in the scale-up methods for an SSF bioreactor; however, these methods have not yet been widely used in practice. Much experimental work is needed before these tools can be used successfully in this scale-up.

References

Ashley VM, Mitchell DA, Howes T. Evaluating strategies for overcoming overheating problems during solid-state fermentation in packed bed bioreactors. Biochem Eng J. 1999;3:141–50.

Barstow LM, Dale BE, Tengerdy RP. Evaporative temperature and moisture control in solid substrate fermentation. Biotechnol Technol. 1988;2:237–42.

Cassel D, Biggar J, Nielsen D. Soil-water movement in response to imposed temperature gradients. Soil Sci Soc Am J. 1969;33:493–500.

Chen HZ, Li ZH. Gas dual-dynamic solid state fermentation technique and apparatus. US Patent 20,030/138,943; 2003.

Chen HZ, Xu J. Principle and application of modern solid-state fermentation. Beijing: Chemical Industry Press; 2004.

Chen HZ, Han YJ, Xu J. Simultaneous saccharification and fermentation of steam exploded wheat straw pretreated with alkaline peroxide. Process Biochem. 2008;43:1462–6.

Chinn MS, Nokes SE. Temperature control of a solid substrate cultivation deep-bed reactor using an internal heat exchanger. Trans Am Soc Agric Eng. 2003;46:1741–9.

Constantz J, Murphy F. The temperature dependence of ponded infiltration under isothermal conditions. J Hydrol. 1991;122:119–28.

Corona A, Sáez D, Agosin E. Effect of water activity on gibberellic acid production by *Gibberella fujikuroi* under solid-state fermentation conditions. Process Biochem. 2005;40:2655–8.

Costa JAV, Alegre RM, Hasan SDM. Packing density and thermal conductivity determination for rice bran solid-state fermentation. Biotechnol Technol. 1998;12:747–50.

De Vries D. Simultaneous transfer of heat and moisture in porous media. Trans Am Geophys Union. 1958;39:909–16.

Duan YY, Chen HZ. Effect of three-phase structure of solid-state fermentation substrates on its transfer properties. CIESC J. 2012;63:1204–10.

Favela-Torres E, Córdova-López J, García-Rivero M, et al. Kinetics of growth of *Aspergillus niger* during submerged, agar surface and solid state fermentations. Process Biochem. 1998;33:103–7.

Gardner W. Solutions of the flow equation for the drying of soils and other porous media. Soil Sci Soc Am J. 1959;23:183–7.

Gervais P, Molin P. The role of water in solid-state fermentation. Biochem Eng J. 2003;13:85–101.

GutierrezRojas M, Hosn SAA, Auria R, et al. Heat transfer in citric acid production by solid state fermentation. Process Biochem. 1996;31:363–9.
Hölker U, Höfer M, Lenz J. Biotechnological advantages of laboratory-scale solid-state fermentation with fungi. Appl Microbiol Biotechnol. 2004;64:175–86.
Ikasari L, Mitchell DA. Oxygen uptake kinetics during solid state fermentation with *Rhizopus oligosporus*. Biotechnol Technol. 1998;12:171–5.
Jensen R, Haridasan M. Effect of temperature on pressure head-water content relationship and conductivity of two soils. Soil Sci Soc Am J. 1972;36:703–8.
Jia SR. Bioreaction engineering principles. Beijing: Science Press; 2003.
Jou RY, Lo CT. Heat and mass transfer measurements for tray-fermented fungal products. Int J Thermophys. 2011;32:523–36.
Liu BC. Study on the heat and mass transfer in soil with phase change and simulation of the growing of plant roots system. Dissertation, Huazhong University of Science and Technology; 2004.
Liu J, Yang J. Process calorimetry on solid-state fermentation of vinegar wastes in bioreactor with air pressure pulsation. Chem Biochem Eng Q. 2006;20:449–55.
Liu W, Fan AW, Huang XM. Theory and application of heat and mass transfer in porous media. Beijing: Science Press; 2006.
Lonsane B, Saucedo-Castaneda G, Raimbault M, et al. Scale-up strategies for solid state fermentation systems. Process Biochem. 1992;27:259–73.
Martynenko OG, Pavlyukevich NV. Heat and mass transfer in porous media. 1998;71:1–13.
Membrillo I, Sánchez C, Meneses M, et al. Particle geometry affects differentially substrate composition and enzyme profiles by *Pleurotus ostreatus* growing on sugar cane bagasse. Bioresour Technol. 2011;102:1581–6.
Mitchell DA, von Meien OF. Mathematical modeling as a tool to investigate the design and operation of the Zymotis packed-bed bioreactor for solid-state fermentation. Biotechnol Bioeng. 2000;68:127–35.
Mitchell DA, Do DD, Greenfield PF, et al. A semimechanistic mathematical model for growth of *Rhizopus oligosporus* in a model solid-state fermentation system. Biotechnol Bioeng. 1991;38:353–62.
Mitchell DA, Pandey A, Sangsurasak P, et al. Scale-up strategies for packed-bed bioreactors for solid-state fermentation. Process Biochem. 1999;35:167–78.
Mitchell DA, Krieger N, Stuart DM, et al. New developments in solid-state fermentation II. Rational approaches to the design, operation and scale-up of bioreactors. Process Biochem. 2000;35:1211–25.
Mitchell DA, von Meien OF, Krieger N. Recent developments in modeling of solid-state fermentation: heat and mass transfer in bioreactors. Biochem Eng J. 2003;13:137–47.
Mitchell DA, Krieger N, Berovic M. Solid-state fermentation bioreactors: fundamentals of design and operation. New York: Springer; 2006.
Nagel FJJI, Tramper J, Bakker MSN, et al. Model for on-line moisture-content control during solid-state fermentation. Biotechnol Bioeng. 2001;72:231–43.
Nagel FJ, Van As H, Tramper J, et al. Water and glucose gradients in the substrate measured with NMR imaging during solid-state fermentation with *Aspergillus oryzae*. Biotechnol Bioeng. 2002;79:653–63.
Oostra J, Le Comte E, Van den Heuvel J, et al. Intra-particle oxygen diffusion limitation in solid-state fermentation. Biotechnol Bioeng. 2001;75:13–24.
Pandey A. Effect of particle size of substrate of enzyme production in solid-state fermentation. Bioresour Technol. 1991;37:169–72.
Philip J, De Vries D. Moisture movement in porous materials under temperature gradients. Trans Am Geophys Union. 1957;38:222–32.
Raghavarao K, Ranganathan T, Karanth N. Some engineering aspects of solid-state fermentation. Biochem Eng J. 2003;13:127–35.
Roukas T. Solid-state fermentation of carob pods for ethanol-production. Appl Microbiol Biotechnol. 1994;41:296–301.

Sabu A, Sarita S, Pandey A, et al. Solid-state fermentation for production of phytase by *Rhizopus oligosporus*. Appl Biochem Biotechnol. 2002;102:251–60.

Sangsurasak P, Nopharatana M, Mitchell DA. Mathematical modeling of the growth of filamentous fungi in solid state fermentation. J Sci Res Ind Res. 1996;55:333–42.

Saucedo-Castaneda G, Lonsane BK, Krishnaiah MM, et al. Maintenance of heat and water balances as a scale-up criterion for the production of ethanol by *Schwanniomyces castellii* in a solid state fermentation system. Process Biochem. 1992;27:97–107.

Schutyser M, Padding J, Weber F, et al. Discrete particle simulations predicting mixing behavior of solid substrate particles in a rotating drum fermenter. Biotechnol Bioeng. 2001;75:666–75.

Schutyser MAI, Weber FJ, Briels WJ, et al. Heat and water transfer in a rotating drum containing solid substrate particles. Biotechnol Bioeng. 2003;82:552–63.

Shao MA, Wang XJ, Huang MB. Soil physics. Beijing: Higher Education Press; 2006.

Stuart D, Mitchell DA, Johns M, et al. Solid-state fermentation in rotating drum bioreactors: operating variables affect performance through their effects on transport phenomena. Biotechnol Bioeng. 1999;63:383–91.

Thibault J, Pouliot K, Agosin E, et al. Reassessment of the estimation of dissolved oxygen concentration profile and KLa in solid-state fermentation. Process Biochem. 2000;36:9–18.

Weber FJ, Oostra J, Tramper J, et al. Validation of a model for process development and scale-up of packed-bed solid-state bioreactors. Biotechnol Bioeng. 2002;77:381–93.

Zambra C, Moraga N, Escudey M. Heat and mass transfer in unsaturated porous media: moisture effects in compost piles self-heating. Int J Heat Mass Trans. 2011;54:2801–10.

Zhang YP, Bai JL. Effect of temperature on soil water potential. Acta Pedol Sin. 1990;27:454–8.

[illegible] Solid-state nanomaterials for [illegible] by [illegible] Spectrosc [illegible]

[illegible]

[illegible]

Chapter 4
Aerobic Solid-State Fermentation

Abstract Based on the nature of biological processes, aerobic solid fermentation can be defined as a biological metabolic process that uses air containing oxygen as the continuous phase. In the natural environment, the majority of microorganisms live under aerobic conditions, so aerobic solid fermentation simulates the natural environmental condition, and it may be more suitable for the growth of microorganisms. Current model simulations of different fermentation technologies describe the fermentation transfer principle. Various bioreactors have been designed, investigated, and scaled up. The large-scale industrial application of aerobic solid-state fermentation concludes the production of antibiotics, organic acids, enzymes, biofeeds, biopesticides, edible fungi, and so on. In this chapter, the physical and biological characteristics of aerobic solid fermentation are introduced; the related fermentation technologies and bioreactors are described and discussed, especially gas double dynamic solid-state fermentation.

Keywords Aerobic solid-state fermentation • Gas double dynamic solid-state fermentation • Tray bioreactor • Packed bed bioreactor • Rotating drum bioreactor • Gas-solid fluidized beds

4.1 Biology and Physics Foundation of Aerobic Solid-State Fermentation

4.1.1 Introduction to Aerobic Solid-State Fermentation

Oxygen is one important factor that affects the process of aerobic solid-state fermentation. Based on the nature of biological processes, aerobic solid fermentation can be defined as a biological metabolic process that uses air containing oxygen as the continuous phase. Solid-state fermentation involves the growth of microorganisms on moist solid particles. There is a continuous gas phase in the space between the particles. The majority of water of the system is absorbed within

H. Chen, *Modern Solid State Fermentation: Theory and Practice*,
DOI 10.1007/978-94-007-6043-1_4,

the moist solid particles, and there are thin water films on the particle surfaces. The interparticle water phase is discontinuous, and most of the interparticle space is filled by the gas phase. In the natural environment, the majority of microorganisms live under aerobic conditions, so the aerobic solid fermentation processes simulate the natural environment, and they may be more suitable for the growth of microorganisms.

With regard to solid-state fermentation equipment, researchers have developed tray-type bioreactors, packed bed bioreactors, rotating drum bioreactors, gas-solid fluidized bed bioreactors, and gas double dynamic bioreactors. In 1929, the British scholar Fleming first discovered that bacteria could not grow in the plate where *Penicillium* had grown and named this antibacterial substance penicillin. This began the era of large-scale study and use of antibiotics. In the initial stage, penicillin was produced using aerobic tray fermentation. Because of the limitations of the production process, the levels of production, extraction, and purification were low. In the 1940s, with the development of submerged liquid fermentation technology, the production of penicillin was scaled up to the industrial level, which opened a new chapter of modern aerobic fermentation (Mitchell et al. 2006).

For different products or fermentation technologies, the processes of an aerobic solid-state fermentation procedure may be different, but the basic flow can be summarized in the following aspects (Fig. 4.1): (1) There is pretreatment of raw materials, such as crushing, cooking, molding, starter propagation, cooling, and so on. (2) Compared to the liquid fermentation process, the flow properties of the solid substrate are poor. Consequently, material handling is an important factor that influences the efficiency of the solid-state fermentation process and should be paid more attention. (3) Microorganisms in aerobic solid-state fermentation include some natural microorganisms and some artificial screening strains. (4) For the process and control of solid-state fermentation with respect to liquid fermentation, the solid substrate environmental conditions are more complex, and the fermentation process is more difficult to control. (5) Compared to anaerobic solid fermentation, besides the transfer of mass and heat, the distribution and transfer of oxygen in a fermentor are other important factors that influence the fermentation process. (6) Solid-state fermentation postprocessing consists of product purification, product drying, sterilization, deployment, repackaging, and so on.

4.1.2 Aerobic Microorganisms and Nutrition

4.1.2.1 Aerobic Microorganisms

According to the different demands for oxygen, microorganisms could initially be divided into two categories, aerobes and anaerobes. Obligate aerobic aerobes have an entire respiratory chain that uses oxygen as the final electron acceptor and can perform complete metabolic processes. Facultative anaerobes can be grown under both aerobic and anaerobic conditions. The microbes will obtain energy from

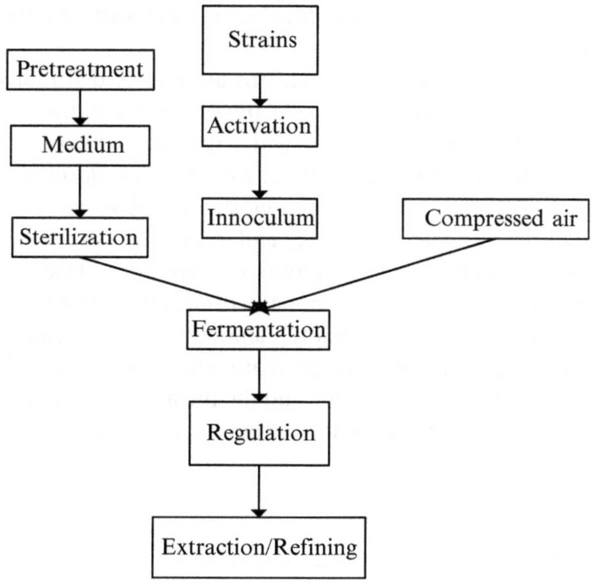

Fig. 4.1 General aerobic solid-state fermentation technical processes (Chen and Xu 2004)

aerobic respiration under the aerobic condition or from anaerobic fermentation under the anaerobic condition. Microaerophilic anaerobes include a complete respiratory chain that can use oxygen as the final electron acceptor, but they only live in an environment with a low oxygen concentration.

Lignocellulosic Enzyme-Producing Microorganisms

Cellulase-producing microorganisms, including bacteria, fungi, and actinomycetes, all can produce cellulase; examples are *Trichoderma reesei*, *Trichoderma viride*, *Trichoderma koningii*, *Aspergillus aculealus*, *Neurospora crassa*, and *Fomes fomentaris*. *Trichoderma* has been widely studied and applied to cellulase production in solid-state fermentation. Some information about biological characteristics of *Trichoderma* follows.

Trichoderma species are mainly distributed in moist soil, and the mycelia grow rapidly. The colonies with a green surface are amorphous floccules or a dense plexus bundle. Mycelia with separated branches produce chlamydospores and conidia. The conidia are mostly ovoid, colorless, or green clustered at the top of the mycelia's small stems. *Trichoderma* growth requires higher humidity; the optimum growth relative humidity is usually higher than 90 %, and the optimum growth temperature is 20–28 °C. *Trichoderma* have a wider range of growth pH values; the pH values are around 1.5–9.0, but the optimal growth pH value is 5–5.5. *Trichoderma* can use a variety of organic compounds as a carbon source, such as glucose or starch. Ammonium is the nitrogen source most available to *Trichoderma*, and other nitrogen sources, such as amino acids and urea, also can maintain normal growth for *Trichoderma*.

Laccase-producing organisms are widely distributed in nature, such as bacteria, fungi, insects, and plants. At present, most laccase-producing strains are derived from fungi, especially the white rot fungi, such as *Bjerkandera adusta*, *Cerrena unicolor*, *Coriolopsis gallica*, *Fomes sclerodermeus*, *Funalia trogii*, *Ganoderma lucidum*, *Irpex lacteus*, *Pycnoporus cinnabarinus*, *Polyporus pinsitus*, *Rigidoporus lignosus*, *Trametes hirsute*, and *Trametes versicolor*. *Trametes versicolor* is the most common laccase-producing strain; a brief description of its biological characteristics follows: annual; coriaceous; sessile or equatorial reflexed semicircle to shell-like; color variety; smooth; narrow with concentric rings; edge thin; incomplete or wavy. Mycelia are white and grow rapidly. The growth temperature is in the range of 5–32 °C, and the most suitable temperatures are around 25–28 °C. The growth pH values range from 3.5 to 7.5, and the optimal pH values are between 5.5 and 6.5.

Antibiotic-Producing Microorganisms

Antibiotics are the major secondary metabolites produced by microorganisms, such as penicillin, cephalosporins, and streptomycin. Here, we use penicillin-producing strains as examples to introduce these kinds of microorganisms. *Penicillium chrysogenum* (asexual) is widely distributed in the soil and air. Colonies grow quickly, densely, and with velvet-like radial grooves, white edges, and blue-green spores. The optimum growth temperatures range from 20 to 30 °C, and the optimum pH is above 9. Glucose and sucrose are both important factors that influence the penicillin biosynthesis process.

For plant pest and disease control, some microbes (e.g., *Bacillus thuringiensis*) could produce insecticidal crystal proteins, vegetative insecticidal protein, hemolysin, and chitinase, which play important roles in agriculture. *Bacillus thuringiensis* is a gram-positive bacterium; the vegetative cells of the rounded ends are rod-like, and the parasporal crystal proteins can be formed in one or both ends of the cell and are square, spherical, cubic, rhombic, or ellipsoidal triangular. *Bacillus thuringiensis* is widely distributed in the soil, dead insects, vegetation, sewage, and dust. *Bacillus thuringiensis* belongs to heterotrophic-type bacteria and could use organic carbon such as starch or oligosaccharide as a carbon source. Nitrogen sources are mainly from organic nitrogen compounds such as fish meal, peptone, and the like (Sanahuja et al. 2011).

Biological Metallurgy

Biometallurgical technology is also known as bioleaching technology and started in the 1960s and 1970s. It usually refers to the oxidation of ore by bacteria or other microorganisms. The microorganisms previously discussed usually rely on pyrite, arsenopyrite, and other metal sulfides, such as chalcopyrite and copper uranium mica, and directly or indirectly leach out metal from ore. Bioleaching microorganisms can be divided into three categories based on their temperature requirements: mesophilic bacteria, 25–35 °C; thermophilic bacteria, 40–55 °C; and extreme thermophilic bacteria, above 60 °C. *Thiobacillus ferrooxidans* and *Thiobacillus thiooxidans* are common bacteria. *Thiobacillus thiooxidans* (Carol and Kelly 2008) widely exists in soil, sulfide ore wastewater, and seawater. The gram-negative, rod-end-born flagella are about 1 μm long and about 0.5 μm wide and gain energy by oxidation of the sulfur.

4.1.2.2 Nutrition

Nutrition is the process by which microorganisms obtain energy and nutrients from the external environment, which also provides basic physiological functions for structural substances, energy metabolism regulation substances, and the necessary physiological environment for metabolism (Zhou 2004). Microbial basic nutritional elements can be divided into six categories: carbon sources, nitrogen sources, energy, minerals, water, and growth factors. The carbon sources are the major nutrients for microorganisms and include organic carbon sources and inorganic carbon. Various sugars, petroleum compounds, and agricultural straw substances are all carbon sources. In the solid-state fermentation process, the carbon source substances often can be used both as nutrients and as inert carrier material that maintains the growth of the microorganisms. Consequently, it is essential to go into the characteristics of the solid substrate during the solid-state fermentation process, especially for amplification.

I have paid much attention to nutritional adsorption carrier solid-state fermentation using steam-exploded straw as carrier. To study the physical properties, cell growth, and metabolic interactions of heat and mass transfer processes, researchers divided the steam-exploded straw into long fibers and small fibers based on the characteristics of the solid substrate. At the same time, researchers explored the effect of fiber length on microbial metabolism and the interaction between the substrates and microbial metabolism during the fermentation process. These studies have enriched solid-state fermentation knowledge. Nitrogen sources mainly provide nitrogen elements for microbial growth, and nitrogen sources are used to synthesize important life protein materials and nucleic acid. Common raw protein materials mainly include bean substances, such as soybean peas, soybean cavings, bran, urea, peptone, cicada chrysalis powder, and more. For example, in the soy sauce brewing process, soybean meal is often used as a raw material (bean cake), and the crude protein content is more than 40 %.

Solar energy mainly provides initial energy sources for nutrition or for the microbial organisms. For autotrophic microbes, energy mainly comes from the metabolic process of the carbon source; several autotrophic microbes also need to use the energy of light as an energy source and synthesize essential nutrients for life activity. For several heterotrophic microbes, energy also comes from the inorganic matter metabolic process, such as NH_4^+, NO^{2-}, Fe^{2+}, and so on. Inorganic salt and growth factor are two other kinds of substances needed for the microbial growth. Their common characteristics are decreased demand and essentialness for the growth of microorganisms, yet they cannot be synthesized by the microbial organisms themselves. They mainly include vitamin base, amine, and small molecule fatty acids. Inorganic salt refers to K, P, S, Ca, Mg, and the like. During the fermentation process, the added carbon and nitrogen sources are often mixtures that include combinations of ingredients. For example, straw is rich in K, P, S, Ca, Cu, Mg, and so on and can provide enough inorganic salt for microbial growth (Yu and Chen 2010).

Water is a necessary nutritional element for microbes and an important part of organisms. For example, bacteria are composed of 80 % water, and for mold, the proportion is as high as 85 %. Water can assist microbes in transferring nutrients for metabolism from outside into the cell. On the other hand, water molecules also play a role in the maintenance of macromolecular stability and provide a relatively stable microenvironment. In aerobic solid-state fermentation, water can be divided into bound water and free water, and water content is an important factor that influences heat and mass transfer. Bound water exists as a thin water film layer and plays a role in the absorption of nutrients and desorption of metabolism substances. Free water, commonly expressed by water activity, can be defined as the ratio of solvent fugacity and pure solvent fugacity (van den Doel et al. 2009). Water transfer in fermentation can be summarized as surface water evaporation and water evaporation from the solid phase. The temperature gradient and the characteristics of the fermentation substrate such as pore degrees, morphology, and the like are all important factors that influence water vapor movement.

4.1.3 Physical Chemistry Foundations of Aerobic Solid Fermentation

In the aerobic solid-state fermentation process, almost all water is absorbed by the solid particles, and it forms a thin water film layer; there is almost no free water. Wet solid particles are filled with continuous gas, and microorganisms can grow in the damp solid particles (Fig. 4.2). There is a continuous gas cycle through the solid substrate, so aerobic solid-state fermentation has many unique properties compared to liquid fermentation and anaerobic solid-state fermentation.

First, in the aerobic solid-state fermentation process, solid substrate dries more easily, especially when it is exposed long term to the rapid flow of gas. Second, heat generated by the microorganisms could cause the uneven distribution of temperature during the fermentation process. Third, oxygen distribution is easy to control when the solid substrate loading is low. Yet, with the increases in the substrate loading coefficient, the gradient of oxygen distribution will appear uneven. Fourth, pH values are relatively constant but are more difficult to detect. Last, there are

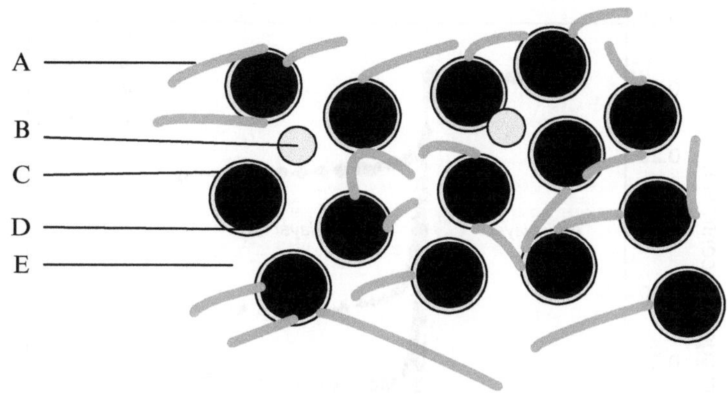

Fig. 4.2 *A* fungal hyphae; *B* droplets of water; *C* water film; *D* solid substrate; *E* continuous gas phase (Mitchell et al. 2006)

complicated interactions between solid substrate and microbes; the dynamic change of solid substrate will cause damage to the microorganisms. At the same time, microbial metabolism modifies the solid substrate.

Oxygen is an important element for aerobic microbes to complete the physiological and biochemical processes. Oxygen transfer is the important factor for aerobic fermentation (Richard et al. 2010). Consequently, here we discuss the transfer process of gas in the solid substrate. In the solid-state fermentation process, the gas transfer process can be divided into both micro and macro transfer procedures. Macro transfer is the processes of oxygen transmission in the material space, which includes the air going into the biological reactor and natural air convective diffusion. The oxygen macro transfer process has two forms. The first is diffusion; the air circulates on top of the materials. This process is relatively simple, and the solid substrate is the uniform system. Another type of oxygen transfer is forced ventilation in the gap, which lets air circulate through the material layer. Under this condition, the oxygen transfer within the material is mainly caused by gas flow. Macroscopic transfer processes are mainly affected by material thickness, the bulk density of the material layer, the particle size, and so on. From the microscopic point of view, oxygen transfers are mainly by transmembrane delivery or within the biofilm. For example, filamentous fungi and single-cell microorganisms that grow on the surface or inside the solid substrate can absorb oxygen from the external environment and discharge carbon dioxide. For gas transfer within the particles, the oxygen circulates between the substrate and microbial cells. The factors that influence the microscopic oxygen transfer processes can be briefly stated as follows: (1) thickness of the layer of wet cells; (2) density of the layer of wet cells; (3) microbial respiration activity of the layer of wet cells; and (4) the oxygen transfer coefficient of the layer of wet cells. Several researchers (Oostra et al. 2001) drew conclusions from experiments: The oxygen variation was relatively gentle in the aerial hyphae layer, and the oxygen concentration changed severely in the layer of wet cells. With increase in layer depth, the oxygen

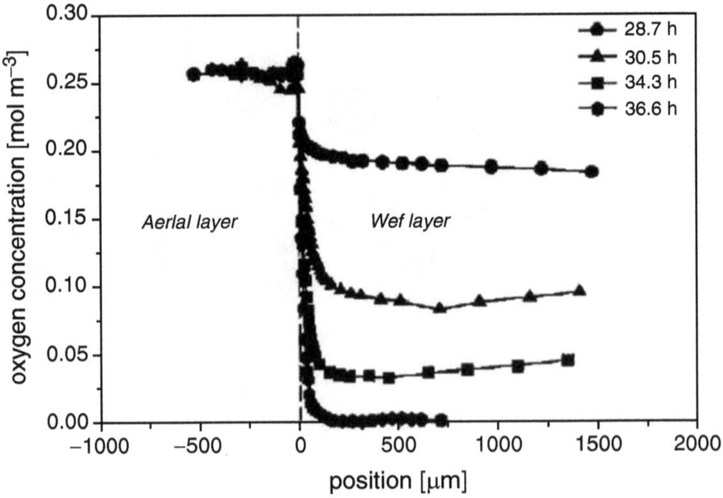

Fig. 4.3 The concentrations of oxygen in the different substrate depths (Gervais and Molin 2003)

concentrations decreased (Fig. 4.3). The internal oxygen concentrations decreased when the fermentation proceeded.

In the process of solid-state fermentation using filamentous fungi, the fungal hyphae not only grow on the particle surface but also extend into the substrate, which results in the formation of agglomerates of fungi and substrate. Within the solid particles, the aerobic mycelia extend to obtain oxygen for their growth. Some studies suggested that the level of dissolved oxygen in the interior of the particle was influenced by particle radius. There was a critical radius; the diffusion distance of oxygen in the solid media particles was very small when the radius was smaller than the critical radius. If the dissolved oxygen was about zero, the aerobic microorganisms could not live. During the fermentation process, the oxygen content is dynamic and is influenced by carbon dioxide discharged by microbial metabolism. The oxygen supply within the particles and hyphae are two important factors that affect the solid-state fermentation process; consequently, research needs to be further strengthened.

In solid-state fermentation, microorganisms can take oxygen directly from the air; yet, many operating factors and the fermentation substrate characteristics will affect the rate of oxygen transfer, such as air pressure, ventilation rate, substrate porosity, material layer thickness, substrate humidity, reactor and mechanical stirring device characteristics, and so on. The transfer efficiency of air from the substrate voids into the substrate is determined by the nature of the substrate itself, such as its porosity, particle size, and moisture content. In the fermentation process, the substrate water content is closely related to the transfer of oxygen in the void. If the water content is too high, the pores are filled with the free water, and the air is excluded from the substrate, which results in an anaerobic microenvironment. If the water content is too low, growth of the microorganisms will be restricted. Agitation and forced ventilation both can enhance oxygen transfer in the pore spaces, and its efficiency is also influenced by the porosity and water content. When the porosity is high, agitation and aeration may not be necessary because the oxygen in the pores has been able to meet the need for the growth of microorganisms. With high porosity, the air in the pores can effectively circulate within the surrounding environment so that the oxygen in the pores can be replenished simultaneously.

Water transfer is another focus of concern. The water not only affects the microbial physiological and biochemical processes but also relates to the temperature variation (Gervais and Molin 2003).

At present, the influence of water on the microbial reaction is studied based on food industry methods. The water activity is defined as the ratio of fugacity of the solution to the fugacity of the pure solvent, which is approximately equal to the ratio of the vapor pressure of water P to a vapor pressure of pure water at the same temperature P_0 in a sealed container (Hu and Xu 2009):

$$\alpha_W = \frac{P}{P_0} \tag{4.1}$$

P = the actual pressure of the water vapor;
P_0 = the actual pressure of the pure water vapor

The different microorganisms have different suitable water activities. The variations of water activity not only have a direct impact on microbial physiology but also influence the nature of the substrate, such as specific surface area and the oxygen transfer rate (Astoreca et al. 2012).

Another important role of water is to reduce the temperature of the fermentation system by evaporation. The heat absorption of water vaporization is relatively large, and the thermal conductivity of the solid substrate is low; consequently, evaporation is one of the effective means of adjusting the temperature in most fermentation systems. Heat removal can be calculated according to the following formula (Hu and Xu 2009):

$$Q_V = \lambda k_W A\left(a_{ws} - a_{ws}^*\right) \tag{4.2}$$

Q_V = the heat removal rate of water vapor;
λ = the heat of water vaporization;
a_{ws} = the actual water activity of substrate;
a_{ws}^* = the actual water activity of substrate that is balanced to the gas phase.

In general, the transfers of water within the substrate can be divided into three levels: within the gaseous phase, on the surface of the substrate, and within the substrate. Because there is nearly no free water in solid-state fermentation and the relatively complex transfer process is within the substrate, current research mostly focuses on the transfer in the water and gas phases of the solid substrate.

The factors that influence the variations of substrate water content are as follows: nonbound water contained in the matrix, the diffusion and convection of liquid water, evaporation (phase transition), the diffusion and convection of water vapor, and reaction metabolism. Diffusive convection of water vapor as well as the phase transition are the major factors that influence the fermentation process.

There are complex interactions between microorganisms and the environment; these can be summarized as follows: Direct response mainly refers to the primary response of the microbial cells to the solid substrate and the surrounding environment, such as various inducers, carbon sources, nitrogen sources, and the morphology of the solid substrate. On the other hand, direct response also refers to the interaction between the microenvironmental conditions and microorganisms. First, in aerobic biological metabolism, oxygen as the final electron acceptor is reduced into water. Oxygen can also be catalyzed by the dioxygenase enzymes to form organic molecules. The microbial growth will be inhibited if the oxygen concentration is low, yet high concentrations of oxygen may be toxic for microbes. In the aerobic solid-state fermentation process, the concentration of available oxygen is an important factor that affects the rate of microbial growth, the change of cell composition, and metabolites. At the same time, the response of the microorganism for oxygen would affect the concentration of oxygen. Second, for several heterotrophic microbial metabolic processes, carbon dioxide will be produced, which will affect the metabolism. However, for some autotrophic microbial metabolic processes, the supply of carbon dioxide will be affected when oxygen is too high. Therefore, in the early stage of solid fermentation, a lot of ventilation is not needed to maintain a suitable carbon dioxide concentration level. Sometimes, the media pH will be affected when the concentration of carbon dioxide is high. A large amount of molecular carbon dioxide enters into the cell, resulting in the generation of hydrogen ions and bicarbonate ions. To maintain the ion balance, microorganisms must pump out hydrogen ions from the cell, avoiding futile circulation. There are complex interactions between the temperature and the microorganisms, and the microorganisms are extremely sensitive to the environment. At a certain temperature, the microorganisms may grow preferably; yet, the DNA, RNA, protein, and other cell components are affected if the temperature changes. When the microbes grow quickly, various forms of energy are transformed into metabolic energy, and much heat is released at the same time. The rapid rise in ambient temperature also affects the metabolic activity of authigenic microbes. Finally, there are complex interactions between the vigor of the water in the environment, the concentration of hydrogen ions, and the microorganisms.

4.2 Mixed Solid-State Fermentation

4.2.1 Overview

As an important kind of biological resource, microorganisms have been studied and utilized for many years; these microorganism studies can be divided into two kinds: those with natural mixed cultures and those with pure cultures. A variety of microorganisms participate in traditional microbial fermentation; for example, primitive tribes use fruit sugar to ferment wine koji. Only since Leeuwenhoek invented the microscope in 1680 and Pasteur and Koch established the microbial purebred cultivation process did fermentation enter the modern pure fermentation stage. Modern fermentation industries usually take a single pure culture for the fermentation process. The researchers can get rid of the unstable traits for a complex situation and the mutual interference that exists among a variety of microorganisms. By studying the pure strains, our understandings of microbial morphology, physiological metabolism, and genetic characteristics are enriched. Humans can separate and utilize products of microbial primary and secondary metabolites. For example, the modern beer and wine industries both use pure culture fermentation technologies (Chen and He 2012).

But, during the long experimental and production practice, people gradually found that a single microorganism often participated in only one or several steps of biochemical metabolic processes, and the efficiencies of metabolic catalysis were very low. Therefore, humans began to look again at the advantage of a mixed culture. In nature, the catalytic processes of microbial metabolism are often completed by two or more microorganisms, which form a close community by interaction with each other. A special microenvironment is formed by the biological interaction of mutual promotion and mutual competition, and the microbes that are adapted to the environment can grow and reproduce. Some parts of the metabolites produced by mixed microorganisms can be used as fermentation flavor products; these advantages cannot be obtained by pure culture fermentation. For example, the koji-making process is accomplished by the symbiotic group of mold, yeasts, and bacteria. In mixed solid-state fermentation, the microorganisms are various, and most of them are unknown. Consequently, people often utilize flora characteristics to realize and control the culture conditions and metabolic process.

4.2.2 Process Principles

Aerobic mixed solid-state fermentation can be divided into coculture fermentation and mixed-culture fermentation (Bader et al. 2010). Coculture fermentation is a process by which a variety of known microorganisms grow under sterile aerobic conditions. Mixed-culture fermentation is a cultivation process by which a variety of known or partially known microorganisms grow under approximately aseptic

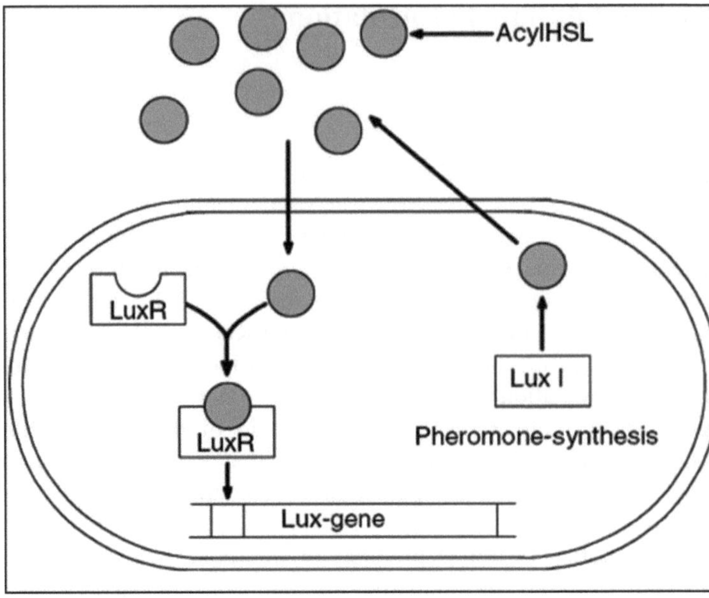

Fig. 4.4 Information transmission between microbes (Fuqua et al. 2001)

conditions. In this process, microbes often manifest in a competitive and symbiotic interaction relationship.

The microbial interaction mechanisms of coculture fermentation mainly refer to the interaction of the microbial metabolites or the biological information during the fermentation processes to complete the metabolic processes. The process of microbial interactions is achieved by some chemical substances. For example, the cellular interaction of a single bacterial cell is achieved by small soluble signaling molecules (Fuqua et al. 2001). As shown in Fig. 4.4, when the concentrations of AcylHSL (acylated homoserine) were low, expression of the other gene could not begin. And, when the concentration of AcylHSL reached a certain value, this gene started to express; thus, biological mutual regulation was realized.

The specificity of mixed-culture fermentation is the complicated synergistic interaction of a variety of microbes. To illustrate the coordination process of natural microbial decomposition, here we take the degradation of plant biomass in nature as an example. The soil microbial community is the most complicated biological community; according to statistics, the microbial number could be up to 10×10^{11}/g soil and includes bacteria, actinomycetes, fungi, viruses, algae, and small protozoans. According to the difference in the utilization of the carbon source, the concerned microorganisms can be divided into early, middle, and late flora. The early flora mainly are microorganisms that could only use the soluble carbohydrate, lipid, amino acid, and pectin as nutrients. The early flora include weak parasitic fungi and some soil inhabitants, such as *Cunninghamella*, *Mucor*, and *Cladosporium*. Wood rot fungi are mainly microorganisms that take part in the middle decomposition of plant biomass; they include small filamentous fungi and large brown rot basidiomycetes that could use cellulose and hemicellulose and some white rot fungi (large white rot basidiomycetes and white rot ascomycetes). White rot fungi can decompose cellulose and lignin simultaneously. Small filamentous fungi are important cellulose decomposition groups, such as the *Trichoderma*, *Aspergillus*, *Penicillium*, *Rhizopus*, *Cladosporium*, and so on. As far as is known, white rot fungi, as an important group of wood rot fungi, are the only microbes that can degrade lignin into CO_2 and H_2O completely. In the case of polypores, these include *Antrodiella*, *Bjerkandera*, and so on. Polypore species are more than 90 % of the total species. Last, the final part of decomposition is organic soil humus; this mainly comes from raw material debris and the condensation product of biomass microbial decomposition. Most humus is the dark acidic polymer compounds of some nitrogen-containing aromatic, such as humic acid, fulvic acid, and humin acid. This metabolic remainder humus soil can be absorbed and transformed by soil inhabitant microbes.

4.2.3 Process Evaluation

4.2.3.1 Advantages of Mixed Solid-State Fermentation

Raw Materials are Cheap and Easy to Obtain

Natural substrate or waste residue is inexpensive and widely distributed. For example, solid-state fermentation research used lignocellulose as a substrate, which not only significantly reduces cost but also is beneficial to environmental protection.

High Raw Material Utilization Efficiency

Raw materials often contain a variety of nutrients. Single microorganisms often only utilize one or two kinds of nutrients, and other substances are usually discarded as fermentation waste residue. Different microorganisms usually choose different substrates as their optimal nutrients, so the utilization of raw materials is improved in the mixed solid-state fermentation process.

High Equipment Utilization Efficiency and Low Energy Consumption

Different desired products can be produced in the same or a similar process in the same fermentation vessel. Consequently, it is beneficial to make full use of the equipment (equipment utilization efficiency), reduce staff time, and improve labor efficiency.

High Substrate Conversion Rate

There are always beneficial interactions between the mixed microorganisms. The related mixed strains benefit each other through their metabolic processes, which can achieve a multigene function in fermentation. By the coupling of different metabolic processes, the complex metabolism that a single microbe has difficulty completing can be achieved. Consequently the substrate conversion rate is high.

Production of a Unique Flavoring Substance

A variety of microbes form an organic unity, which could generate different products such as unique flavoring substances through the metabolism of various nutrients.

Low Cultivating Condition Requirements

Compared to pure fermentation, there are antagonism effects existing in mixed culture fermentation; thus, it is more resistant to contamination. Examples of low cultivating condition requirements are as fermented manure and feed production.

4.2.3.2 Problems of Mixed Solid-State Fermentation

Although mixed solid-state fermentation has been widely used in many fields, there are still some problems with its application.

1. Microbial fermentation strains are unclear. Only about 1 % of the microbes can be cultured; thus, most of them have not been recognized. The recognition of complicated mixed flora is limited; therefore, a stable artificial mixed flora cannot be made to serve industrial production.
2. The interactions among the fermentation microorganisms are not clear, and the relationships between mixed culture systems cannot be effectively coordinated to reach the best ecological effect level. This limits the development and application of mixed culture fermentation.
3. The lack of awareness of the interaction between the microbes causes problems. The synergy among the strains is also random, so effective theoretical guidance for industrial production cannot be obtained.

4.2.4 Key Technologies

Because fermentations are completed by different microorganism types, it is crucial to choose microbes. Compared with animals and plants, the ability of microorganisms to adapt to the environment is stronger. There are strong interactions between microorganisms and the environment. To adapt to the external environment, the microbial genes will be gradually altered. A coculture fermentation system usually involves more than two microorganisms; it is organic and coupled by physiological metabolic characteristics and genetic information. It is not necessary to separate and purify the microbe from the mixed cultures. Researchers usually design special culture conditions to achieve directional control of the microbial flora.

4.2.4.1 Naturally Enriched Mixed Culture Process

The naturally enriched process of mixed culture mainly refers to cultivating microorganisms by controlling different culture conditions based on the various genetic traits of the microbes, such as different temperatures, oxygen supply conditions, and the pH values of the medium, so the selected microbes could grow, inhibit, and outnumber other microorganisms if the culture conditions are suitable, with the selected microbes forming the dominant flora. At the same time, the other microorganisms beneficial to the flora could also grow and reproduce. So, a stable and efficient mixed solid-state fermentation system could ultimately be established. With further research, several scholars have put forward methods to strengthen the mixed solid-state fermentation process. For example, researchers inoculated pure strains to the cultures to inhibit the growth of other microorganisms; as a result, a series of certain microbes could be enriched. In the silage feed process, scholars add lactic acid strains to the media to enhance the effect of silage. The growth of other spoilage microorganisms can be inhibited by antagonistic action from the lactic acid strains used; at the same time, the flavor and quality of silage are well preserved.

4.2.4.2 Control of the Mixed Solid-State Fermentation Process

Mixed solid-state fermentation is accomplished by a variety of microorganisms, and the conditions of the microorganism varieties are various. Creation of suitable fermentation conditions for continuous growth and reproduction of microbes in the fermentation system is important.

4.2.5 Application of Mixed Solid-State Fermentation

4.2.5.1 Utilization in Lignocellulosic Enzyme Production

Mixed solid-state fermentation has a long history, is familiar to most people, and has been applied in many fields. At present, research mostly focuses on anaerobic mixed solid-state fermentation, such as traditional Chinese liquor brewing, kimchi making, tobacco fermentation, tea fermentation, silage, and compost fermentation.

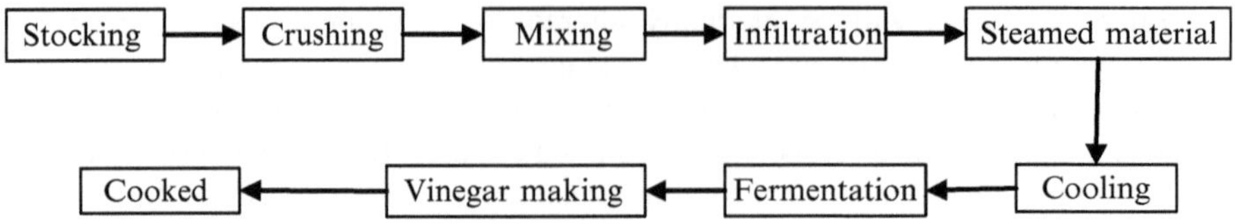

Fig. 4.5 The process of mixed culture solid fermentation of vinegar

Lignocellulose resources are abundant in nature. The catabolic process is accomplished by the synergy of various microbes. These microorganisms form a microecological system of lignocellulose degradation through interaction (alternate, symbiotic, parasitic, antagonistic, and predatory). Each microbe plays an important role in the particular microecosystem through its metabolic cycle. The proportion of various enzymes is appropriate, and the microecosystem is in an optimal state. Once the related microbes are separated from the microecosystem, the cellulose hydrolysis function caused by coexisting microorganisms will subsequently be lost. Some scholars cocultivated *Aspergillus niger* and *Trichoderma* to produce cellulase and xylanase in solid-state fermentation using wheat straw, wheat bran, and corncobs as the solid substrate. During the process, both of the microbes secreted enzymes and degraded substrate to service each other for their own growth and reproduction and provided nutrients for the other's growth and reproduction; as a result, a mutually beneficial situation was formed, and the production of two enzymes improved (Hu et al. 2007). Vadlani cocultivated *Trichoderma* and *Aspergillus oryzae* to produce cellulase enzymes in tray solid-state fermentation using soybean hull and bran as a nutrient medium. The research results showed that both the filter paper activity and glucosidase production were significantly improved (Brijwani et al. 2010).

4.2.5.2 Mixed-Culture Solid-State Fermentation of Vinegar

As shown in Fig. 4.5, the mixed solid-state vinegar fermentation process can be summarized as follows (Wu and Han 2012):

1. Preparation: Commonly used raw materials include sorghum, sweet potato, and smashed rice. The starch content of the raw materials should be around 14–16 %, and the produced alcohol content is up to 7 %. The growth and reproduction of acetic acid bacteria will be inhibited when the alcohol content is too high.
2. Crushing: The raw materials need to be properly pretreated to enhance the utilization of raw materials.
3. Mixing: The crushed raw materials need to be mixed together to certain proportions.
4. Wetting: Water is added to the materials in an appropriate proportion. The water content is usually controlled around 62–66 %.

5. Steaming: The materials are cooked at atmospheric pressure for about 2 h and cooled rapidly.
6. Fermenting: Koji and yeast are added to the materials. The temperatures are controlled at around 30–40 °C, and the moisture content is about 60 %.
7. Vinegar making: Bran and acetic acid bacteria are added to produce vinegar.
8. Cooking: At the end of fermentation, salt and vinegar are added to inhibit the activities of the acetic acid bacteria.

4.2.5.3 Production of Polyhydroxyalkanoates by Mixed Culture

Previous studies showed that polyhydroxyalkanoates (PHAs) could be produced by active sludge under microaerobic conditions, and PHBs poly-β-hydroxybutyrate could be produced under aerobic conditions (Salehizadeh and Van Loosdrecht 2004). Under microaerobic conditions, the microorganisms obtained energy and accumulated organic matter by oxidation degradation of minority organics. Under sufficient oxygen, microorganisms were able to obtain enough energy to accumulate protein, glycogen, and other organic materials, yet if oxygen conditions became the limiting factor, absorption vitality would decrease, and PHAs were accumulated by microbes. Under this condition, the PHA accumulation function can be selectively started. Whether the microflora had the function to accumulate protein or glycogen, they almost did not synthesize protein and glycogen. The microflora were a highly active specific dynamic combination; consequently, scientists proposed using the cycle stimulation theory and method to promote the PHA-producing ability of the microflora. By setting the circular changes of nutrient content and establishing the two states of surplus-hunger, the uneven growth of microorganisms was achieved. In this process, the growth of microflora and PHA accumulation proceeded; alternately, when there was excess nutrient, PHAs could accumulate. When nutrients were lacking, PHA was used as a nutrient and for energy. The synthesis of polymer did not affect the growth and metabolism of microbes. The physiological adaptability became important; if there were excess nutrients in the culture conditions for a long time, accumulation of polymers was a priority for the growth of microbes. In different states of surplus-hunger, more than 60 % of the PHA could be accumulated in the hunger state (Fig. 4.6).

4.3 Static Closed Solid-State Fermentation

Static closed solid-state fermentation is the traditional solid-state fermentation technology; in these processes, the substrate maintains a relatively quiescent state. The most typical static solid-state fermentation reactor is the tray bioreactor, which has a long history. For example, Chinese traditional liquor brewing and Western cheese making both can be attributed to the static solid-state fermentation process. Static closed solid-state fermentation reactors mainly include tray bioreactors and packed bed bioreactors. Compared to submerged liquid fermentation, the most

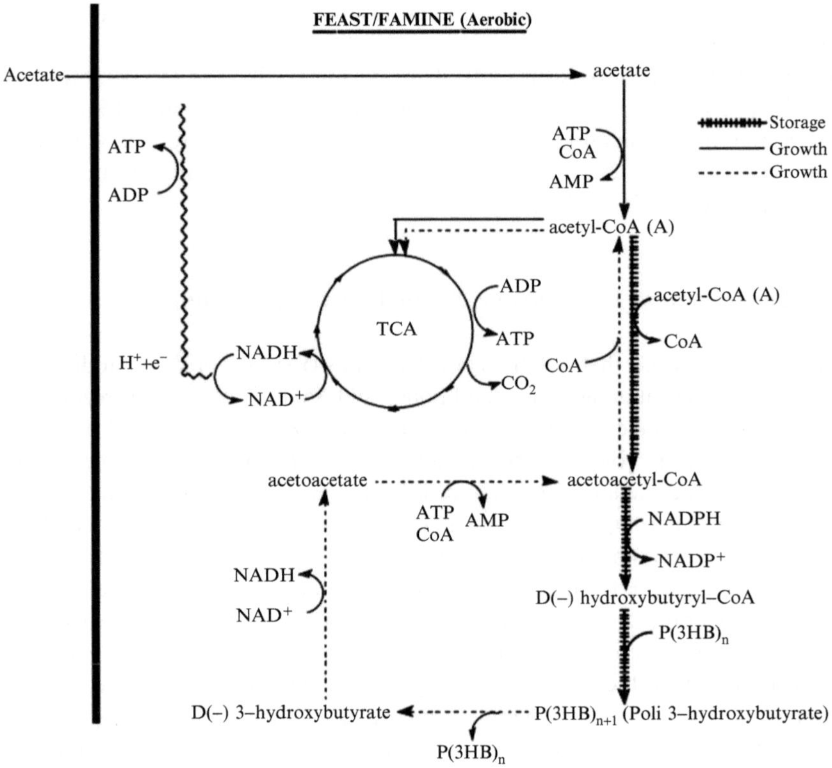

Fig. 4.6 The model of surplus hunger in anaerobic microorganisms (Salehizadeh and Van Loosdrecht 2004) ATP: adenosine-triphosphate; NADH: nicotinamide adenine dinucleotide phosphate; NaD+: nicotinamide adenine dinucleotide; AMP: adenosine monophosphate

significant characteristic of these bioreactors is that they are without mixing equipment. The heat and mass transfer efficiency are low, and the fermentation processes are more easily to be contaminated (Singhania et al. 2009).

4.3.1 *Tray Solid-State Fermentation Technology*

4.3.1.1 Introduction to Tray Bioreactors

Tray bioreactors have been studied for many years. During the fermentation process, there is no mandatory ventilation and mechanical agitation, which is true static aerobic solid-state fermentation. Tray bioreactors include a large empty chamber, in which the temperature and humidity should be purposefully controlled. Second, there should be a series of shallow trays (Fig. 4.7a, b, c), and these trays should be successively placed in the empty chamber. Third, each tray bioreactor is an independent small space; it may be a small reactor or a larger special room. The

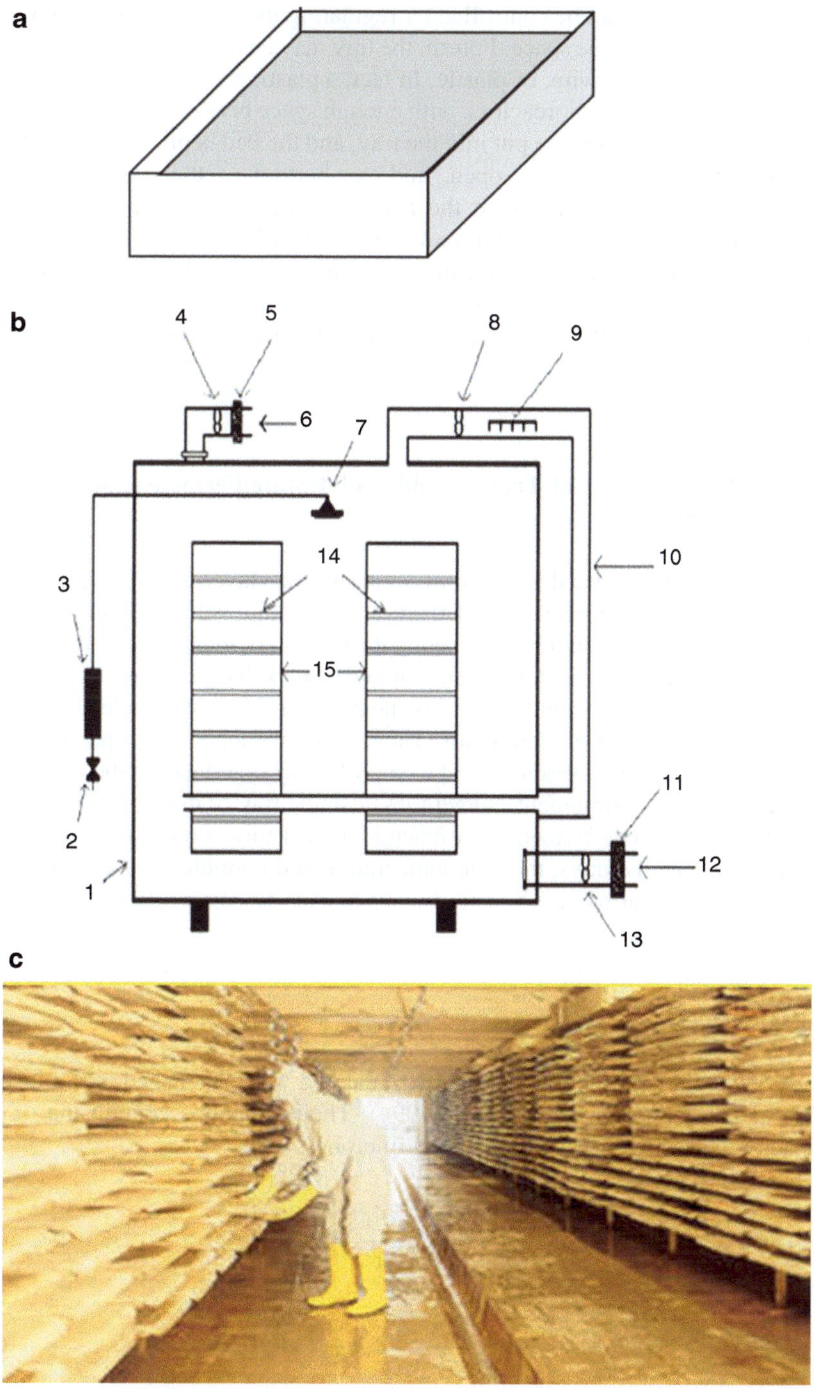

Fig. 4.7 (**a**) Tray model (Mitchell et al. 2006); (**b**) tray bioreactor model (Chen and Xu 2004); (**c**) industrial application level of tray solid-state fermentation (Chen and Xu 2004)

fermentation process can be controlled by regulating the temperature and humidity of the air that goes into the space. Fourth, the tray may be made of various materials, such as wood, bamboo, wire, or plastic. In fact, a plastic bag might be used instead of a rigid tray. Fifth, tray bioreactors, with enough space between each other, are put into the chamber. Cultures are put into the tray, and the bed depth is about 5–15 cm. The upper part of the tray is open, and the bottom of the tray may also be appropriately opened according to the actual situation to facilitate gas transfer. The trays filled with culture medium are sterilized and are pushed into the related room or fermentor. Sixth, the overall temperature of the bioreactor should be controlled by the integration of a water cooling-heating system. The moist air should pass into the fermentor periodically during the large-scale industrial application.

4.3.1.2 Characteristics of Tray Aerobic Solid-State Fermentation Technology

Many trays are filled with a thin, uniform, solid substrate layer, and the layer height of the material is severely restricted. If there are no additional ventilation measures, numerous small holes should be punched in the trays to promote air circulation in the trays and in the entire fermentation bed. Air and heat exchange always exist between the fermentor and the surrounding environment during large-scale industrial production. The temperature of the fermentor changes with the ambient temperature variation, and the internal temperature can be controlled by coordinating the temperature of the surrounding environment. Sometimes, a single tray or the entire chamber can be deemed a solid-state fermentation reactor. If each tray is open and there is heat exchange between the trays, then the temperature and humidity of each tray can be controlled in a similar range.

Oxygen Balance

The oxygen balance equation of a tray stromal bed was established by Smits et al. (1999) and Rajagopalan and Modak (1994). Their equations contain the oxygen diffusion in space and oxygen uptake by microbes.

$$\frac{\partial C_{O_2}^{b}}{\partial t} = D_{O_2}^{b} \frac{\partial^2 C_{O_2}^{b}}{\partial z^2} - r_{O_2} \tag{4.3}$$

$$\frac{\varepsilon \partial C_{O_2}}{\partial t} = D_{O_2}^{b} \frac{\partial^2 C_{O_2}}{\partial z^2} - K_{a} a x \left(C_{O_2} - H C_{O_2}^{f} \right) \tag{4.4}$$

t = time;
$C^{b}_{O_2}$ = unit volume oxygen concentration within the bed;
C_{O_2} = oxygen concentration of the space;
$C^{f}_{O_2}$ C_o = oxygen concentration within the biofilm;
ε = porosity;
z = vertical coordinates;
r_{O_2} = microbial uptake rate of oxygen;
$D^{b}_{O_2}$ = diffusion coefficient;
K_a = transfer coefficient of oxygen on the air/biofilm interface;
a_X = air/biofilm interface area;
H = Henry's law constant.

The first term of the right-hand side of the equation represents the amount of oxygen diffusion in the pores. Obviously, these two formulas differ in the oxygen transfer within the biofilm. Factually, the latter assumes that there is no oxygen accumulation in the biofilm; the concentration of oxygen is kept constant, and the oxygen going into the biofilm is used immediately by microorganisms.

In the model established by Rajagopalan and Modak (1995), the porosity is not constant and it is the growth function of the biomass. Here, we assume that the biofilm is uniformly mixed. This can make the oxygen concentration of the biofilm surface clearly be expressed as the transfer of oxygen at the gas/biofilm interface:

$$\frac{\partial C_{O_2}\varepsilon(t)}{\partial t} = D^{b}_{O_2}\frac{\partial^2 C_{O_2}}{\partial z^2} - K_a a_X\left(C_{O_2} - HC^{f}_{O_2}\big|_{R(t)}\right) \tag{4.5}$$

The applicability of these oxygen transfer modeling methods depends on the purpose of modeling as well as the available experimental information. In contrast, the method established by Smith et al. (1999; Mitchell et al. 2003) may be simpler, and in most cases, this simplification is allowed. The model proposed by Rajagopalan and Modak (1995) that involved the growth of microbes on the microparticle biofilm may be more in line with the actual situation. The aim of this model is to study whether the diffusion of oxygen in the pores, oxygen delivery into the biofilm, and diffusion within the biofilm are the rate-limiting steps, but the model's form and computational procedure are much more complex.

Water Balance

The water balance equation in the tray bioreactors is established by considering water vapor diffusion and evaporation from the liquid phase of the material layer to the gas phase (Mitchell et al. 2003).

$$\frac{\partial C_W}{\partial t} = r_{H_2O} - \left[\frac{\partial C_{VAP}}{\partial t} - D^*_{VAP}\frac{\partial^2 C_{VAP}}{\partial z^2}\right] \tag{4.6}$$

C_w = stromal bed unit volume water content;
C_{vap} = stromal bed unit volume steam content;
D^*_{VAP} = stromal bed water vapor effective diffusion coefficient;
r_{H_2O} = reaction water biological metabolic rate.

The first item of the right-side of Eq. 4.6 represents the increase or decrease in the water content caused by microbial metabolism. The first term in the brackets of the equation represents variations of moisture caused by water evaporation, and the second term in the square brackets represents the diffusion of water vapor in the voids in the vertical direction.

Energy Balance

According to the basic principles of heat transfer balance (Cha and Chen 1997), the enthalpy change rate of an element is equal to the heat transfer rate plus the heat reaction rate. The heat balance equation for tray bioreactors can be stated as follows (Rajagopalan and Modak 1994, 1995):

$$\rho_s C_{ps}\frac{\partial T}{\partial t} = k_b\frac{\partial^2 T}{\partial z^2} + r_Q \tag{4.7}$$

ρ_s = stromal bed density;
C_{ps} = stromal bed heat capacity;
T = stromal bed temperature;
K_b = stromal bed thermal conductivity;
r_Q = microbial tissue metabolism heat production rate.

The solid-state fermentation heat removal is achieved mainly by water evaporation from the materials; consequently, Eq. 4.7 heat balance is usually followed by a water evaporation phase (last item on the right side of the equation) (Mitchell et al. 2003):

$$\rho_s C_{ps}\frac{\partial T}{\partial t} = k_b\frac{\partial^2 T}{\partial z^2} + r_Q + \lambda D^*_{VAP}\frac{\partial^2 C_{VAP}}{\partial z^2} \tag{4.8}$$

where λ is the enthalpy of water vaporization.

The third term Eq. 4.8 represents the heat dissipation of water diffusion and evaporation; it is assumed that the water and heat of the gas and solid are balanced.

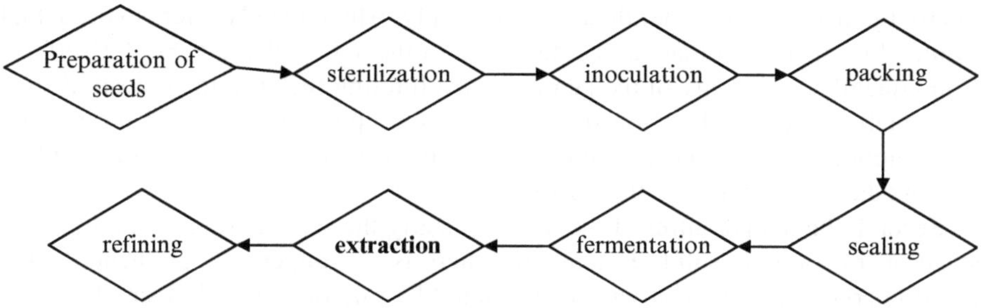

Fig. 4.8 The tray solid-state fermentation process

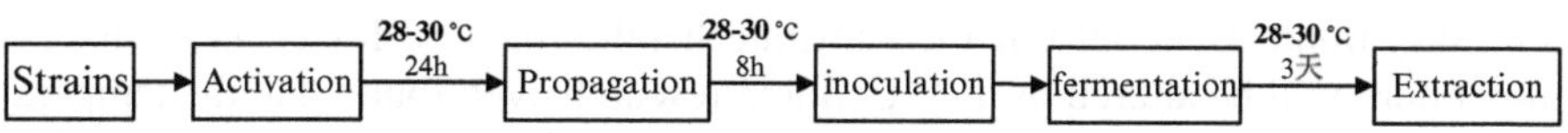

Fig. 4.9 Schematic diagram of production of biopesticides by *B. thuringiensis*

The simulation model established by Mitchell et al. (2003) demonstrated that the contribution of evaporated cooling should be ignored when the air humidity is more than 98 %.

4.3.1.3 Applications

The general procedure of tray solid-state fermentation is shown in Fig. 4.8. The seaming or sealing chamber mainly refers to the fermentation that would proceed in a closed tank or room, where the temperature and humidity can be adjusted.

Biopesticide production in tray solid-state fermentation using *B. thuringiensis* was briefly described (Fig. 4.9). Strains commonly used included AS1.949 and AS1.1013. The carbon sources were starch and polysaccharide substance. The nitrogen sources were soybean meal, cottonseed cake, and chaff. During the fermentation process, the inoculum size should be greater than 50 %, and the fermentation temperature should not exceed 35 °C. For the massive ventilation pool, the medium should be covered with a layer of sterile chaff, which plays an important role in water retention and sterilization.

4.3.1.4 Tray Solid-State Fermentation Technology Bottlenecks

Regarding the tray solid-state fermentation technology bottlenecks, first, there are no forced ventilation measures; the transfer of oxygen and carbon dioxide are completely dependent on diffusion, which results in a huge problem of heat and mass transfer during the process. The oxygen consumption of the aerobic microorganisms is far higher than oxygen supply, which mainly refers to the actual availability of oxygen that can be dissolved in the solid substrate biofilm surface. Consequently, the supply of oxygen is often a limiting factor in the tray solid-state fermentation process. Because of oxygen transfer process limits and the oxygen consumption of microbial metabolism, a gradient of oxygen concentration often appears during the fermentation process.

Second, the temperatures of the tray are nearly the same during the preliminary stage of tray aerobic solid fermentation; there is no temperature gradient. As the reaction continues, the heat will be generated by microbial metabolism. The poor thermal conductivity of the solid substrate results in the difficult diffusion of heat and the gradient temperature of the entire tray.

Studies have shown that the temperatures of various heights of solid substrate are not the same, and with the increase in the packing height, the temperature shows an increasing trend. Generally, if the substrate height changes, for every 1 cm, the temperature will be altered by 1.7 °C. Some research even found that the temperature difference could reach 50 °C when the height of the substrate was increased up to 5 cm. The solid substrate will be transformed by the effect of microbial metabolism, which may hamper heat transfer and divide the entire system into a high-temperature zone and a low-temperature region.

Sometimes, the influence of the temperature gradient is significant, which leads to microbial growth, and the production of the desired substance is affected. A high temperature would influence microbial growth, spore germination, fruiting body growth, and metabolite formation. However, a lower temperature is not beneficial for microbial growth and biochemical reactions. At the same time, the temperature gradient of the substrate will result in the generation of natural air convection, which affects not only the transfer of heat but also the transfer of oxygen and carbon dioxide and moisture evaporation. Some research showed that the material layer tray height should be restricted to only a few centimeters to maintain the rapid growth of microorganisms. Researchers promoted the evaporation of water by lowering the humidity of the air circulating in the fermentor. The evaporation promotes the cooling of the solid substrate, thereby reducing substrate temperature. However, during this process, the culture substrate surface would dry quickly, which is not beneficial for the growth of the microorganisms. During the fermentation process, the trays should be kept artificially flipped. Because of the intrinsic characteristics of tray bioreactors, the mechanization of operation is difficult to achieve. This technology is a labor-intensive industry.

4.3.1.5 Current State of Tray Solid-State Fermentation Technology

At present, the tray solid-state fermentation process plays an important role in human life; it is the technology used the most widely. Tray fermentation also plays an important role in laboratory studies. For example, the petri dish, Erlenmeyer flasks, and plastic pots are all simple tray bioreactors. Some researchers used castor bean as a substrate to culture *Penicillium simplicissimum* for lipase production (Godoy et al. 2011). Some researchers used straw as a substrate to culture *Bacillus* sp. for amylase production (Hashemi et al. 2011). Compared to other fermentation means, the tray solid-state fermentation process is a simple operation widely applied in strain selection and optimization of fermentation conditions. In industrial applications, tray solid-state fermentation bioreactors are simple, and the operation technology requirement is not high. After several years of further research, tray solid-state fermentation bioreactors have successfully completed the stages from laboratory, to pilot, to industrialized production. Now, this technology has been widely used in liquor production. On the other hand, there may be interactions between the microorganisms and other microbes, which results in the introduction of some flavor compounds in the fermentation process. The process also has its unique value, especially for some of the low-cost fermentation products. The development of tray solid-state fermentation is still important for Third World countries because of the labor-intensive and less staff technical requirements.

However, tray solid-state fermentation reactors need a large room and require more manpower in industrialized production than other solid state fermentation bioreactors. The height of the loading substrate must be strictly controlled to maintain the transfer of heat and mass. A low loading substrate height will result in a lower yield and lower utilization of the fermentor. Yet, a high loading substrate height will lead to problems of heat and mass transfer that hinder the fermentation process. In the fermentation process, the microbial growth is susceptible to external factors, the heat transfer is poor, amplification is difficult, and labor intensity is high; these are all the factors that limit its widespread application. Therefore, the design and improvement of tray solid-state fermentation reactors need further study.

With the development of modern science and technology, new materials have been applied on a large scale to tray solid-state fermentation, such as for bag solid-state fermentation bioreactors. The bag can be made of plastic, paper, or a special fabric for facility ventilation. The fermentation substrate is encased by the special bags; the transfer of oxygen and carbon dioxide are promoted. Meanwhile, water cannot evaporate freely, thus keeping the humidity of the entire environment consistent.

Ngo designed a new type of sponge tray bioreactor for the removal of organic pollutants in sewage (Nguyen et al. 2011). Large size of cylindrical urethane resin foam was prepared, and there were a large number of mesh holes in it. These conditions were ideal for the growth of microorganisms.

4.3.2 Packed Bed Aerobic Solid-State Fermentation Technology

4.3.2.1 Introduction to Packed Bed Aerobic Solid-State Bioreactors

Typically, the packed bed reactor is a cylindrical tube filled with solid substrate, and the gas can freely pass from the bottom. The solid substrates are held by a plate

(Fig. 4.10a). The control of temperature and humidity are achieved by gas circulation through the fermentor. In addition to the cylindrical shape, the bioreactors can be a crate, vertical or inclined chamber, and so on. The fermentor may be aerated from either end. For a vertical column, the air may enter the bed from either the top or the bottom (Fig. 4.10b).

The column may have a perforated inserted tube along its central axis, allowing an extra air supply in addition to end-to-end aeration (Fig. 4.10c). However, this will only be effective for bioreactors with very small diameters. The column may be water jacketed, or heat transfer plates may be inserted into the bed.

4.3.2.2 Characteristics of Packed Bed Aerobic Solid-State Fermentation Technology

The packed bed bioreactor is generally a high and thin column, and there are intake and outlet ports in the upper and lower ends of the column. The air goes into the fermentor and leaves from the other end. During the actual operation, the solid-state substrates remain relatively static; the transfer of heat and mass is achieved by airflow. Consequently, packed bed solid-state fermentation is suitable for aerobic microorganisms that are more sensitive to shear forces. The column may lie horizontally or at any angle. This alters the relative directions of the forces because of gravity and air pressure. Usually, the materials are placed on the plate of the reactor, and air is blown from the bottom and is discharged from the top.

The main design and operation parameters of the bioreactor include the height of the reactor, the airflow, and the temperature of the inlet air. The temperature and humidity of the entire reactor can be controlled by forced ventilation or water jacketing. The packed bed bioreactor is often designed as a thin cylinder, which is beneficial for the increase in surface and heat transfer areas. The advantage of the packed bed reactor is the simple design requirements, especially for the control of temperature and humidity.

At present, research into packed bed aerobic solid-state fermentation technology can be briefly stated as exploring the following areas: (1) control of the axial and radial temperature gradients of the entire reactor; (2) control of water evaporation in the reactor to avoid drying of the media; (3) increased ventilation pressure because of the increase in height of the reactor and the growth of filamentous fungi; (4) no need to consider oxygen supply for a small packed bed bioreactor. The aim of forced ventilation is only to promote the transfer of heat and mass, yet for a large-scale packed bed reactor, the oxygen supply must be considered.

Energy Balance

The basic form of energy balance for the packed bed bioreactor is as follows (Sangsurasak and Mitchell 2000; Vaziri and Fanael 2008):

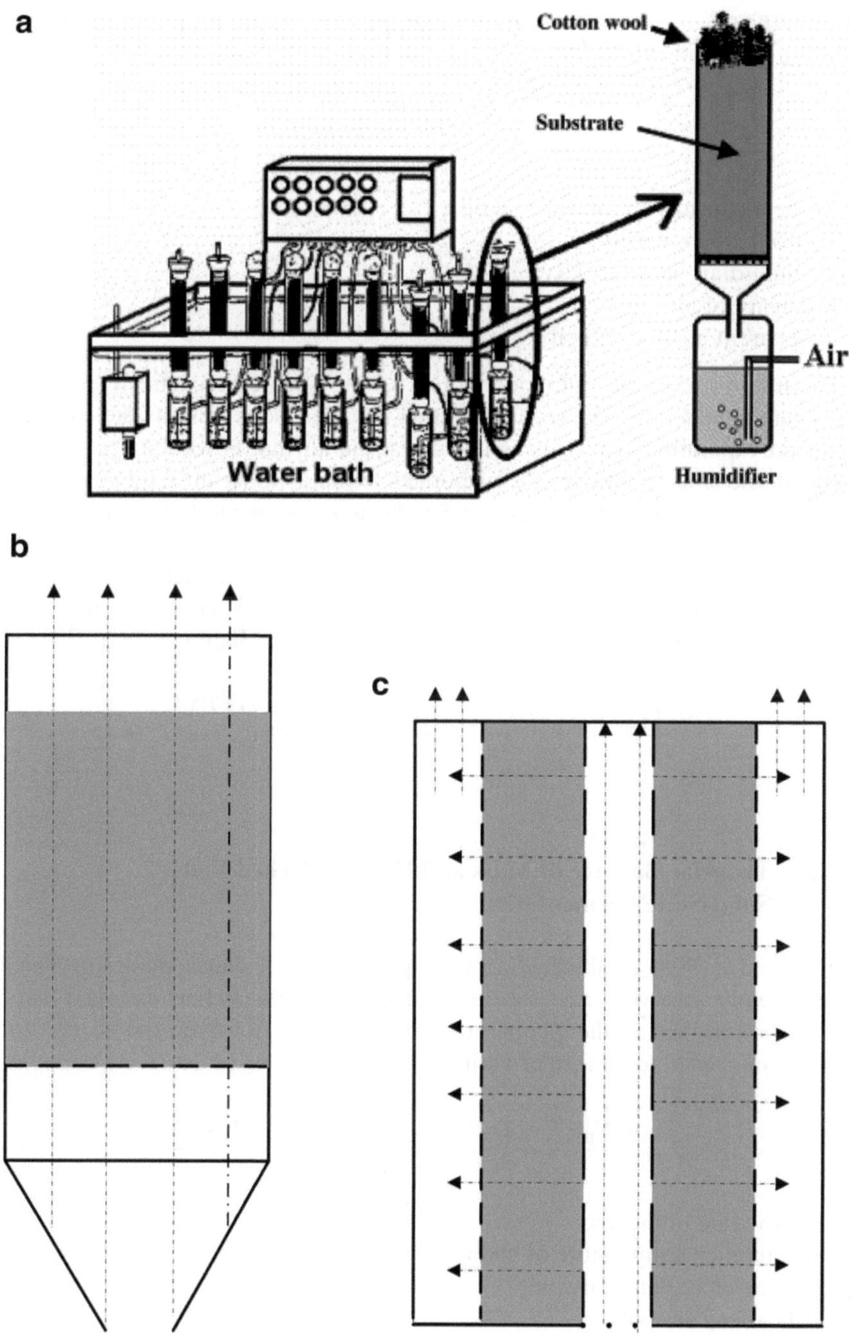

Fig. 4.10 (**a**) Packed bed solid-state fermentation bioreactor; (**b**) traditional packed bed bioreactor; (**c**) radial flow packed bed bioreactor

$$\rho_b C_{pb}\left(\frac{\partial T}{\partial t}\right)+\rho_a(C_{pa}+f\lambda)V_z\left(\frac{\partial T}{\partial z}\right)=\left[\frac{k_b}{r}\left(\frac{\partial T}{\partial r}\right)+k_b\left(\frac{\partial^2 T}{\partial r^2}\right)\right]+k_b\left(\frac{\partial^2 T}{\partial z^2}\right)+r_Q \tag{4.9}$$

C_{pb} = heat capacity of the stromal bed;
ρ_b = stromal bed density;
C_{pa} = humid air heat capacity;
ρ_a = density of air;
V_z = apparent airflow velocity.

The first term on the right side of Eq. 4.9 (in square brackets) represents the radial heat conduction. The second term on the left side represents the convection heat and evaporation heat. It is assumed that the air that flows through the bed is saturated, the water continuously evaporates to maintain steam saturation, and the air has a higher apparent heat capacity. So, the association coefficient appears in the equation.

The axial heat conduction can also be ignored if the diameter of the column is very small. The main heat dissipation is achieved by forced ventilation. In this case, Eq. 4.9 can be modified as follows (Ashley et al. 1999; Membrillo et al. 2011):

$$\rho C_{pb}\left(\frac{\partial T}{\partial t}\right)+\rho_a(C_{pa}+f\lambda)V_z\left(\frac{\partial T}{\partial z}\right)=k_b\left(\frac{\partial^2 T}{\partial z^2}\right)+r_Q \tag{4.10}$$

4.3.2.3 Transfer Balance of Mass and Heat in Packed Bed Solid-State Fermentation

Weber et al. (1999) established an energy and water balance model in which a pseudo steady state was used. At the same time, the researchers assumed that the water vapor content in the gas phase varied linearly. However, the scope of this assumption needed to be further verified:

$$0=r_Q+F_{air}\frac{d\left(C_{pg}(T-T_{ref})+y_{VAP}\left(C_{pVAP}(T-T_{ref})+\lambda\right)\right)}{dz} \tag{4.11}$$

F_{air} = flow rate of the air;
T_{ref} = reference temperature of enthalpy value;
y_{VAP} = gas phase humidity;
C_{pg} = heat capacity of air;
C_{pVAP} = heat capacity of vapor.

Another formula for external water balance is represented as follows:

$$(1-\varepsilon)C_S\frac{dX_{WS}}{dt} = r_{H_2Oext} - (1-\varepsilon)X_{WS}\frac{dC_S}{dt} - F_{air}\frac{y_{out}-y_{in}}{H} \tag{4.12}$$

C_s = weight of dry materials per volume of the bioreactor;
X_{ws} = mass ratio of water to dry matter;
r_{H_2Oext} = extracellular water produced by the growth of microorganisms;
H = height;
y_{in} and y_{out} = inlet and outlet humidity, respectively.

Application of Packed Bed Solid-State Fermentation

Sella et al. (2009) studied spore production in a packed bed bioreactor using *Bacillus atrophaeus*. A column bioreactor was used; the diameter was 4 cm, and height was 20 cm. The fermentation temperature was maintained at around 36 °C by water bath. The moist air passed into the column from the bottom. The fermentation proceeded for 9 days. The results showed that during the fermentation process, if the water content was more than 88 % of the maximum water content, cell growth would be affected significantly. Weber et al. (2002) established a mathematical model of an industrial-scale packed bed solid-state fermentation bioreactor. Using *Coniothyrium minitans* and *Aspergillus oryzae* as strains, he compared changes in physical characteristics of marijuana, oats, sugarcane bagasse, and perlite substrate, such as scalability and permeability in the fermentation process. The process and the optimum operating conditions were tested.

4.3.2.4 Packed Bed Solid-State Fermentation Technology Bottlenecks

Compared to the oxygen gradient, the temperature gradient in a packed bed solid-state bioreactor is more damaging to microbes. Consequently, the temperature gradient of a packed bed solid-state fermentation reactor is the first bottleneck that needs to be overcome. With the increase in the packing height and the decrease in the aeration rate, the bioreactor temperature gradient gradually increases (Sangsurasak and Mitchell 1998). When the temperature exceeds a certain value, the growth of microbes will be suppressed, and microbial death may occur. The highest temperature that the microbe can stand is called the critical temperature, which determines the substrate packing height of the fermentor. The critical height is influenced by its own characteristics and the cultivation conditions. With respect to these problems, researchers mainly select forced ventilation. The evaporation of water is strengthened by the air circulation, thus achieving cooling and reducing the axial temperature gradient of the entire bioreactor. Evaporation plays an important role in heat transfer in the packed bed bioreactor. In practice, approximately

65–78 % of the heat is taken away by water evaporation. Although maintaining the temperature by water evaporation is good, at the same time the solid substrate would dry quickly. The excessive loss of water is harmful to solid-state fermentation. So, the use of saturated vapor may be a better alternative.

On the other hand, reducing the temperature of the inlet air is a good alternative. In practice, the temperature of the air inlet is 10–15 °C lower than the microbial optimum temperature. The growth of microbes near the inlet will be inhibited because of the lower temperature. The diameter of the bioreactor is usually reduced to strengthen the ventilation effect. In the packed bed bioreactor with a smaller diameter, the temperature at the bottom of the reactor is low, which is suitable for microbe growth. Consequently, the microbial metabolic activity is enhanced, and because of the low efficiency of heat radial conduction, the upper temperature of the solid substrate would be high. The upper microbial growth is affected by the increase in temperature, which results in the reduction of metabolic heat, and the temperature is gradually decreased. For the small-diameter packed bed solid-state fermentation bioreactor, this results in a low packing coefficient and product separation difficulties. Therefore, large-scale production applications are limited. In large-scale production, water jackets are usually used to control the temperature. However, the effect of water jackets will be not very obvious if the heights are more than 20 cm.

4.3.2.5 Current State of Packed Bed Solid-State Fermentation Technology

After nearly a decade of research, the experimental and mathematical models of the packed bed bioreactor have been studied in depth. Many models describing the gradient of temperature, humidity, and oxygen concentration in the fermentation process have been established. In Asia, many industrial examples showed that these bioreactors can be successfully used to produce low-value-added products. However, a packed bed solid-state fermentation bioreactor has its own shortcomings that are limiting factors for large-scale applications, such as difficulties of product separation, low heat transfer efficiency, and amplification difficulties. This kind of bioreactor should be researched further.

The latest studies of packed bed solid-state fermentation technology mainly focused on the optimization of the fermentation process by using various solid substrates, through enhancing the heat and mass transfer, and by promoting the flow of oxygen. Baños et al. (2009) studied a new type of packing material as a solid substrate. Polyurethane foam was used as a packing solid substrate, and *Aspergillus terreus* was cultivated to produce lovastatin under packed bed solid-state fermentation. The artificial polyurethane foam was cut into small pieces (1–3 cm^3); the diameter of the column was 0.021 m, and the length was 0.15 m. The moist air passed into the column. The results showed that the production of lovastatin could be significantly improved by controlling the packing quantity of the solid substrate and the rate of ventilation. Compared to traditional solid-state fermentation using bagasse as the solid substrate and liquid fermentation, the production of lovastatin could be increased by nearly 2-fold and 16-fold, respectively. The products may be different from liquid fermentation products, indicating the unique advantage of solid-state fermentation. The same results also were found by Minjares-Carranco et al. (1997); the production of pectinase by a mutant strain of *Aspergillus niger* in solid-state fermentation and liquid fermentation was studied. The diameter of the bioreactor was 2 cm, and the height was 15 cm. The result showed that the heat resistance of pectinase from solid-state fermentation was significantly superior to that from liquid fermentation.

Roussos et al. (1993) designed a Zymotis packed bed solid-state fermentation bioreactor made of acrylic plastic. The length of the box was 40 cm; the width was 15 cm, and the height was 65 cm. The working volume was about 100 L. A rectangular cover was buckled in the bioreactor to prevent the exchange of mass between the internal and external reactor. There was a gas circulation system in the right side of the reactor. Ten stainless steel heat exchange plates were placed parallel along the bioreactor. The distance between the heat exchange plates could be controlled. The results showed that the homogeneity of the entire reaction process was good when this distance was less than 5 cm. Air could go into the nine gas flow pipes after being degreased, sterilized, and humidified. The reaction temperature was controlled by a cold-hot water circulation plate. The concentrations of oxygen and carbon dioxide were monitored online. Mitchell and von Meien (2000) studied the energy balance of the growth process of *A. niger* in a Zymotis packed bed bioreactor (Fig. 4.11). The established mathematical model laid a solid foundation for condition and amplification optimization. The study showed that the optimal fermentation results could be obtained when the distance between the filler plates was about 5 cm.

4.4 Dynamic Solid-State Fermentation

Dynamic solid-state fermentation bioreactors have the advantage of simplicity in mass transfer and heat transfer, but several studies also have shown that they have a negative impact on fermentation. For example, shearing force may change the characteristics of the solid substrate, which is harmful to the fermentation process. Therefore, the requirements to strengthen measures are as follows: First, the measures should ensure the fermentation process is aseptic; second, the damaging effects of shearing force should be minimized to keep the integrity of the solid substrate; third, the temperature should be controlled consistently by a water jacket.

The performance of the rotating drum and stirred drum bioreactors will depend strongly on the effectiveness of the exchange of water and energy between the bed and the headspace gases. The effectiveness of this exchange will be affected by the flow patterns within the bed and headspace. It is likely that rotating or stirred drum bioreactors will be well mixed, and there is no need to pay specific attention to the promotion of mixing in the design stage. The flow patterns within the bed and the headspace of these bioreactors have only recently started to be explored.

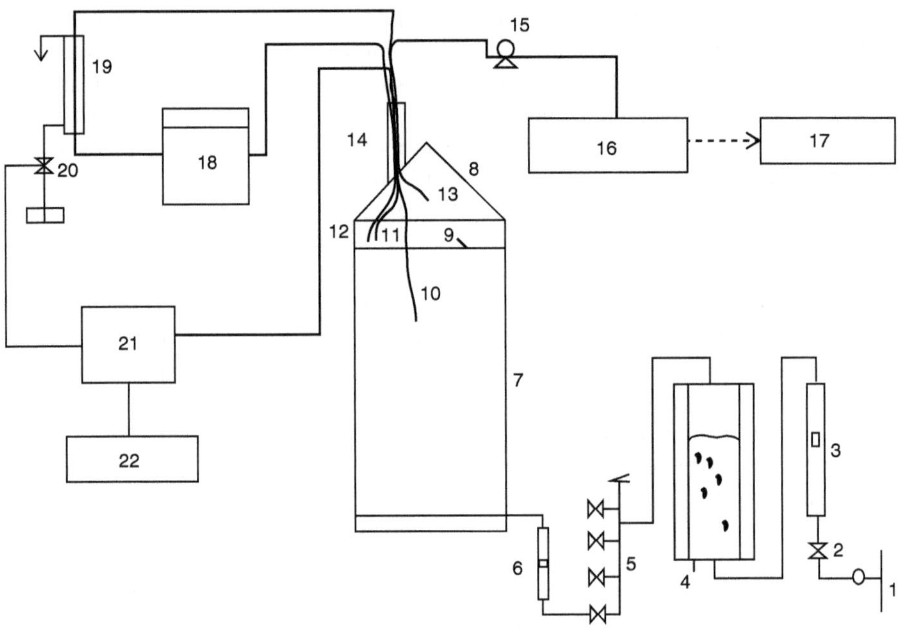

Fig. 4.11 Diagram of Zymotis packed bed solid-state fermentation. *1* Air compressor. *2* Pneumatic valves. *3* Speed monitor. *4* Humidified column. *5* Airflow detector. *6* Speed display. *7* Fermentor. *8* Cover. *9* Heat exchange plate. *10* Temperature probe. *11* Water inlet. *12* Water outlet. *13* Airflow outlet. *14* Line. *15* Air pump. *16* Gas detection system. *17* Recording system. *18* Temperature control system. *19* Heat exchange column. *20* Valves. *21* Temperature control system. *22* Temperature recorder

4.4.1 Rotating Drum Aerobic Solid-State Fermentation Technology

4.4.1.1 Introduction

Takamine (1914) first developed tray bioreactors and then invented rotating drum bioreactors; he utilized *Aspergillus oryzae* to produce the amylase in solid-state fermentation using wheat bran as a substrate. In the early 1940s, the equipment was further improved and was applied in the commercial-scale production of penicillin. There were 40 rotating drum bioreactors of 1.22 m diameter and 11.28 m length, meaning that each bioreactor had a total volume of 13 m^3.

The main body of the bioreactor is a horizontal or inclined cylinder; the cylinder rotates along its axis. The rotating drum bioreactor usually contains a stromal bed, gas circulation space, and the drum wall. Several bioreactors also contain a baffle system. The air goes into the fermentor from the top of the bioreactor, and there is no forced ventilation. The direction of rotation is changed periodically. The solid substrate should be a large amount of wet small particles, and the volume is about

10–40 % of the entire volume of the bioreactor. The speeds of different drum bioreactors are various, typically around 1–15 rpm/min. The rotating drum solid-state fermentation reactor only has a short research history, and there are few application reports.

4.4.1.2 Characteristics

The design requirements for a rotating drum solid-state fermentation bioreactor can be briefly stated as follows:

1. The inclination of the central axis of the bioreactor is usually horizontal.
2. The shapes of the stirrer are different in the various devices.
3. The design of the intake and exhaust ports will affect the working process of the whole device.
4. The temperature is controlled by a jacket, and the jacket pipe should rotate with the stirrer simultaneously.
5. The design of the system is for the addition of water or other additives to the bed during the process.
6. The size and shape of the mixing device within a stirred drum and the number, size, and shape of baffles in a baffled rotating drum are various.

In the rotating drum solid-state fermentation process, the loading coefficient is determined at the start of the fermentation and cannot be arbitrarily changed. With fermentation, the substrate will be reduced gradually. The heat produced by the microbial metabolism determines the temperature, humidity, and flow rate of the flowing air. In practice, the substrate is wet by interval spraying replenishment. The stirring speed of the fermentation process is an important factor that influences fermentation efficiency. With the stirring speed increase, the efficiency of fermentation is enhanced, and then the fermentation efficiency begins to decrease. On the one hand, a fixed substrate structure is formed by the stirring rotation, which facilitates the transfer of oxygen, carbon dioxide, and heat. On the other hand, shear force may be harmful to the growth of microbes.

The heat transfer between the substrate and the reactor space is a critical factor that determines fermentation efficiency. The stirring method is the most important factor that affects industrial applications. Schutyser et al. (2002) simulated the mixing process of the solid particles in the solid-state fermentation process by the three-dimensional (3D) model. Three different mixing strategies were created: (1) without a stirring blade, (2) with a vertical stirring blade, and (3) with a curved stirring blade. The experimental results showed that method using the curved stirring blade was the best and could effectively promote heat transfer in the longitudinal and axial directions. The mathematical model of industry amplification was established. The amplification process of a 28-L stirring drum bioreactor was studied, and the fermentation process was characterized using a two-phase model. These results showed that the model can represent changes in the temperature gradient well.

Energy Balance

It is difficult to describe the dynamic process of every point by using the previous microelement balance method because the bioreactor is more complex (Stuart 2000; Mitchell 2002). Thus, the commonly used method is overall balance. That is, for a system, only the states of the inlet and outlet need to be considered, regardless of their specific intermediate process. The model established by Stuart divided the rotating drum bioreactor into three subsystems: the stromal bed, headspace, and the wall of the bioreactor. Then, the equilibrium equation was established for each system (Hardin et al. 2000; Costa et al. 1998).

The energy balance equation for the stromal bed is as follows:

$$\frac{\mathrm{d}\left[T_{\mathrm{s}}M\left(C_{\mathrm{pm}}+C_{\mathrm{pw}}W\right)\right]}{\mathrm{d}t}=r_{\mathrm{Q}}-h_{\mathrm{sa}}A_{\mathrm{sf}}\left(T_{\mathrm{s}}-T_{\mathrm{f}}\right)-h_{\mathrm{sa}}A_{\mathrm{sa}}\left(T_{\mathrm{s}}-T_{\mathrm{a}}\right) \\ -kA_{\mathrm{sa}}\left(C_{1}-C_{\mathrm{B}}\right)\left(T_{\mathrm{s}}C_{\mathrm{pw}}+\lambda-\left(T_{\mathrm{s}}-T_{\mathrm{a}}\right)C_{\mathrm{pVAP}}\right) \tag{4.13}$$

T_{s} = stromal bed temperature;
M = dry weight;
C_{pm} = dry substrate heat capacity;
W = stromal bed moisture content;
C_{pw} = heat capacity of water;
h_{sf} = heat transfer coefficient;
A_{sf} = area of wall;
T_{f} = wall temperature;
H_{as} = heat transfer coefficient;
A_{sa} = area of stromal bed;
T_{a} = air temperature at the top;
K = mass transfer coefficient;
C_{1} = vapor concentration;
C_{B} = vapor concentration at the top.

The second right-hand term of Eq. 4.13 describes the heat transfer from the stromal bed to the reactor wall; the third term describes the convective heat transfer from the stromal bed to the top space of the fermentor, and the fourth term describes the heat dissipation by water evaporation.

The energy balance of the top space is presented as follows:

$$\frac{\mathrm{d}\left[T_{\mathrm{a}}G\left(C_{\mathrm{pg}}+C_{\mathrm{pVAP}}H\right)\right]}{\mathrm{d}t}=T_{\mathrm{i}}F_{\mathrm{i}}\left(C_{\mathrm{pg}}+C_{\mathrm{pVAP}}H\right)-T_{\mathrm{a}}F_{0}\left(C_{\mathrm{pg}}+C_{\mathrm{pVAP}}H\right) \\ +kA_{\mathrm{sa}}\left(C_{1}-C_{\mathrm{B}}\right)T_{\mathrm{a}}C_{\mathrm{pVAP}}+h_{\mathrm{sa}}A_{\mathrm{sa}}\left(T_{\mathrm{s}}-T_{\mathrm{a}}\right)+h_{\mathrm{fa}}A_{\mathrm{fa}}\left(T_{\mathrm{f}}-T_{\mathrm{a}}\right) \tag{4.14}$$

G = weight;
H = humidity of the space at the top;
T_{i} = inlet air temperature;

F_i = inlet air flow rate;
H_i = inlet air humidity;
C_{pg} = dry air heat capacity;
F_o = outlet air flow rate;
h_{fa} = heat transfer coefficient;
A_{fa} = contact area.

The energy balance of the wall is presented as follows:

$$\frac{d(T_f V_f \rho_f C_{pf})}{dt} = h_{sf}A_{sf}(T_s - T_f) - h_{fa}A_{fa}(T_f - T_a) - h_{fe}A_{fe}(T_f - T_e) \qquad (4.15)$$

V_f = overall reactor volume of metal;
ρ_{pf} = metal density;
C_{pf} = metal heat capacity;
h_{fe} = heat transfer coefficient;
A_{fe} = contact area;
T_e = external air temperature.

The first term on the right of Eq. 4.15 describes the transfer of heat between the stromal bed and the wall of the reactor; the second term describes the transfer of heat between the wall surface of the reactor and the top space. The third term describes the transfer of heat between the wall surface and the outside air.

The mass balance of matrix bed moisture is presented as follows:

$$\frac{dMW}{dt} = -kA_{sa}(C_1 - C_B) + r_{H_2O} \qquad (4.16)$$

The first term on the right of Eq. 4.16 describes moisture loss from the stromal bed caused by evaporation; the second term represents the water content produced by the microbial metabolites. Another mass balance equation is based on the moisture of the space at the top:

$$\frac{dGH}{dt} = F_i H_i - F_0 H + k_w A_{sa}(C_i - C_b) \qquad (4.17)$$

The third term on the right side of Eq. 4.17 describes the water content of the inlet air and the outlet air that evaporated from the stromal bed.

Intermittently Stirred Solid-State Fermentation

Under the stationary state, the intermittently stirred solid-state fermentation bioreactor is similar to a tray solid-state bioreactor. However, when it is under the stirring state, the intermittently stirred solid-state fermentation bioreactor is similar to the continuously rotating drum bioreactor. Because of the presence of the quiescent

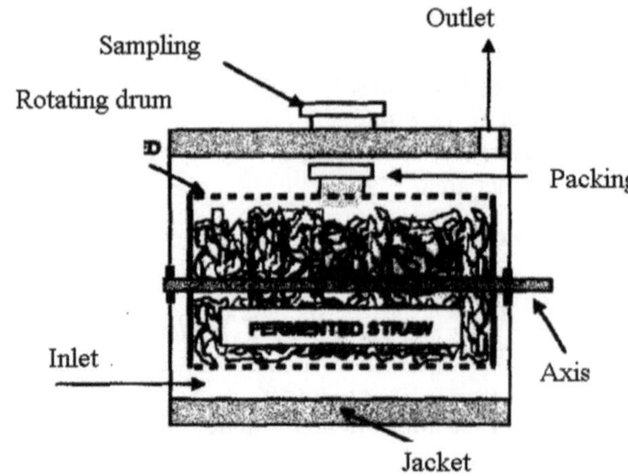

Fig. 4.12 Schematic diagram of rotating drum solid-state fermentation bioreactor (Kalogeris et al. 1999)

period, the packing height is affected to some extent. Kalogeris et al. (2003) self-designed a new batch drum bioreactor (Fig. 4.12) for the production of cellulase and hemicellulase that was successful for scale-up. The bioreactor consisted of a stainless steel cylinder that was wrapped by a water jacket for temperature control and had a rotatable stainless steel drum that was connected to a motor. The diameter of the drum was 0.15 m, and the length was 0.59 m. Many pores with a diameter of 1 mm were distributed on the surface. The entire volume of the drum was 1 L. The entire temperature of the fermentation tank was controlled by water circulation in the jacket. The heat exchanger and humidification were controlled by gas circulation in the fermentor. The gas left the fermentor in the opposite direction from the way it entered. Water vapor was condensed and collected by a peristaltic pump. Thermal-resistant strains of *Thermoascus aurantiacus* were used, and wheat straw was used as a solid substrate. The temperature was controlled at about 49 °C; the gas flow rate was about 5 L/min/kg dry substrate. The results showed that the production of cellulase and hemicellulase was higher than the control group through controlling the moisture content, fermentation temperature, and air velocity of the fermentation process.

4.4.1.3 Rotating Drum Solid-State Fermentation Technology Bottlenecks

During the rotating drum solid-state fermentation process, small media particles form groups of knots, which affects the heat and mass transfer in the entire fermentation process. Second, the growth of filamentous fungi is affected by shearing forces during the rotation process. Finally, there are complex interactions between the stromal bed and gas phases within the solid substrates. The rotational speed of the fermentor is an important factor that affects the fermentation process. On the other hand, when the speed exceeds more than 10 % of the critical rate, the energy consumption will become the limiting factor for large-scale application. Consequently, researchers usually take measures that have a low speed yet multiple stirring blades to complete the heat and mass transfer process. The stirring blades

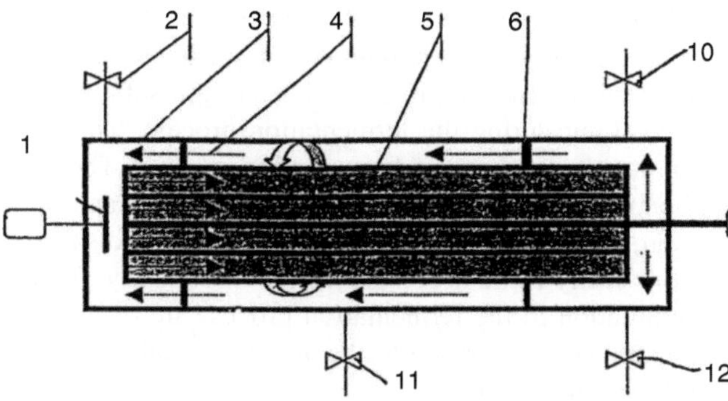

Fig. 4.13 Semicontinuous extraction solid-state fermentation reactor (Chen and Xu 2004). *1* Circulating fan. *2* Intake valve. *3* Horizontal fermentation tank. *4* Circulation air duct. *5* Fermentor. *6* Stent. *7* Electric machine. *8* Gas distribution plate. *9* Hole. *10* Leaching fluid valve inlet. *11* Exhaust valve. *12* Leaching fluid valve outlet

are sometimes designed with a curved shape to promote substrate mixing efficiency at the end of the fermentor.

4.4.1.4 Current State of Rotating Drum Solid-State Fermentation Technology

Compared to other fermentor devices, rotating drum bioreactors have been applied in many fields. The fermentor plays an important role in modern large-scale solid-state fermentation, which represents one of the important directions for future solid-state fermentation development.

With respect to the long period of the traditional solid-state fermentation process, I designed a semicontinuous extraction solid-state fermentation bioreactor (Fig. 4.13) to solve the difficulties of product separation.

The specific steps are as follows: sterilization, inoculation, installation of the gas distribution plate, and sealing of the tank. The circulating fan is opened, the fermentation starts, and the fermentation product is generated. The leaching fluid inlet valve is opened when the product reaches its peak. After leaching for 20 min, the fermentation cylinder is rotated by 180°, and the fermentation product in the other half of the fermentation tank is leached for 20 min; then, the extract is discharged.

4.4.2 Gas-Solid Fluidized Bed Fermentation

4.4.2.1 Introduction

Gas-solid fluidized beds consist of a vertical chamber with a perforated baseplate. The air or some other gas with sufficient velocity that fluidizes the substrate particles is blown from the perforated baseplate into the fermentor, and a large amount of air rapidly leaves from the top. We say that this bed is fluidized. The height of the fermentation tank is an important design parameter and is determined by multiple factors. There are usually stirring paddles in gas-solid fluidized beds to avoid solid substrate caking during the fermentation process. To save gas costs, circulating air is commonly used in the fermentation process. The concentration of oxygen and carbon dioxide gas should be maintained at an appropriate range. In the fermentation process, the heat exchange between the solid substrate and the surroundings are more easily to be accomplished. Consequently, the problems of metabolic heat accumulation in the fermentation process are overcome. The gas-solid fluidized bed also could be applied to the anaerobic solid-state fermentation process by using nitrogen instead of air.

In the 1980s, Rottenbacher first designed the gas-solid fluidized bed bioreactor using nitrogen as the cycle gas. Ethanol was produced under anaerobic fermentation through continuous circulation of the airflow to reduce product inhibition and promote ethanol fermentation. According to the actual needs, the gas stream sometimes goes into the fermentor along the central axis; only a part of the solid substrate is in a somersault state by the airflow. There is continuous particle circulation in the bottom of the fermentor bed. In 1993, Matsuno designed a gas-solid fluidized bed fermentor with a diameter of 0.2 m and a height of 2 m. At the same time, the fermentor was successfully scaled up to 1,600 L. The research results showed that the production of protease and amylase was significantly higher than production in the liquid fermentation process. (1) The condition was suitable for the growth of aerobic microorganisms because of the good ventilation. (2) The metabolic heat was completely removed, and the phenomenon of high temperatures in local media could be avoided. (3) Volatile metabolites could be quickly removed, so the feedback inhibition could be reduced. (4) The effect of mixing was good; the temperature and humidity gradient in the fermentation process could be avoided, which was conducive to the control of the fermentation parameters. (5) Compared to traditional solid-state fermentation technology, the production efficiency was improved significantly.

4.4.2.2 Technology Characteristics

For the gas-solid fluidized bed bioreactor, the fermentation conditions are easier to control, and the axial and radial temperatures still are consistent when the diameter of the bed is greater than 10 cm. The heat transfer efficiency in gas-solid fluidized beds is good, so it does not need to be considered.

Foong et al. (2009) studied feed production in a gas-solid fluidized bed bioreactor using palm oil cake as the substrate (Fig. 4.14). The length of entire reactor was 1 m, and the inner diameter was 0.046 m. There was an automatic drip system at the top of the fermentor, which was quantitatively regulated in a timely manner by the humidity of the reactor. There was a perforated plate at the bottom, which was used for gas distribution. The gas aeration rate was 0.6 m/s, and palm oil cake was crushed into 855-μm particles. Heat and mass transfer in the reaction process were

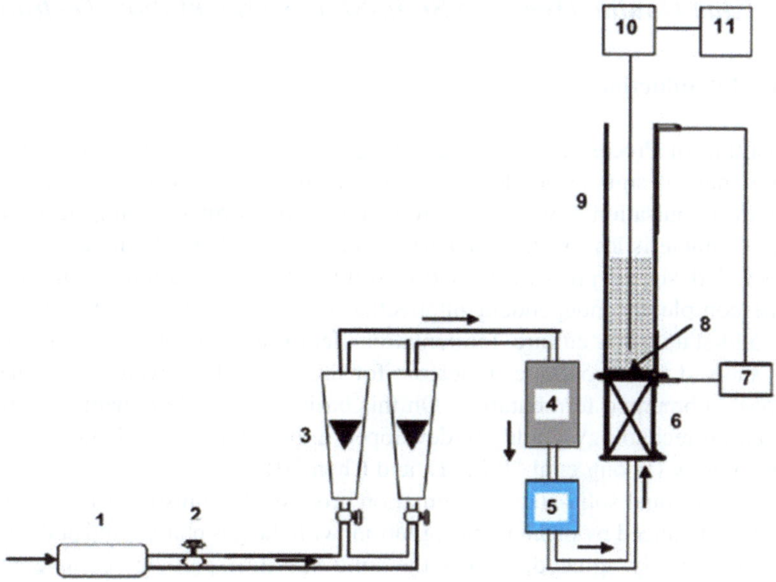

Fig. 4.14 The gas-solid fluidized bed (Li and Chen 2010). *1* Compressor. *2* Pressure controller. *3* Speed measurement instrument. *4* Humidifier. *5* Humidity controller. *6* Glass beads. *7* Divider. *8* Gas distribution plate. *9* Fluidized bed column. *10* Thermocouples. *11* Data logger

promoted by regulating the airflow changes. The water content of the fermentation process was maintained by controlling the dripping speed. The research results showed that the transformation of biomass can be achieved under the gas-solid fluidized bed fermentation bioreactor using nutritional adsorptive carriers as the substrate. This study laid the foundation for the high-value utilization of biomass.

4.4.2.3 Gas-Solid Fluidized Bed Solid-State Fermentation Technology Bottlenecks

The characteristic of the solid substrate is an important factor that affects the design, development, and applications of a gas-solid fluidized bed reactor. Sometimes, there will be large agglomeration phenomena because of the low viscosity of the fermentation substrate. The fermentation process will be influenced if the sticky group cannot be broken up by airflow. The size of the solid substrate particles is also an important factor that influences fermentation. The inconsistent size of the fermentation particles would result in the suspending heterogeneity of the particles in the fermentation process. The characteristics of the solid substrate would change when the microbial metabolism proceeded. For example, the weight and the shape of the substrate both can result in low-efficiency fermentation.

4.4.3 Gas Double Dynamic Solid-State Fermentation Technology

4.4.3.1 Introduction

The Institute of Process Engineering, Chinese Academy of Sciences, researchers proposed new design principles for a bioreactor using normal pressure as the outside cycle pulsation power source to stimulate the fermentation process. Based on the characteristics of raw materials and the biological characteristics of microbes, I designed pressure pulsation solid-state fermentation technology and own the completely independent intellectual property rights. In 1998, the large-scale solid-state pure culture fermentation demonstration plant was built. The results showed that economic indicators for this technology were better than for traditional submerged fermentation. On this basis, gas double dynamic solid-state fermentation technology gradually developed into a modern solid-state fermentation technology (Foong et al. 2009; Li and Chen 2010).

In the traditional solid-state fermentation process, the transfer of heat and mass are usually enhanced by mechanical agitation, with the gas phase fixed and the solid phase continuously agitated, to mix the solid substrate particles completely and strengthen the contact between the particles or gas molecules. During the agitating solid-state fermentation process, the growth of microbes will be damaged by the shearing force. Second, the equipment is difficult to seal, and the energy consumption is high. Third, the sticky wet materials are in contact with fermentation tanks, which easily cause the appearance of a dead angle that is difficult to be sterilized in the fermentor. If the agglomeration of media cannot be completely avoided, the efficiency of heat and mass transfer will be influenced. These shortcomings of traditional solid-state fermentation all can be overcome by gas double dynamic solid-state fermentation. Mass and heat transfer can be improved, and the concentration gradients of temperature, O_2, and CO_2 can be reduced. At the same time, the microbial metabolism activity phase can be promoted by circular high-pressure pulse.

4.4.3.2 Characteristics

The gas double dynamic solid-state fermentation bioreactor consists of a horizontal solid-state fermentation cylinder, built-in circular duct, cooling pipes, blowing devices, and an air circulation system. The solid-state fermentation cylinder can be divided into two kinds: binocular body and monocular body solid-state fermentation tanks. The characteristics of gas double dynamic solid-state fermentation can be summarized as follows: (1) There is no mechanical agitation device. The transfers of mass and heat are achieved by air circulation. (2) The bioreactor structure is simple and easy to seal. (3) The fermentation tank is a pressure-resistant container that can be sterilized by steam pressure. (4) During the fermentation process, the pressure of the fermentor is always maintained at a positive stage, which is easy to keep the environment sterile. (5) Microbial metabolism can be enhanced by cycle stimulation. (6) The temperature and humidity of the bioreactor are easy to control. (7) The fermentation process can be automated (Chen et al. 2007).

4.4.3.3 Gas Double Dynamic Solid-State Fermentation Bottlenecks

A periodic pressure pulse is conducive to the transfer of heat and mass in the fermentation process and to the growth of microorganisms. However, the high frequency of the pressure pulse will accelerate water loss from the solid substrate, which leads to a decrease in water activity, which affects the growth of microorganisms. Thus, the cycle of the pressure pulse should be properly optimized. During actual operation, the temperature changes of the solid substrate are detected by the temperature probes. The relationship between the temperature change curve and cell growth is established; the pressure pulse cycle is optimized by considering the curve and the actual situation. Air circulation in the fermentor is always in the convection-diffusion state. The air circulation rate should be increased with the intensification of the microbial metabolic activities. But, when the air convection-diffusion is too strong, the surface of the material layer will be blown on, which could affect the fermentation process.

4.4.3.4 Gas Double Dynamic Solid-State Fermentation Process

Gas double dynamic solid-state fermentation technology developed from tray solid-state fermentation. Pressure pulsation in the process is accomplished by supercharging and decompression of sterile air. One cycle of pressure pulsation consists of the stamping, decompression, maintenance, and valley stages. The supercharging stage is long, and the curve rises gently. The decompression time is as short as possible, generally from a few seconds to 1 min. The solid substrate could suddenly be expanded. The time of the high-pressure stage and the atmospheric stage can be set freely according to different fermentation processes. Usually, in the microbial logarithmic growth period, circulation is frequent. Yet, in the delay growth and stable periods, the cycle is infrequent. The circle time ranges from 15 to 150 min. The wet solid particles are rapidly loosened by the rapid expansion of gas, which enhances heat and mass transfer (He and Chen 2002; Selinheimo et al. 2006).

4.4.3.5 Current State of Gas Double Dynamic Solid-State Fermentation

Gas double dynamic solid-state technology has groundbreaking significance in terms of both theory and industry production applications. Based on experimental and practice results, the technology can be applied to a wide range of microorganisms, such as bacteria, fungi, or actinomycetes.

Gas double dynamic solid-state fermentation technology breaks the monopoly of submerged fermentation technology in the modern fermentation industry. Because of the unique advantages of this fermentation, much liquid fermentation technology

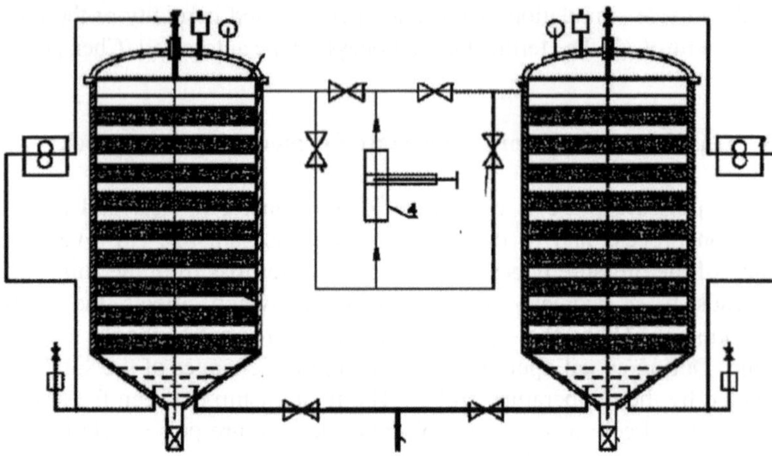

Fig. 4.15 Breathing solid-state fermentation bioreactor (Chen and Li 2011)

could be replaced by gas double dynamic solid-state fermentation technology, such as for production of pesticides, cellulase, pectinase, and riboflavin. Many new products can be produced by gas double dynamic solid-state fermentation; more important, the biotransformation of lignocellulosic substrate can be achieved, such as for cellulose ethanol or bioorganic fertilizer. Compared to traditional solid-state fermentation technology, the fermentation time tends to be shortened by one-third. In addition, it could also play an important role in mixed culture fermentation, such as for Chinese traditional liquor brewing and food flavor production.

Based on the laboratory level of the gas double dynamic solid-state fermentation process combined with bionics knowledge, I designed and established a breathing solid-state fermentation bioreactor. The whole fermentation system consists of two fermentation tanks. There is a reciprocating pump between the two fermentation tanks. The air passes from one fermentor into the other. The negative pressure tank sucks fresh air and forms an atmospheric pressure tank. At the same time, high-pressure tank discharges exhaust gas and forms atmospheric pressure tank. Circulation of negative pressure, atmospheric pressure, and high pressure proceeds until the end of fermentation (Fig. 4.15). CO_2 can be discharged and heat can be removed by "breathing" and "sucking" repeating cycles in the two parallel fermentors.

4.5 Numerical Simulation of the Fermentation Process Under Different Operating Conditions

Here, the characteristics of heat and mass transfer are compared under three different operations: tray solid-state fermentation, forced ventilation solid-state fermentation, and gas double dynamic solid-state fermentation. Based on the quality of the three operations, the equations and numerical simulation of the fermentation process using a nutritional fermentation substrate were modified. The mechanism of the gas double dynamic solid-state fermentation process was quantitatively studied, the critical control points of the fermentation process were clarified, and the operating conditions optimized. So that the process was regulated well and the efficiency of industrial fermentation could be enhanced, the pace of its industrial application was accelerated.

4.5.1 Boundary Conditions and Control Equations for Tray Solid-State Fermentation

The main features of tray solid-state fermentation are slow heat exchange and gas exchange between the solid substrate and the environment, yet the natural convection is weak. Water evaporation is serious under the temperature gradient of the substrate, which causes the heat conductivity of the substrate to decrease gradually during the fermentation process. In the logarithmic period of fermentation, the substrate is dried, and the growth ability of the microorganisms is limited the boundary conditions and control equations for tray solid state fermentation process were studied.

In tray solid-state fermentation heat transfer equations, evaporation becomes the main cooling item. The outstanding feature is the decrease in substrate thermal conductivity. Therefore, the water evaporation equation and the thermal conductivity reduction equation are added to the tray solid-state fermentation transfer equation:

$$\frac{\mathrm{d}\rho_1 V_1}{\mathrm{d}t} = m + Y_{1/\mathrm{X}}\frac{\mathrm{d}X}{\mathrm{d}t} \tag{4.18}$$

$$k = 1.166 - 4.343e^{(-\mathrm{TPSI}/21.1)}$$

$$\mathrm{TPSI} = 0.157V_\mathrm{s} + 0.562V_1 + 0.326V_\mathrm{g}$$

In Eq. 4.18, m is the water evaporation rate (g/s), and TPSI is the index for the three-phase structure of the substrate.

4.5.2 Boundary Conditions and Control Equations for Forced Ventilation Solid-State Fermentation

The features of forced ventilation are that the gas passes into the fermentor from the bottom and discharges from the top. The substrate is loosened in the vertical direction. The heat transfer is also strongly enhanced.

The difference between the equation of forced ventilation and static solid fermentation is that air fills into the bottom of the fermentor. The speed of gas movement is high (1 m/s), and the migration rate is 0.0005–0.001 m/s in the 8-cm high substrate. Under the physical case described, the corresponding transfer control was added to the equation:

The air diffusion and migration at the bottom of the space:

$$\frac{\partial C_{O_2}^{x}}{\partial t} = D_{O_2}^{x} \frac{\partial^2 C_{O_2}}{\partial x^2} - v_x \nabla C_{O_2} \tag{4.19}$$

The changes of the oxygen flow speed within the substrate:

$$\frac{V_g \partial C_{O_2}}{\partial t} = D_{O_2} \frac{\partial^2 C_{O_2}}{\partial y^2} - K_{gas}(C_{O_2} - HC_{O_2}^{f}) - (v_y + v'_y) \frac{\partial C_{O_2}}{\partial y} \tag{4.20}$$

The movement of substrate in the wind direction:

$$\rho_s Cp \frac{V_s \partial T}{\partial t} = k \frac{\partial^2 T}{\partial y^2} + \Delta H * \frac{dX}{dt} - \rho_s Cp v_{sy} \nabla T \tag{4.21}$$

4.5.3 Boundary Conditions and Control Equations of Gas Double Dynamic Solid-State Fermentation

Compared to tray solid-state fermentation, the main feature of gas double dynamic solid-state fermentation is the two-way dynamic movement of the gas within the reactor: On one hand, the air goes into the fermentor through the side wall of the reactor channel from the direction perpendicular to the x direction into the reactor by external air compressor transportation; on the other hand, the gas transfer along the y and z directions is helped by an inner fan. The direction of air movement changes from the z direction to the y direction because of the gas distribution plate. Therefore, in terms of outside space of the solid substrate, the air movement includes both the x and y directions, which causes turbulent flow gas within the solid substrate. The convection within the solid substrate is enhanced by the process of supercharging and decompression. This is the main feature of gas double dynamic solid-state fermentation that is different from the forced ventilation operation. Therefore, based on the point of heat and mass transfer, the convection within the substrate layer is enhanced by gas double dynamic solid-state fermentation. Previous studies showed that the fluctuations of ambient humidity are small by the humidifying device, such as a water tray placed inside the reactor. Therefore, similar to forced ventilation, here we assumed that in the process of gas double dynamic solid-state fermentation, the air of the inlet and outlet was saturated. The water activity substrate layer was maintained at a high relative level. Consequently, compared with forced ventilation, the characteristics of the equation of gas double dynamic solid-state includes the following:

The movement directions of the bottom of the gas includes the x and y directions.

$$\frac{\partial C_{O_2}^{x}}{\partial t} = D_{O_2}^{x} \frac{\partial^2 C_{O_2}}{\partial x^2} - \left(v_x \frac{\partial C_{O_2}^{x}}{\partial x} + v_y \frac{\partial C_{O_2}^{y}}{\partial y} \right) \tag{4.22}$$

The change of gas migration speed within the substrate layer:

$$\frac{V_g \partial C_{O_2}}{\partial t} = D_{O_2} \frac{\partial^2 C_{O_2}}{\partial y^2} - K_{gas} \left(C_{O_2} - H C_{O_2}^{f} \right) - (v_y + v'_y) \frac{\partial C_{O_2}}{\partial y} \tag{4.23}$$

4.5.4 Numerical Simulation of Cellulase Production in Different Solid-State Fermentation Operating Modes

The mass thermal coupling model was solved by multiphysics simulation software (COMSOL Multiphysics 3.4) using the finite element method (FEM). The physical model is shown in Fig. 4.16; the physical area was divided into the substrate area with a height of 8 cm and the bottom of the space area with a height of 2 cm. For static solid-state fermentation, air zone was not active; yet, for the other two modes of operation, air zone was active, and the corresponding boundary conditions were set. The solving triangular linear unit was also different; the size of the linear elements of the substrate zone was 4e-3, while the size of the linear elements of the air zone was 2e-2. The unit of physical area was divided only once; the total grid number was 1,484, and the number of grid nodes was 769. The controlling equation was solved by direct analyzer; the step length was 1 s. The result of the step length was 12 h. The absolute deviation was 0.001, and the relative deviation was 0.01.

4.5.4.1 Process Simulation of Cellulase Production in Static Solid-State Fermentation Culturing *Penicillium decumbens*

The result of static solid-state fermentation simulation of cellulase production culturing *Penicillium decumbens* is shown in Figs. 4.17, 4.18, and 4.19; the

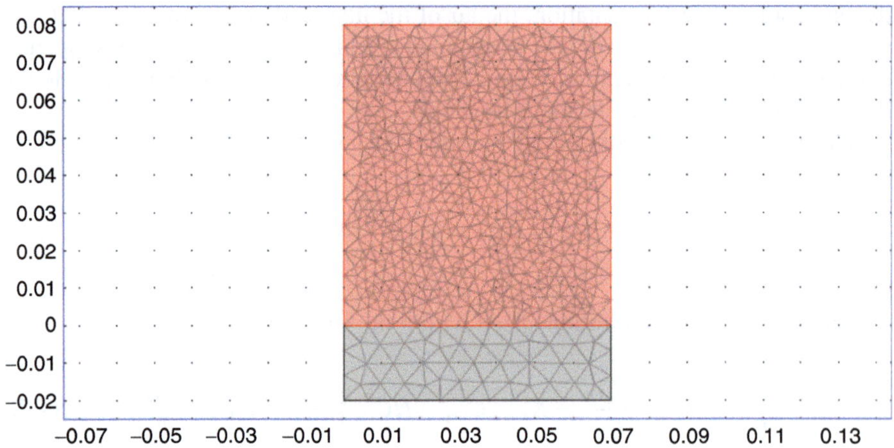

Fig. 4.16 The physical model and its linear grid of the solid matrix

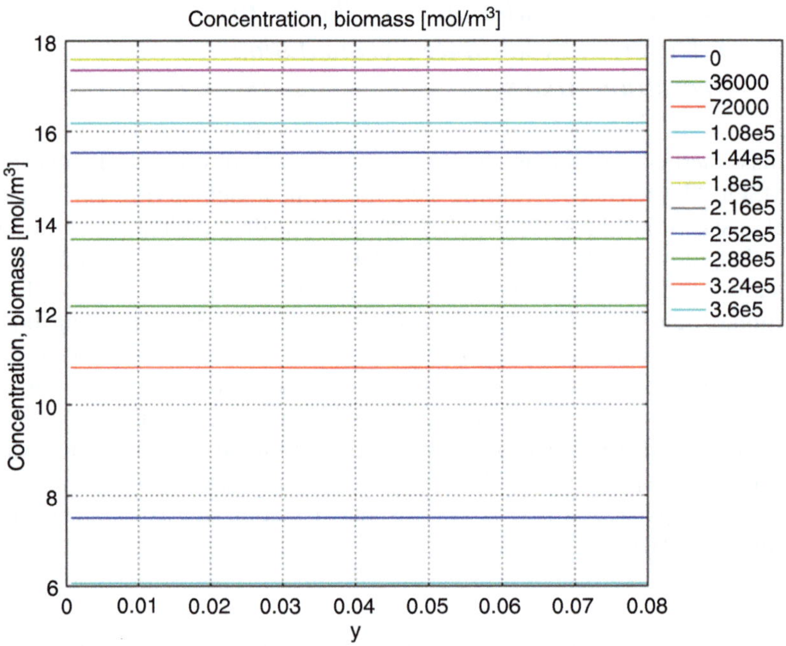

Fig. 4.17 The variations of biomass in static solid-state fermentation using *P. decumbens*

assumption was that cell growth was still in line with the logistic equation. The maximum temperature of the substrate layer could reach 43 °C in the metabolism period of cell growth and be maintained for 72 h. The temperature did not decrease until the late period of cell growth.

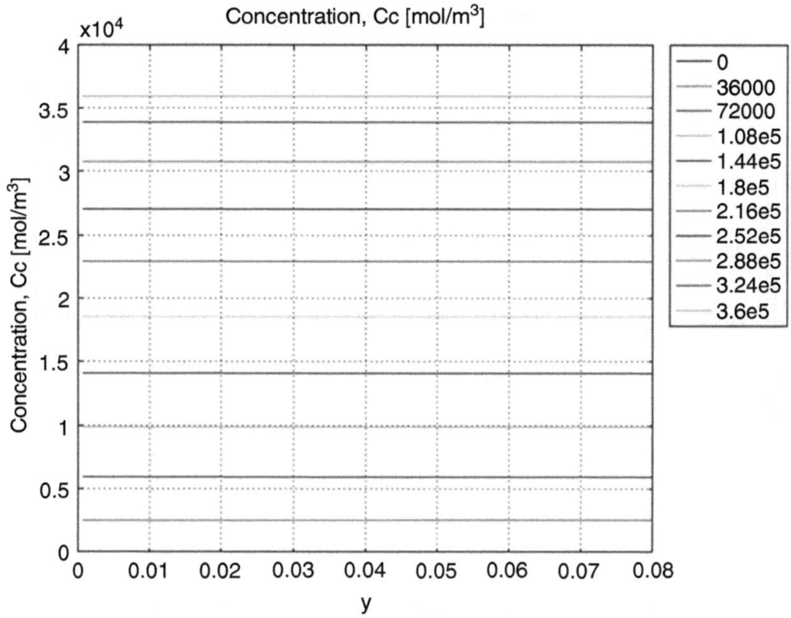

Fig. 4.18 The variations of cellulose production in static solid-state fermentation using *P. decumbens*

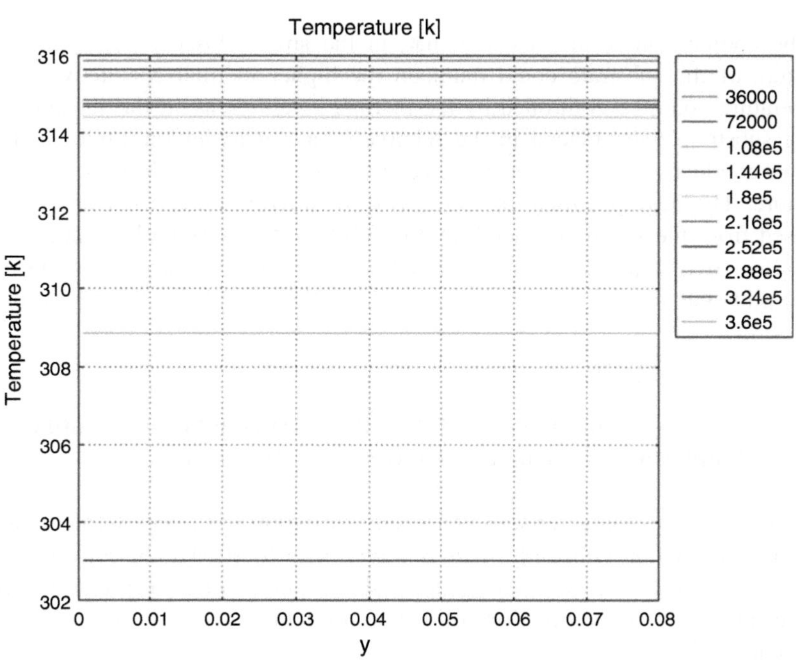

Fig. 4.19 The variations of temperature in static solid-state fermentation using *P. decumbens*

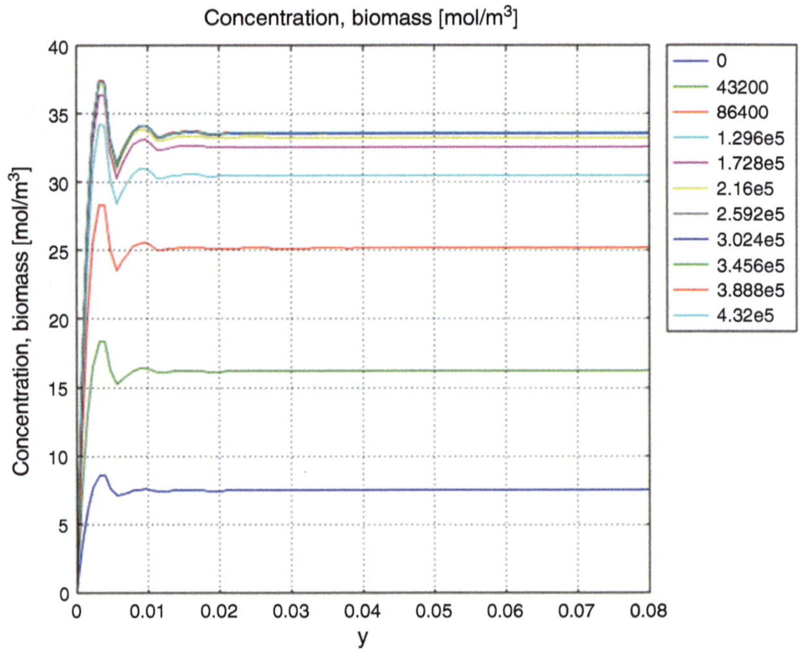

Fig. 4.20 The variations of biomass in forced ventilation solid-state fermentation using *P. decumbens*

The comparison results showed that, in the small closed devices, the higher moisture of the substrate could maintain the higher thermal conductivity, so the temperature could be controlled by the thermal conductivity of the substrate itself. If the substrate were placed in the relatively easy dehydration environment, the thermal conductivity of substrate would decrease because of the decrease in substrate moisture. The heat generated by the growth of microbes could not be effectively released from the substrate; therefore, the substrate temperature built up, which in turn affected cell growth. Water evaporation was effective for cooling, but in the static solid-state fermentation case, evaporation did not justify the full release of heat.

4.5.4.2 Process Simulation of Cellulase Production in Forced Ventilation Solid-State Fermentation Culturing *P. decumbens*

As mentioned, the characteristics of forced ventilation solid-state fermentation for cellulase production culturing *P. decumbens* were to strengthen convective heat transfer in a certain direction. Without consideration of water evaporation, simulation results for forced ventilation solid-state fermentation are shown in Figs. 4.20, 4.21, and 4.22.

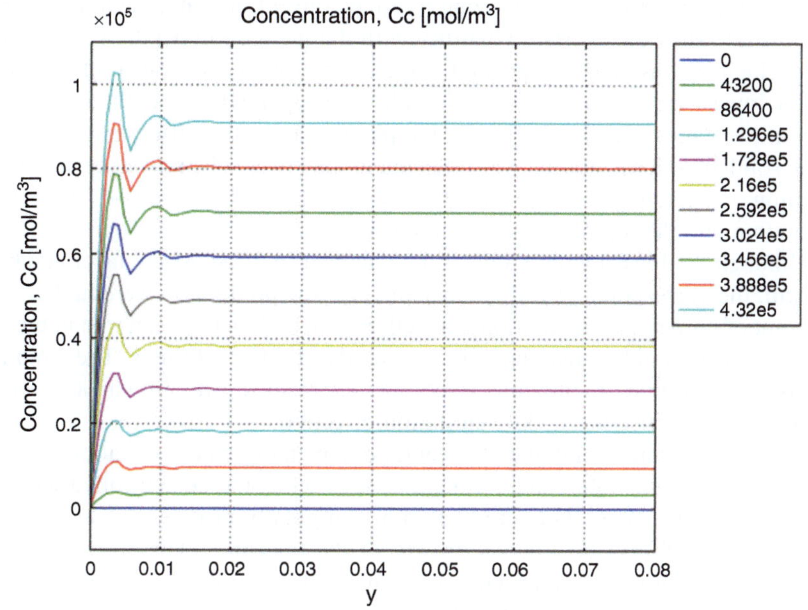

Fig. 4.21 The variations of cellulose production in forced ventilation solid-state fermentation using *P. decumbens*

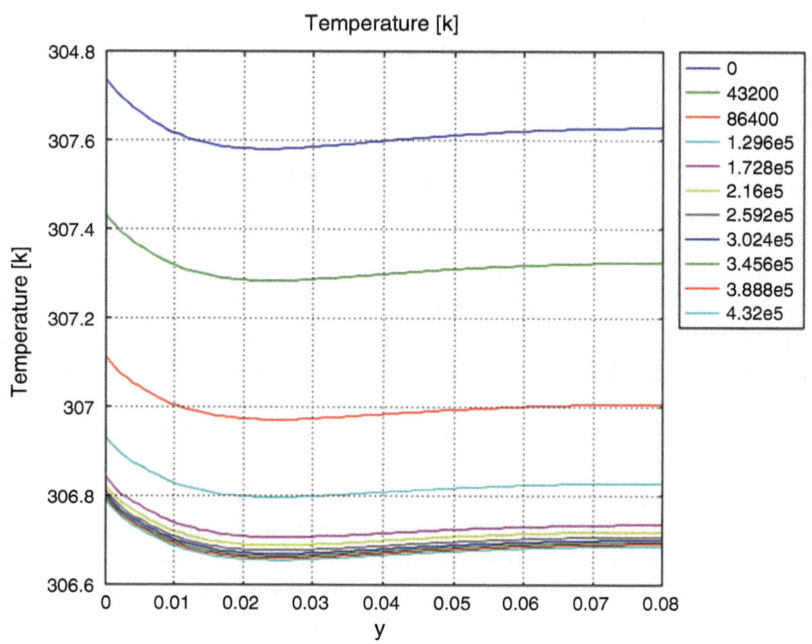

Fig. 4.22 The variations of temperature in forced ventilation solid-state fermentation using *P. decumbens*

When the substrate was in a strong internal convection state, the temperature of the substrate showed a certain volatility with the increase in depth of the substrate layer. The heat was mostly deposited in the bottom of the substrate (below 0.5 cm), and the lowest-temperature point of the substrate appeared at a height of 2–3 cm for the various stages. Consequently, the temperature reduction effect in forced ventilation solid-state fermentation was reflected in the lower substrate layer. In the axial direction of the substrate, there was still a temperature difference (1–2 °C) between the bottom and the top of the substrate layer.

Although the lowest temperature did not appear in the substrate layer at a height of 1 cm, the corresponding cell growth and enzyme production were higher. From the point of numerical distribution, the cell growth and metabolism rate in the same substrate layer could be maintained consistently at different times. The simulation results were similar to previous experimental results; cell growth and enzyme production weakened along the ventilation direction.

Compared to static solid-state fermentation, the temperature variation was small in forced ventilation solid-state fermentation, which could be controlled within the range from 33 to 34 °C. Forced ventilation can greatly promote substrate cooling and maintain uniform continuous growth of the microbes. Therefore, the visible water-holding ability and thermal conductivity of substrate have a crucial role in microbial growth. Previous studies showed that even when the humidity was maintained in a saturated state in the interior of the reactor, forced ventilation still caused substrate dehydration, and the absolute dehydration percentage was about 2 % (Gowthaman et al. 1993). Thus, in the actual production process, measures should be taken to ensure that the substrate maintains a high thermal conductivity to reduce the fermentation temperature gradients when the substrate is subjected to forced convection.

4.5.4.3 Process Simulation of Cellulase Production in Gas Double Dynamic Solid-State Fermentation Culturing *P. decumbens*

The main features of gas double dynamic solid-state fermentation of cellulase production culturing *P. decumbens* were the two-way airflow in the top or bottom of the substrate, and the structure of the substrate could be loosened under fluctuating pressure. Thus, the forced convection in the substrate layer was significantly enhanced.

The simulation results showed that the two-way airflow in the bottom space of the substrate was able to strengthen the heat transfer in the substrate layer along the direction of the substrate. Compared to forced ventilation, the heat in gas double dynamic solid-state fermentation no longer accumulated in the bottom of the layer but was distributed uniformly from the bottom to the top of the layer. The temperature was maintained at about 31–32 °C, and the temperature difference was no more than 0.5 °C during the fermentation process. Thus, although the airflow within the matrix was only unidirectional, the two-way airflow at the bottom could further strengthen its internal forced convection and promote cooling. Compared to forced ventilation, the cell growth and enzyme production were enhanced by the gas

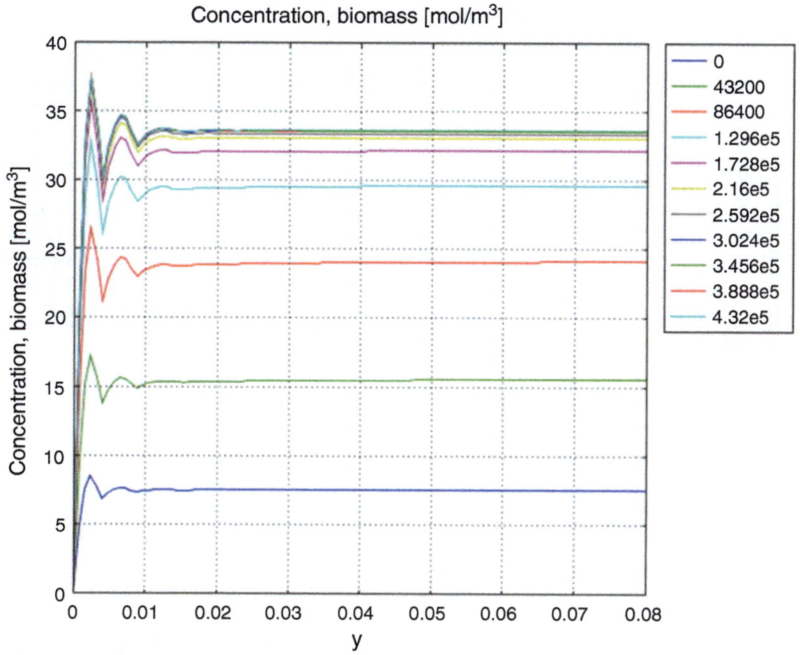

Fig. 4.23 The variations of biomass in gas double dynamic solid-state fermentation using *P. decumbens*

double dynamic method (Figs. 4.23, 4.24, and 4.25). The results further indicated that it was an effective measure that enhanced convection for controlling the temperature of the substrate if the substrate had high water activity.

4.5.4.4 Comparisons of Biomass, Cellulase Production, and Temperature Under Three Different Operations

Compared to the other two operations, gas double dynamic solid-state fermentation showed obvious advantages in promoting cell growth (Figs. 4.26, 4.27, and 4.28); the growth of microbes in the various stages of fermentation was superior. The growth in gas double double dynamic fermentation process could reach a high value after fermentation for 60 h, which is 15 % and 34 % higher than that in forced aeration and static fermentation process. For static solid-state fermentation, the heat generated by microbes in the logarithmic phase affected the cell growth rate. The previous study also showed that the respiration rate and growth vitality in forced ventilation solid-state fermentation were weaker than the results from gas double dynamic solid-state fermentation.

The cellulase production in the two ventilation fermentation operations was higher than for tray solid-state fermentation. However, the cellulase production in

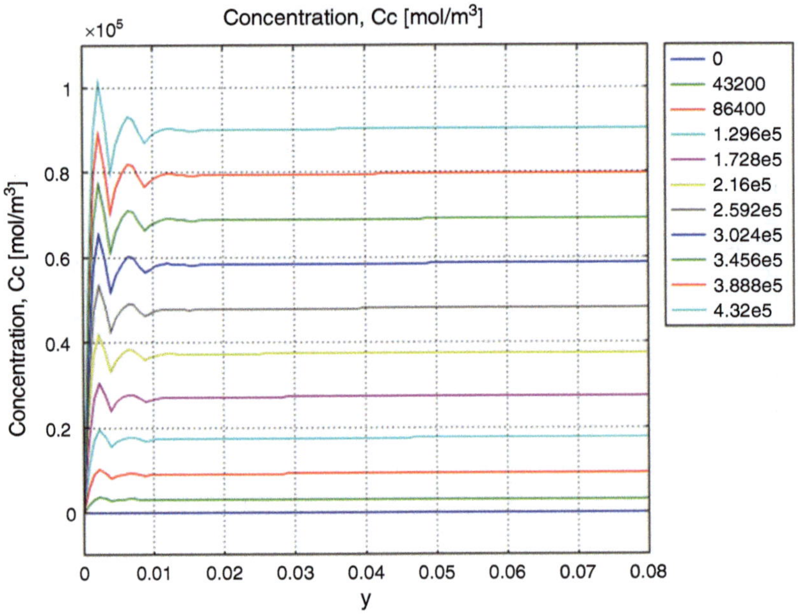

Fig. 4.24 The variations of cellulose production in gas double dynamic solid-state fermentation using *P. decumbens*

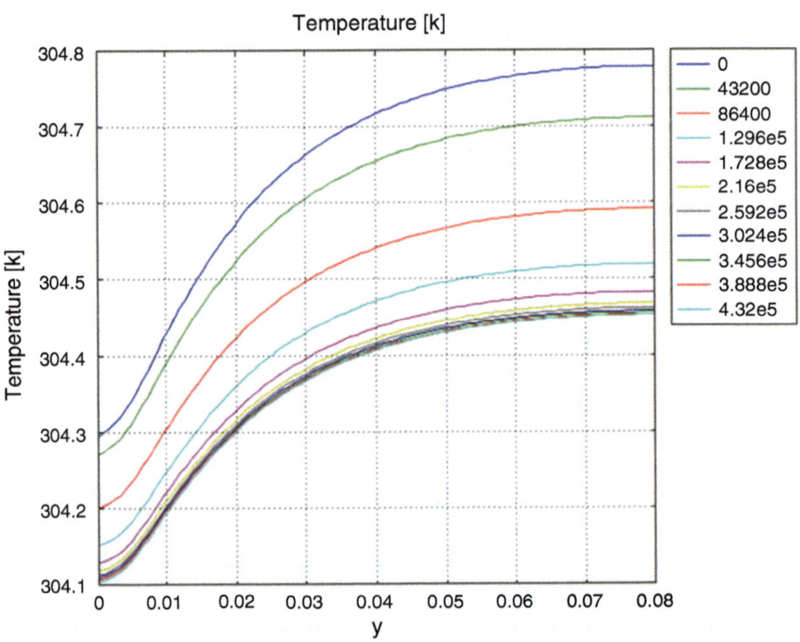

Fig. 4.25 The variations of temperature in gas double dynamic solid-state fermentation using *P. decumbens*

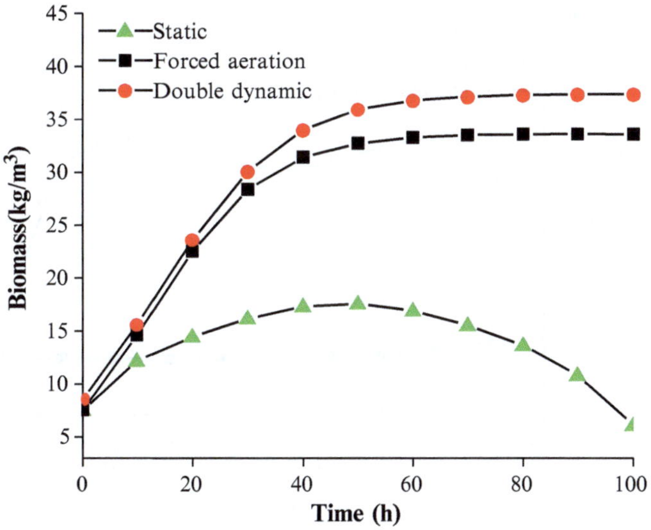

Fig. 4.26 Simulation results for growth curve of *P. decumbens* in three different solid-state fermentation processes

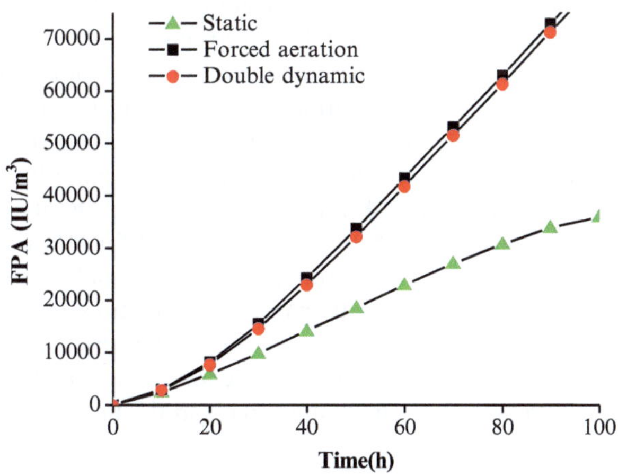

Fig. 4.27 Simulation results for cellulase production of *P. decumbens* in three different solid-state fermentation processes

gas double dynamic solid-state fermentation was not superior to the cellulase production in forced ventilation solid-state fermentation, and there was no significant difference between the two. The reason was that the cell growth in gas double dynamic solid-state fermentation reached the top value after fermentation for 60 h because of the effective control of the substrate temperature, yet the cellulose production was not enhanced. Because of influence by various factors, there was nearly no significant difference in cellulase production between forced ventilation solid-state fermentation and gas double dynamic solid-state fermentation. The

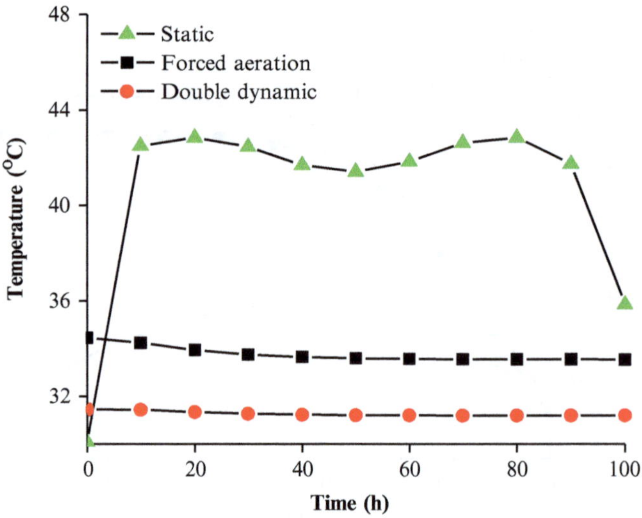

Fig. 4.28 Simulation results for temperature of *P. decumbens* in three different solid-state fermentation processes

substrate temperature was the most prominent difference. In the tray solid-state fermentation process, the temperature was elevated continuously from the starting fermentation at 20 h, and the temperature gradient of the substrate layer could reach about 15 °C. The temperature gradient of the substrate layer could reach about 6 °C in forced ventilation fermentation, yet the temperature gradient of the substrate layer reached only 2 °C in gas double dynamic solid-state fermentation. The simulation results were consistent with the experimental results.

From the physical point of the equation, the thermal conductivity and thermal dispersion of the substrate were essential for nutritional carrier solid-state fermentation. Therefore, maintaining the high thermal conductivity was the critical control point in the nutritional carrier fermentation process. Besides the heat transfer capabilities, the forced convection within the substrate layer played an important role. The efficiency of forced convection can be controlled by regulating the temperature, the water saturation, and the flow rate of the gas.

References

Ashley VM, Mitchell DA, Howes T. Evaluating strategies for overcoming overheating problems during solid-state fermentation in packed bed bioreactors. Biochem Eng J. 1999;3:141–50.

Astoreca A, Vaamonde G, Dalcero A, Ramos AJ, Marín S. Modelling the effect of temperature and water activity of *Aspergillus flavus* isolates from corn. Int J Food Microbiol. 2012;156:60–7.

Bader J, Mast-Gerlach E, Popović M, Bajpai R, Stahl U. Relevance of microbial coculture fermentations in biotechnology. J Appl Microbiol. 2010;109:371–87.

Baños JG, Tomasini A, Szakács G, Barrios-González J. High lovastatin production by *Aspergillus terreus* in solid-state fermentation on polyurethane foam: an artificial inert support. J Biosci Bioeng. 2009;108:105–10.

Brijwani K, Oberoi HS, Vadlani PV. Production of a cellulolytic enzyme system in mixed-culture solid-state fermentation of soybean hulls supplemented with wheat bran. Process Biochem. 2010;45:120–8.

Carol AJ, Kelly DP. Growth of *Thiobacillus ferrooxidans* on ferrous iron in chemostat culture: influence of product and substrate inhibition. J Chem Technol Biotechnol. 2008;33:241–61.

Cha JR, Chen JY. Transfer process principle. Beijing: Metallurgical Industry Press; 1997.

Chen HZ, He Q. Value-added bioconversion of biomass by solid-state fermentation. J Chem Technol Biotechnol. 2012. doi:10.1002/jctb.3901.

Chen HZ, Li HQ. Respiratory solid-state fermentation method and fermentor. Chinese Patent CN201110107956.8; 2011

Chen HZ, Xu J. Modern solid state fermentation: theory and practice. Beijing: Chemical Industry Press; 2004.

Chen HZ, Xu J, Li ZH. Temperature cycling to improve the ethanol production with solid state simultaneous saccharification and fermentation. Appl Biochem Microbiol. 2007;43:57–60.

Costa JAV, Alegre RM, Hasan SDM. Packing density and thermal conductivity determination for rice bran solid-state fermentation. Biotechnol Tech. 1998;12:747–50.

Foong C, Janaun J, Krishnaiah K, Prabhakar A. Effect of superficial air velocity on solid state fermentation of palm kernel cake in a lab scale fermenter using locally isolated fungal strain. Ind Crops Prod. 2009;30:114–8.

Fuqua C, Parsek MR, Greenberg EP. Regulation of gene expression by cell-to-cell communication: acyl-homoserine lactone quorum sensing. Annu Rev Genet. 2001;35:439–68.

Gervais P, Molin P. The role of water in solid-state fermentation. Biochem Eng J. 2003;13:85–101.

Godoy MG, Gutarra MLE, Castro AM, Machado OLT, Freire DMG. Adding value to a toxic residue from the biodiesel industry: production of two distinct pool of lipases from *Penicillium simplicissimum* in castor bean waste. J Ind Microbiol Biotechnol. 2011;38:945–53.

Gowthaman MK, Ghildyal NP, Rao K, Karanth NG. Interaction of transport resistances with biochemical reaction in packed-bed solid-state fermentors—the effect of gaseous concentration gradients. J Chem Technol Biotechnol. 1993;56:233–9.

Hardin MT, Mitchell DA, Howes T. Approach to designing rotating drum bioreactors for solid-state fermentation on the basis of dimensionless design factors. Biotechnol Bioeng. 2000;67:274–82.

Hashemi M, Mousavi S, Razavi S, Shojaosadati S. Mathematical modeling of biomass and α-amylase production kinetics by *Bacillus* sp. in solid-state fermentation based on solid dry weight variation. Biochem Eng J. 2011;53:159–64.

He Q, Chen HZ. Pilot-scale gas double-dynamic solid-state fermentation for the production of industrial enzymes. Food Bioprocess Technol. 2002. doi:10.1007/s11947-012-0956-9.

Hu WF, Xu GR. Solid-state fermentation principle, devices and applications. Beijing: Chemical Industry Press; 2009.

Hu KJ, Wu K, Pan RR, Liu B, Cai JM. Solid-state mixed fermentation increases activities of xylanase and cellulase. Chin J Mycosyst. 2007;26:273–8.

Kalogeris E, Fountoukides G, Kekos D, Macris BJ. Design of a solid-state bioreactor for thermophilic microorganisms. Bioresour Technol. 1999;67:313–5.

Kalogeris E, Iniotaki F, Topakas E, Christakopoulos P, Kekos D, Macris BJ. Performance of an intermittent agitation rotating drum type bioreactor for solid-state fermentation of wheat straw. Bioresour Technol. 2003;86:207–13.

Li ZH, Chen HZ. A peristaltic cycle aerobic solid state fermentation reactor using stimulate as power source. Chinese Patent CN201010113685.2; 2010

Membrillo I, Sánchez C, Meneses M, Favela E, Loera O. Particle geometry affects differentially substrate composition and enzyme profiles by *Pleurotus ostreatus* growing on sugar cane bagasse. Bioresour Technol. 2011;102(2):1581–6.

Minjares-Carranco A, Trejo-Aguilar BA, Aguilar G, Viniegra-González G. Physiological comparison between pectinase-producing mutants of *Aspergillus niger* adapted either to solid-state fermentation or submerged fermentation. Enzyme Microb Technol. 1997;21:25–31.

Mitchell DA. The potential for establishment of axial temperature profiles during solid-state fermentation in rotating drum bioreactors. Biotechnol Bioeng. 2002;80:114–22.

Mitchell DA, von Meien OF. Mathematical modeling as a tool to investigate the design and operation of the Zymotis packed-bed bioreactor for solid-state fermentation. Biotechnol Bioeng. 2000;68:127–35.

Mitchell DA, von Meien OF, Krieger N. Recent developments in modeling of solid-state fermentation: heat and mass transfer in bioreactors. Biochem Eng J. 2003;13:137–47.

Mitchell DA, Krieger N, Berovic M. Solid-state fermentation bioreactors: fundamentals of design and operation. New York: Springer; 2006.

Nguyen TT, Ngo HH, Guo W, Phuntsho S, Li J. A new sponge tray bioreactor in primary treated sewage effluent treatment. Bioresour Technol. 2011;102:5444–7.

Oostra J, Le Comte E, Van den Heuvel J, Tramper J, Rinzema A. Intra-particle oxygen diffusion limitation in solid-state fermentation. Biotechnol Bioeng. 2001;75(1):13–24.

Rajagopalan S, Modak JM. Heat and mass transfer simulation studies for solid-state fermentation processes. Chem Eng Sci. 1994;49:2187–93.

Rajagopalan S, Modak JM. Modeling of heat and mass transfer for solid state fermentation process in tray bioreactor. Bioprocess Biosyst Eng. 1995;13:161–9.

Richard TL, Walker LP, Gossett JM. Effects of oxygen on aerobic solid-state biodegradation kinetics. Biotechnol Prog. 2010;22:60–9.

Roussos S, Raimbault M, Prebois JP, Lonsane BK. Zymotis, a large-scale solid-state fermenter—design and evaluation. Appl Biochem Biotechnol. 1993;42:37–52.

Salehizadeh H, Van Loosdrecht M. Production of polyhydroxyalkanoates by mixed culture: recent trends and biotechnological importance. Biotechnol Adv. 2004;22:261–79.

Sanahuja G, Banakar R, Twyman RM, Capell T, Christou P. Research advances on *Bacillus thuringiensis*. Plant Biotechnol J. 2011;9:283–300.

Sangsurasak P, Mitchell DA. Validation of a model describing two-dimensional heat transfer during solid-state fermentation in packed bed bioreactors. Biotechnol Bioeng. 1998;60:739–49.

Sangsurasak P, Mitchell DA. Validation of a model describing two-dimensional heat transfer during solid-state fermentation in packed bed bioreactors. Biotechnol Bioeng. 2000;60:739–49.

Schutyser MAI, Weber FJ, Briels WJ, Boom RM, Rinzema A. Three-dimensional simulation of grain mixing in three different rotating drum designs for solid-state fermentation. Biotechnol Bioeng. 2002;79:284–94.

Selinheimo E, Kruus K, Buchert J, Hopia A, Autio K. Effects of laccase, xylanase and their combination on the rheological properties of wheat doughs. J Cereal Sci. 2006;43:152–9.

Sella SRBR, Guizelini BP, Vandenberghe LPS, Medeiros ABP, Soccol CR. Lab-scale production of *Bacillus atrophaeus*' spores by solid state fermentation in different types of bioreactors. Braz Arch Biol Technol. 2009;52:159–70.

Singhania RR, Patel AK, Soccol CR, Pandey A. Recent advances in solid-state fermentation. Biochem Eng J. 2009;44:13–8.

Smits JP, van Sonsbeek HM, Tramper J, Knol W, Geelhoed W, Peeters M, et al. Modelling fungal solid-state fermentation: the role of inactivation kinetics. Bioprocess Biosyst Eng. 1999;20:391–404.

Stuart DM. Solid-state fermentation in rotating drum bioreactors: operating variables affect performance through their effects on transport phenomena. Biotechnol Bioeng. 2000;63:383–91.

Takamine J. Enzymes of *Aspergillus oryzae* and the application of its amyloclastic enzyme to the fermentation industry. Ind Eng Chem Res. 1914; 6:824–8.

van den Doel LR, Mohoric A, Vergeldt FJ, van Duynhoven J, Blonk H, van Dalen G, et al. Mathematical modeling of water uptake through diffusion in 3D inhomogeneous swelling substrates. AIChE J. 2009;55:1834–48.

Vaziri BM, Fanael MA. Temperature control in packed-bed solid-state bioreactors. In: Paper presented at 5th international chemical engineering congress and exhibition; 2008 Jan 2–5; Kish Island; 2008.

Weber FJ, Tramper J, Rinzema A. A simplified material and energy balance approach for process development and scale-up of *Coniothyrium minitans* conidia production by solid-state cultivation in a packed-bed reactor. Biotechnol Bioeng. 1999;65:447–58.

Weber FJ, Oostra J, Tramper J, Rinzema A. Validation of a model for process development and scale-up of packed-bed solid-state bioreactors. Biotechnol Bioeng. 2002;77:381–93.

Wu F, Han CP. Cereal science and biotechnology. Beijing: Chemical Industry Press; 2012.

Yu B, Chen HZ. Effect of the ash on enzymatic hydrolysis of steam-exploded rice straw. Bioresour Technol. 2010;101:9114–9.

Zhou DQ. Text book of microbiology. Beijing: Higher Education Press; 2004.

Chapter 5
Anaerobic Solid-State Fermentation

Abstract Anaerobic solid-state fermentation has unique advantages in that it is water and energy saving and provides environmental protection; it will be the future direction of the fermentation industry. Anaerobic solid-state fermentation has shown tremendous application potential in the fermentation industry, agriculture, and treatment of organic solid waste. This chapter first analyzes the biological and physics basis of anaerobic solid-state fermentation. The typical case is reported in more detail, including principles, processes, and trends. The breakthroughs of reactors for production of ethanol and biogas are based on anaerobic solid-state fermentation. Anaerobic solid-state fermentation can be used in to produce such as things as bio-based energy, chemicals, traditional food, and agricultural feed and assist with environmental protection, especially for silage and solid waste treatment.

Keywords Anaerobic solid-state fermentation • Mixed fermentation • Ethanol fermentation • Biogas dry fermentation • Silage • Composting

5.1 Biology and Physics Basis of Anaerobic Solid-State Fermentation

The fermentation industry originated in China. A large-scale fermentation industrial system has been established in recent decades. The essence of the fermentation industry is the deep processing industry for agricultural products, which is an extension of the industrialization of agriculture. But, it causes problems, such as serious pollution and high energy consumption in modern industry. The main environmental pollution caused by the fermentation industry is water pollution, mostly from the residue after processing of the raw materials, such as bagasse or beet pulp. The waste liquor is also from separation and extraction of products, such as waste liquor in monosodium glutamate fermentation and the washing and cooling water from the production process. Although the fermentation industry has made great progress in cleaner production and pollution prevention, because of

H. Chen, *Modern Solid State Fermentation: Theory and Practice*,
DOI 10.1007/978-94-007-6043-1_5,

rapid growth in the yield of fermentation products in China, wastewater and the total emission of pollutants still evidence a growth trend. The outstanding problems of the fermentation industry are still how to comprehensively utilize resources; solve the problems of grain saving, energy saving, water saving, and environmental pollution; and realize clean production.

Solid-state fermentation is the fermentation process completed by one or more microbes on a wet solid-state substrate with little or no free flow of water. From the nature of the biological reactions, solid-state fermentation is the gas continuous phase bioreactor process. The water content of the solid substrate can be effectively controlled at between 12 and 80 %.

Anaerobic fermentation is carried out in sealed conditions and does not need to aerate in confined conditions. The fermentation equipment is relatively simple and has low energy consumption.

In short, anaerobic solid-state fermentation has the unique advantages in that it is water saving and energy saving and protects the environment; it will be the future direction of the fermentation industry. People need to rerecognize anaerobic solid-state fermentation to guide cleaner production by the fermentation industry.

5.1.1 Similarities and Differences of Anaerobic and Aerobic Solid-State Fermentation

Solid-state fermentation is a bioreaction process that uses gas as the continuous phase; it can be divided into anaerobic and aerobic solid-state fermentation, depending on oxygen utilization. The pathway of microbial carbohydrate metabolism in anaerobic conditions is different from that in aerobic conditions, as is the product obtained. In addition, there are still many differences between anaerobic and aerobic solid-state fermentation (Table 5.1).

Anaerobic solid-state fermentation can facilitate the microbes to produce rich flavor substances. Many of the fermentation processes are a combination of aerobic and anaerobic fermentation, such as traditional soy sauce brewing. It is noteworthy that anaerobic solid-state fermentation provides a unique flavor that is irreplaceable in food fermentation and liquor production. Anaerobic solid-state fermentation has made full use of this advantage for developing fermented products with more flavor variety that are rich in nutrients.

5.1.2 Biological Basis of Anaerobic Solid-State Fermentation

Microorganisms that can live under anaerobic conditions are the first need for anaerobic fermentation. The vast majority of anaerobic microorganisms are

Table 5.1 Comparison of anaerobic solid-state fermentation and aerobic solid-state fermentation

Species	Aerobic solid-state fermentation	Anaerobic solid-state fermentation
Fermentation conditions	Maintain ventilation oxygen; strict control of temperature and humidity in the gas supply	Without ventilation (the strict anaerobic fermentation required to drive oxygen), but requires large doses of inoculation
Fermentation microorganisms	Most aerobic bacteria; a wider range of bacteria sources	Usually anaerobes or facultative anaerobes
Fermentation characteristics	Microbial growth fast; short fermentation period	Poor growth microorganisms; long fermentation period; can form the unique flavor of the product
Application	Enzymes, antibiotics, etc	Liquor, biogas, and fuel ethanol

bacteria; a few are actinomycetes and mycoplasma. Anaerobic fungi are also seen in individual reports. With the development of anaerobic culture techniques, some new anaerobes continue to be found. Some studies have advanced in classification or physiology of anaerobic bacteria. The relationships become increasingly important between humans and anaerobes.

5.1.2.1 Anaerobic Fermentation Microorganisms

According to the demand for oxygen, the anaerobic microorganisms can be divided into facultative anaerobes and obligate anaerobes.

Facultative Anaerobes

The facultative anaerobes are able to grow and reproduce in an aerobic or anaerobic environment. They can gain energy by oxidative phosphorylation under aerobic or anaerobic conditions. The respiratory systems of the bacterial cytochrome and other components are reduced or lose energy in anaerobic fermentation. Yeast and *Escherichia coli* are typical facultative anaerobes. The former is an important industrial microbe for production of single-cell protein and alcohol. The latter is an important engineering bacterium in biological engineering research. The growth of facultative anaerobes does not necessarily need oxygen, but if the oxygen is supplied to the culture, there is better growth for microbes such as yeast, which conducts aerobic respiration in the aerobic environment or generates alcohol from fermentation of glucose in the anaerobic environment. So, in ethanol fermentation, the control of dissolved oxygen is divided into two stages: the initial high dissolved oxygen for the microbial expanding stage, then strict control of dissolved oxygen for anaerobic fermentation in the late stage.

Obligate Anaerobes

Obligate anaerobes can only survive in the environment without the presence of free oxygen. Obligate anaerobes have a series of physiological characteristics, such as lack of the intracellular respiratory enzyme system, superoxide dismutase, catalase, and cytochrome oxidase, and thus show high sensitivity to oxygen, which means they can only live under strictly anaerobic conditions and an environment with low oxidation reduction potential. The ability to limit the dissolved oxygen at a lower value is often the key to successful fermentation.

5.1.2.2 Anaerobic Solid-State Fermentation and Metabolism of Fermentative Microorganisms

Fermentation is the only biooxidation process in which reducing power [H] comes from the substrate after dehydrogenation directly transfers it to endogenous oxidative metabolic intermediates rather than the electron transport chain under anaerobic conditions (Zhou 2002).

After the conversion of glucose to pyruvate, obligate anaerobes and facultative anaerobes under anaerobic conditions can transform pyruvate into a variety of fermentation products by different means. Pyruvate can be restored to lactate by lactic acid bacteria. The pyruvate decarboxylate is reduced to acetaldehyde and then to ethanol by yeast. The acetyl-coenzyme A (CoA) comes from pyruvate decarboxylate in *Butyrivibrio* or eubacterium *Clostridium*, then the acetoacetyl-CoA was obtained by the condensation reaction of two acetyl-CoA and then the butyric acid is formed by a serious steps. Intestinal bacteria can ferment pyruvate into a variety of products, including formic acid, acetic acid, lactic acid, succinic acid, ethanol, glycerol, 2,3-butanediol, and other mixed organic acids and alcohols (Fig. 5.1).

Ethanol Fermentation

Ethanol fermentation by yeast was studied early, and its fermentation mechanism is clear. The glucose is converted into pyruvate by the yeast in the Embden-Meyerhof-Parnas pathway (EMP) pathway, and pyruvate is obtained via catalysis of pyruvate decarboxylase to form acetaldehyde, which is restored to ethanol with the help of alcohol dehydrogenase. The fermentation conditions significantly affect the fermentation process and product in ethanol fermentation, such as ventilation conditions, medium composition, and pH control. Ethanol fermentation is anaerobic; when conditions are changed to aerobic, glucose decomposition decreases, and ethanol generation stops. When returning to anaerobic conditions, the glucose decomposition rate is accelerated, accompanied by a large amount of ethanol. Pasteur first discovered this phenomenon, known as the Pasteur effect.

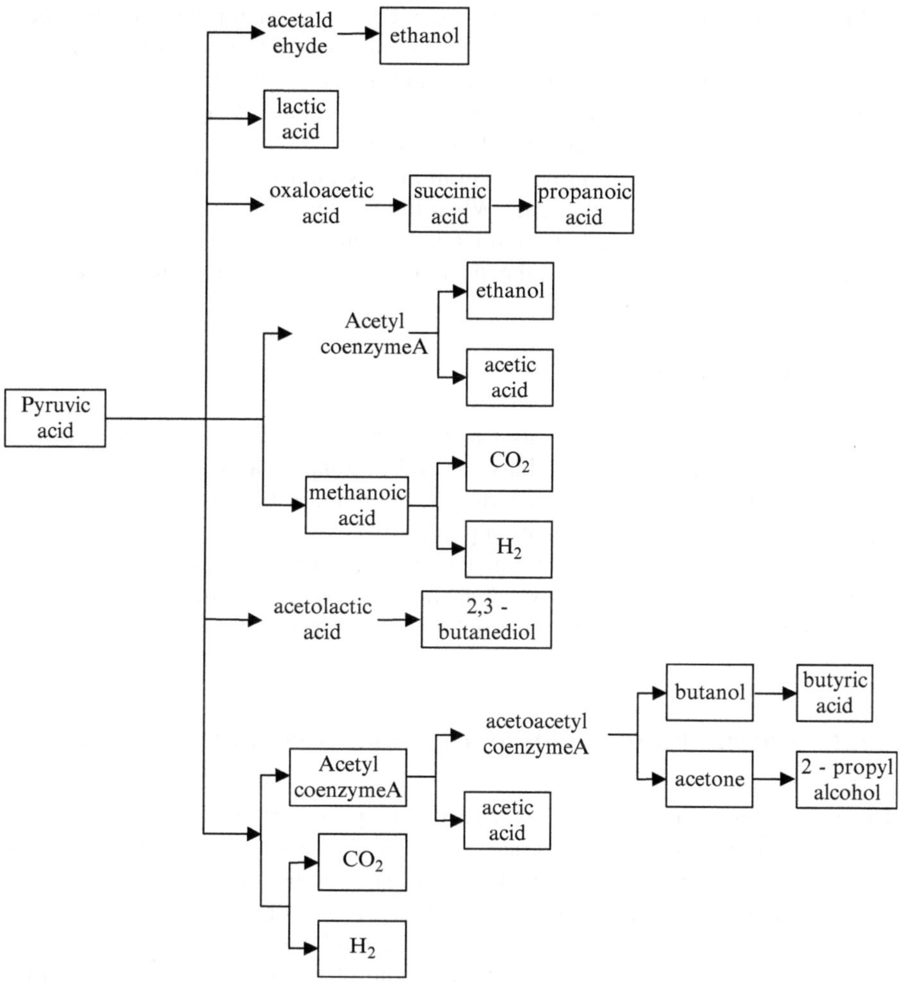

Fig. 5.1 Pyruvate conversion fermentation products

Lactic Acid Fermentation

Lactic acid fermentation is a process of some bacteria using glucose to produce lactic acid and a small amount of other products under anaerobic conditions. Bacteria used for lactic acid fermentation are known as lactic acid bacteria. Common lactic acid bacteria are *Lactobacillus*, *Streptococcus lactis*, *Leuconostoc*, and *Bifidobacterium*. Although most of the lactic acid bacteria are facultatively anaerobic, lactic acid fermentation is completed under strict anaerobic conditions. Lactic acid bacteria lack the ability to synthesize many growth factors; they show complex nutritional requirements for culture, so a certain amount of yeast extract liquid is added for their cultivation.

Lactic acid bacteria conduct lactic acid fermentation through the Embden-Meyerhof-Parnas pathway (EMP) and Phosphoketolase Pathway (PK) pathways. Glucose would be transformed into lactic acid by the EMP pathway in lactic acid bacteria fermentation. When lactic acid is the sole product of lactic acid fermentation, the process is called homolactic fermentation.

Accordingly, lactic acid fermentation in the PK pathway ferments glucose into lactic acid, ethanol, and acetic acid. When the fermentation product is lactic acid and other substances, it is called heterolactic fermentation. Lactic acid fermentation by *Leuconostoc mesenteroides* and *Bifidobacterium* is heterolactic fermentation.

Acetone Butanol Fermentation

Some obligate anaerobes, such as *Clostridium*, *Butyrivibrio*, *Eubacterium*, and *Fusobacterium*, can be used for butyrate and acetone-butanol fermentation. Glucose is degraded into pyruvate, acetyl-coA, butyrate, or butanol and acetone by a series of reactions in the fermentation process.

It should be noted that, in the fermentation pathway mentioned, acetone-butanol fermentation conducted by *Clostridium acetobutyricum* is the only large-scale production of fermented products by obligate anaerobes so far.

5.1.3 Physics Foundations of Anaerobic Solid-State Fermentation

The three phases of solid, liquid, and gas may coexist in the anaerobic solid fermentation process. The substrate is called the solid phase. The gas phase necessarily has trace O_2 and CH_4, CO_2, and H_2. Liquid water includes adsorbed water, pellicular water, capillary water, and gravity water.

Anaerobic microorganisms need to enter the solid matrix porosity, and the mass is transfered on the interface of the solid–liquid phase by liquid. The exothermic and external temperatures of microbial metabolism have a significant impact on mass transfer across the interface of the solid–liquid phase.

The porous medium is a volume that is divided into many tiny volume-containing solids and fluids. The solid parts in the tiny volume are called the backbone, and the fluid-filled portion is referred to as the porosity.

Here, we take production of biogas from straw as an example. Steam explosion pretreatment can increase the straw pore size, volume, and radius. The small changes of the pores will have a big impact on permeability. The pore channels, cellulase, and microbial attachment points increase with the increase in the strength and pressuring time by steam explosion pretreatment. Much lignocellulose is transformed into monosaccharides that can be used to produce biogas. Regarding the critical porosity of seepage, if the porosity is greater than 40 %, the area of seepage expands rapidly. It can be observed that the steam explosion intensity and time affect the increase of porosity, which establishes a functional relation between the porosity and the enzymatic efficiency prediction.

Pretreatment can promote the degradation of hemicellulose to monosaccharides in straw. The nutrient concentration distribution gradient is formed inside the porous medium because of the different ratios of solid to washed water. The injected bacteria from the bacterial solution move to a high concentration of nutrients for chemotaxis. The ideal grid seepage model was used for the percolation theory. It assumed randomness of porous media channels and bacterial chemotaxis forces to provide the bacterial flow driving force following the Markov process.

The fluid group rapidly penetrates both sides of the large pore under the force of chemotaxis. Many pores are filled by the bacterial stream, with an increase of the entering bacteria in porous media flow. Percolation appears until the bacteria flow throughout the media channel. We can design experiments with material for easy microbial attachment, such as glass fiber, nylon fiber, or polyurethane foam used for a microbial carrier. The fermentation carriers have an impact on biogas production as well as microbial distribution in the carrier surface.

The straw fiber is connected by fiber bundles. The reason for water diffusion and convection is the humidity gradients present in fiber groups (water concentration difference). The presence of the pores helps the water move through the percolation process; a variety of physical and chemical phenomena will appear in the anaerobic digestion process, such as mass transfer, adsorption, and phase change.

Porous media have a strong adsorption capacity because of a huge specific surface, so the fluid can exist in adsorbed and free states. Because of selective adsorption, the composition of the fluid is different between adsorbed and free states. With biogas production, the adsorption of the fluid will be changed on the porous medium surface. Thus, the composition of the fluid (adsorbed and free state) is changed, which affects a range of physical parameters, leading to the dynamic changes of the CH_4 content in the methane.

5.2 Types of Anaerobic Solid-State Fermentation

5.2.1 Mixed Fermentation

The traditional solid-state fermentation type includes naturally enriched microorganisms fermentation, strengthened mixed microorganisms solid statefermentation, and limited mixed microorganisms solid state fermentation.

5.2.1.1 Liquor Fermentation

Alcoholic beverage production, such as of beer and wine, is generally liquid fermentation. Moreover, production of brandy, whiskey, and other distilled spirits also uses liquid fermentation. Liquor is a collective term for various colorless transparent distilled alcoholic beverages. Chinese liquor made using a traditional

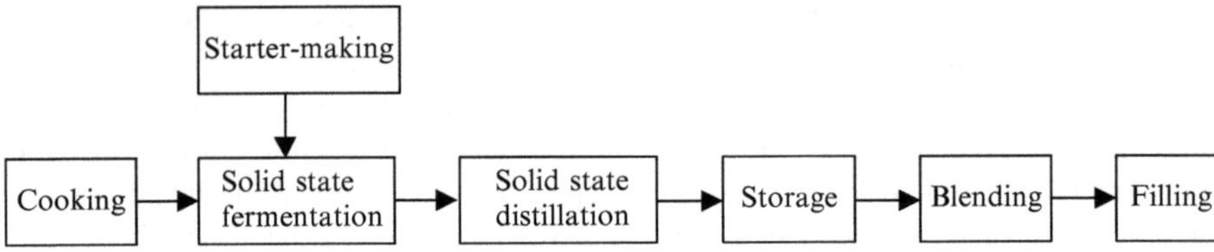

Fig. 5.2 Liquor solid-state fermentation process

operation, which involves solid-state fermentation and solid-state distillation, is unique in the world of liquor-making technology. Solid-state fermentation liquor is a representative of the traditional Chinese food industry, a wine species endemic to China. A special brewing process determines its unique flavor.

Overview of Liquor Solid-State Fermentation

Liquor can be divided into two categories according to the production methods (Fig. 5.2).

- Solid-state fermentation liquor: The fermented feedstock is in the solid state; the moisture content of the fermented grains is about 60 %.
- Semisolid-state fermentation liquor: There is solid saccharification before liquid fermentation or liquid saccharification before solid-state fermentation.

The concern is about the solid-state fermentation liquor, the pure grain as feedstock, solid-state fermentation, solid-state distillation, and blending after storage to produce Chinese liquor. The liquor quality and yield of solid-state fermentation production are affected by the production process, feedstock, distillation methods, equipment, and other factors.

Solid-state fermentation is a traditional technique for Chinese liquor production. Food is used as a raw material. Distiller's yeast is added, and fermenting occurs naturally in the mud pool over a long period. Then, distillation takes place at a high temperature.

The brewing process is characterized by simultaneous saccharification and fermentation (SSF) and solid-state distillation. After distillation of fermented grains, the feedstock needs to have the brew microorganisms mixed again, then saccharification and fermentation occur; this is repeated many times.

Cooking

The starch of the wine material has a protective film on the external layer of the particle shape under the microscope. The protective film must be destroyed before utilization to expose starch (Lv et al. 2003). There are many means to destroy the protective film, such as breaking, acidification, heating, use of water, and addition of biological enzymes. To break the protective film, two or more means are usually used simultaneously in liquor production, such as a combination of crushing, dampening, and cooking.

Starter Making

The mold use for starter making is usually made from mold culture, moldy bran, and glucoamylase in liquor production. The strains are chosen for strong saccharification and adaptability to culture. The essence of starter making is to strengthen and domesticate strains, not merely to ensure pure culture under a specific environment. The actual obtained product is one with highly concentrated mycelia or a high-strength enzyme preparation.

Gelatinization

The purpose of gelatinization is to convert the starch into glucose from wine feedstock. The starch from sorghum and other grains cannot be directly utilized by most microorganisms. It must be transformed into disaccharides or monosaccharides after gelatinization saccharification for microbes to obtain more ethanol and more flavor substances by direct use. Starch granules can be imbibed after absorbing sufficient water; they are easily ruptured and completely gelatinized after heat cooking. Therefore, it is critical to wet the crushing material before cooking.

Saccharification

Saccharification is the process that converts dextrin to sucrose and maltose and then turns them to glucose. From formula analysis, water directly participates in the chemical reaction. If there is no water and the other conditions are fully provided, dextrin cannot be degraded. At present, the liquor brewing process in China are all simultaneous saccharification fermentation.

Fermentation

Pyruvate is generated from starch by the EMP pathway, which converts it to glucose that needs to be degraded. There are two ways to perform continued pyruvate degradation. One is anaerobic degradation, and its product is an organic compound, such as acetic acid, lactic acid, butyric acid, or caproic acid. These compounds produce skeleton components (acid, ester, ketone) and trace components of the liquor after addition, esterification, condensation, cracking, rearrangement, and other biochemical reactions. Another is aerobic degradation. The pyruvate is completely oxidized to CO_2 and H_2O, and large amounts of energy are released because of access directly into the tricarboxylic acid cycle (TCA). Better wine relies on anaerobic glycolysis.

Fermentation in liquor brewing means loading the fermented grains into the wine cellar and fermenting for a period of time. The environment is suitable for the yeast cells to bud and multiply, so the yeast quickly enters the logarithmic phase. As the fermentation time reaches the main fermentation period, the alcohol in the environment continues to accumulate. The temperature of the fermented grains gradually elevates, acidity gradually increases, and adverse factors gradually accumulate; finally, the yeast no longer grows and is subject to mass death.

Saccharification and starch utilization are difficult in solid-state fermentation for the water is encompassed in the raw materials of brewing. The distilled fermented grains continue to be fermented, and residual starch continues to be reused. This is unique to alcoholic fermentation in China, termed continued grain fermentation. The fermented grains are repeatedly fermented, which will accumulate an abundance of precursor flavor component substance. The precursor substance is converted into flavoring substances after fermentation again by facultative anaerobes.

The interface has an obvious impact on the growth of the microorganisms. The solid, liquid, and gaseous states coexist in the solid-state fermentation cellar. The growth and metabolites of the same kind of microorganisms are significantly different from those living in the homogeneous phase or on the interface of the two different states. Solid fermented grains produce rich flavor substances (acids, esters), which have more interfaces in the gas-solid and liquid–solid phases.

Water is distributed in the fermented grains as molecular forms. This can help the microorganisms transfer energy and nutrients, reproduce, and perform metabolic and other physiological and biochemical activities. Water can also be used as an effective solvent for its microbes' metabolism, such as alcohols, aldehydes, acids, and esters. There is a small part of the water that sinks gradually toward the bottom of the wine cellar and, in the sinking process, dissolves the cooked starch, dextrin, amino acid, polypeptide, yeast residues, and ethanol.

Liquor fermentation is an essentially anaerobic mixed solid-state fermentation process. Static fermentation has less effect on heat and mass transfer. The fermentation process lacks regulation of effective means, so the entire reaction process is lengthy, and production is unstable. It has a marked difference on the quality and yield of the wine produced from the different wine cellars or batches in the same wine cellar.

Pure Grain Solid-State Fermentation Wuliangye Liquor Production Process

Wuliangye liquor fermentation uses sorghum (36 %), rice (22 %), glutinous rice (18 %), wheat (16 %), and corn (8 %) as the main raw materials; brewing to produce the liquor is by the processes of gelatinization, saccharification, fermentation, and distillation. The production process is discussed next (Fig. 5.3).

Fermentation is initiated at 72 h with the increase of temperature. Temperature in the cellar after 30 days of fermentation might be 13 °C higher than that after 24 h of fermentation. It was shown that if fermentation is normal in the wine cellar, the quality and liquor yield of the wine may meet the requirements.

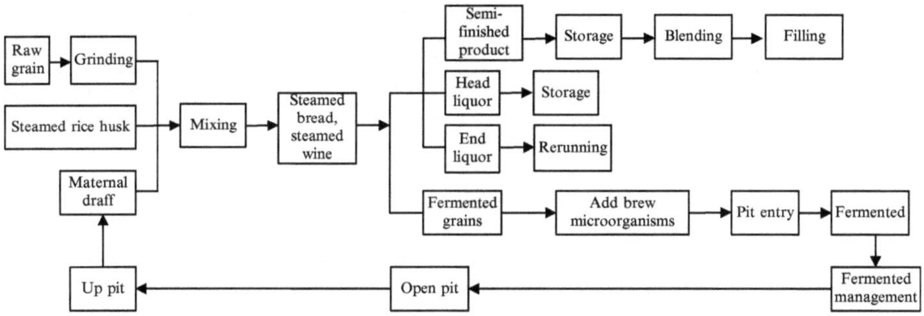

Fig. 5.3 Pure grain solid-state fermentation production process flow of Wuliangye liquor

The heating actually reflects the biochemistry process in the wine cellar. The starch generates glucose after saccharification and fermentation; glucose and water generate the alcohol, which is converted into an acid. The late acid gradually esterifies for the ester formation.

Problems and Prospects in Liquor Solid-State Fermentation

The conventional processing of Chinese liquor is to wet grain first with subsequent stewing. The process aims to facilitate the stewing grain in favor of the saccharification and fermentation through water absorption and gelatinization and to increase the contact area of the starch and enzymes.

High-temperature atmospheric pressure stewing has obvious defects. First, it is time consuming; generally, stewing must be for 1.5–2 h, which does not include soaking in hot water, and therefore consumes a lot of energy. Second, the stewing process generates some substances harmful to bacterial growth and affects the quality of the material of the final product. Third, because of nutrient losses, the nutrition of the starchy feedstock is damaged and lost in the long stewing and soaking process. Fourth, the raw grain cleaning and stewing process inevitably produces large amounts of sewage.

I used the steam explosion technology to replace brewing. The moisture content of the grain increased to 30–60 % from about 10 % of the natural moisture content through adding water to soak the grain before steam explosion. The water can pass into the gas in the steam explosion tank when feeding steam and when pressure is relieved suddenly to make cereal grains puffed and to meet the need for traditional brewing requirements. Steam explosion technology can improve the rate of grain utilization, ahead of the Maillard reaction; improve liquor yield; reduce work intensity; and increase the overall efficiency of the enterprise.

Steam explosion technology is the first step toward mechanization. A pilot study has been completed. With further research, steam explosion applied in the liquor-brewing industry will generate immeasurable economic and social benefits.

Table 5.2 Contrast of different feedstock characteristics in the process of fuel ethanol production

Kind	Starch	Saccharides	Lignocellulose
Pretreatment	Crushing, cooking, gelatinization, acid or enzymatic saccharification	Squeezed, adjustment	Grinding, physical, chemical, biological treatment
Hydrolysis	Easy hydrolysis, single product, no fermentation inhibitors	No hydrolysis, no fermentation inhibitors	Difficulty in hydrolysis; product complex; increased fermentation inhibitor
Fermentation	Produces amylase; yeast ferment hexose to ethanol	Resistance to ethanol; yeast ferment hexose to ethanol	Screening of yeast or bacteria; fermentation of hexose and pentose for ethanol
Extraction purification	Distillation, rectification, purification	Distillation, rectification, purification	Distillation, rectification, purification
Integrated utilization	Feed, methane, CO_2	Feed, methane, CO_2	Lignin (fuel), methane, CO_2

Source: From Zhuang et al. (2009)

5.2.1.2 Fuel Ethanol Production by Anaerobic Solid-State Fermentation

Ethanol is an important industrial raw material, widely used in the chemical, food and beverage, industrial, military, household chemical, pharmaceutical, and health fields. Fuel ethanol, referred to as 99.5 % water-free ethanol, is clean-burning fuel with a high octane value; it is the most promising alternative to petroleum for renewable energy. Therefore, it has broad prospects for future use (Mustafa and Havva 2009). Fuel ethanol production by solid-state fermentation has received increasing attention because of the pressure on the environment and the potential energy crisis. It has advantages of less investment and low energy consumption for vinasse disposal. By learning the essence of Chinese traditional fermentation technology, it will become the new processing trend for energy conservation and low pollution (Dong and Liu 2008).

Overview of Fuel Ethanol

The raw materials for fuel ethanol production are mainly three kinds: starch crops such as corn, wheat, cassava, and sweet potato; sugar crops such as sugar beets, sugarcane, sweet sorghum; and straw feedstocks such as lignocellulose (Table 5.2).

With the energy crisis and environment problems, countries around the world have paid much attention to the non-fossil energy. The problems of the tuber crops for ethanol production are the diverse variety of raw materials, planting, harvesting, and storage. Particularly in terms of storage, because of the need for guarantee of year-round use, comprehensive matching is needed, and this will result in increased costs and affect the large-scale application of tuber crops. Energy crops

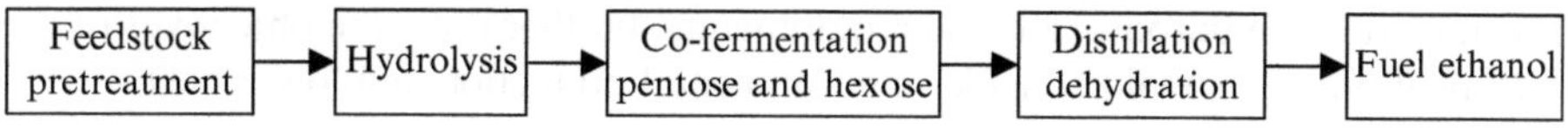

Fig. 5.4 Lignocellulosic ethanol fermentation process

like sweet sorghum for ethanol production are more seasonal. A harvest of about 2 months and storage of only about a month present difficult problems for continuous production (Han et al. 2010). Lignocellulose is the most abundant resource on Earth and has the minimum utilization of resources for this process. Therefore, use of lignocellulosic ethanol is the inevitable trend for future fuel.

Fermentation of Lignocellulosic Ethanol

In the fermentation of lignocellulosic ethanol (Fig. 5.4), the process of changing cellulosic biomass to fuel ethanol involves several unit operations, including feedstock pretreatment, hydrolysis, fermentation, and product recovery. Lignocellulose pretreatment by chemical or physical methods separates the cellulose, hemicellulose, and lignin; cellulose can be hydrolyzed to glucose, while the hemicellulose can be hydrolyzed into monosaccharides such as xylose and arabinose. The pentose and hexose can be converted into mash after fermentation and then become fuel ethanol through distillation and dehydration.

Pretreatment

The pretreatment step is used for separating the biomass into cellulose, hemicellulose, and lignin. In this step, lignin can be removed, and some hemicellulose can be hydrolyzed to soluble sugars. Complexity of the biological structure and chemical composition of lignocellulosic materials results in inefficient direct degradation. The purpose of pretreatment is to break the compact structure of straw and to increase the effective ratio of the enzymatic reaction. Moreover, promotion of degradative efficiency, decrease of inhibitors beneficial for the following conversion steps, and comprehensive utilization of separated components of lignocellulose are also the main purposes.

The current pretreatment can be carried out through dilute acid, steam explosion, alkali, hot water, microwaves, ammonia fiber explosion, and pretreatment with the white rot fungi. Steam explosion is considered to be a relatively ideal method for its features of short treatment time, less utilization of drugs, energy saving, and environmental protection (Chen and Qiu 2010).

Because of the shortcoming of each method, increasingly researchers prefer combined methods to decrease the inhibitors that affect later utilization. These methods could resolve all the defects and make the reaction exert the maximum effect. The specific pretreatment method should be selected according to the type of raw material, the purpose, and the requirements. Furthermore, the objectives of the technical process should consider environmental protection and lower cost.

Hydrolysis

In the energy production process through lignocellulosic conversion, hydrolysis is a limiting step. Lignocellulose should be converted to fermentable monosaccharide through acid or enzyme hydrolysis. The yield of acid hydrolysis is commonly below 60 %. Many inhibitors are produced in the acid hydrolysis process; the huge investment and environmental burden make it difficult to achieve large-scale industrialization. Compared with acid hydrolysis, enzyme hydrolysis is characterized by moderate reaction conditions, environmental protection, unique products, high saccharide yield (conversion rate over 90 %), and low equipment input. Therefore, it has become a focus in research (Alvira et al. 2010).

Cellulase is the primary enzyme in lignocellulosic degradation. The cost of cellulase is the main restraint for enzyme hydrolysis. As a matter of fact, the factor that leads to high cost is the large amount of cellulase in industrial production and its low enzymolysis efficiency. To figure out this problem, extensive research has been carried out centered on cellulase and degradation of lignocellulose. Metabolic engineering of ethanol fermentation was improved in addition to an explosion of activity in the natural world, which enlarged the available range of the bacterial substrate and ethanol yield.

Fermentation

Lignocellulose could be converted to energy products, such as fuel ethanol, methane, and hydrogen, through microorganism fermentation. A good fermentation process needs not only an efficient bacterium but also the optimal conditions and equipment to exert sufficient potential for production.

Fuel ethanol fermentation processes using lignocellulose as raw materials are very different from those for starch and sugar. For example, the biomass hydrolysate often contains components that are harmful to fermentation microorganisms. Also, hydrolysate contains more xylose (Chen 2009a).

To promote conversion efficiency, various new types of fermentation process have been developed by researchers recently. These types include separate hydrolysis and fermentation (SHF), SSF, simultaneous saccharification and cofermentation (SSCF), and consolidated bioprocessing (CBP) (Li and Chen 2010).

Separate hydrolysis and fermentation (SHF): Lignocellulose needs to be hydrolyzed and fermented after pretreatment to obtain ethanol and other bio-based products. The first process is hydrolysis, and then fermentation is used, which is called separate saccharification and fermentation. The advantages of this process are the rapid fermentation rate and absence of solids in the reactor. Moreover, comparatively pure lignin is obtained for heating, and the cellulase can also be recovered. However, the main problem is the inhibition of end product to cellulase.

Simultaneous saccharification and fermentation (SSF): Since SSF was proposed by Takagi in 1977, it has displayed significant economic advantages compared with SHF. The integration of the two steps can reduce the equipment investment and the risk of contamination and increase the concentration of ethanol. Microorganisms can ferment glucose to ethanol to reduce the accumulation of glucose, which can avoid the hydrolysis product inhibition of enzymes in SSF. However, because of lignin residue, it is difficult to guarantee the full use of fermentable sugars for yeast. In addition, the optimum temperatures of yeast fermentation and cellulase hydrolysis are different. So, it will be difficult to achieve optimized conditions for both yeast and cellulase in the SSF process. The development of high-temperature yeast is a hot research topic.

There are some problems in Simultaneous saccharification and fermentation such as low sugar and ethanol concentrations, a large amount of cellulase, high water content of the fermentation residues, and difficult comprehensive utilization in the popular SSF process. Chen and Qiu (2007) proposed that solid-phase hydrolysis coupled with liquid phase fermentation technology could effectively increase the efficiency of hydrolysis of cellulose and ethanol fermentation, which would solve the problem of high water consumption and uncoordinated temperature of enzymatic hydrolysis and fermentation in SHF and lower the process cost. High-intensity ethanol fermentation-separation coupling with gas stripping, which integrates an air loop bioreactor, vacuum reflux, CO_2 stripping, and activated carbon adsorption, achieves the triple coupling of hydrolysis saccharification, ethanol fermentation, and adsorption separation.

Simultaneous saccharification and cofermentation (SSCF): SSCF is now considered to be a new process that shows more economical attraction compared with SSF in the short term. Cofermentation of hexose and pentose is the research goal of lignocellulosic ethanol production. By changing the xylose metabolic pathway of the yeast, cloning the related genes to control metabolic pathways, and altering the metabolism flow of the carbon source, an engineering strain of *Saccharomyces cerevisiae* can be obtained. The specific strain not only can utilize xylose and glucose alone but also can achieve the cofermentation of xylose and glucose. Cofermentation can increase the final ethanol concentration in the fermentation broth and thereby considerably reduce the ethanol production cost. The process is shown in Fig. 5.5 (Vane 2005).

Consolidated bioprocessing (CBP): Among the processes, consolidated bioprocessing only utilizes microorganisms to produce cellulase and hemicellulase; the sugar produced from enzymolysis is still provided by the same bacterium. This is significant for the total utilization of lignocellulose for its simple process and short duration. The great significance of the full use of lignocellulose has impassioned more researchers' concern (Salimi et al. 2010). However, for complexity and heterogeneity of the lignocellulose structure, so far the achievement of directly converted lignocellulose resources to fuel ethanol are rarely reported in industry.

Distillation and Dehydration (Product Recovery)

The main components of fermenting mash are water and ethanol, and the concentration of ethanol is generally 5–10 % (w). The concentration of ethanol in the vapor

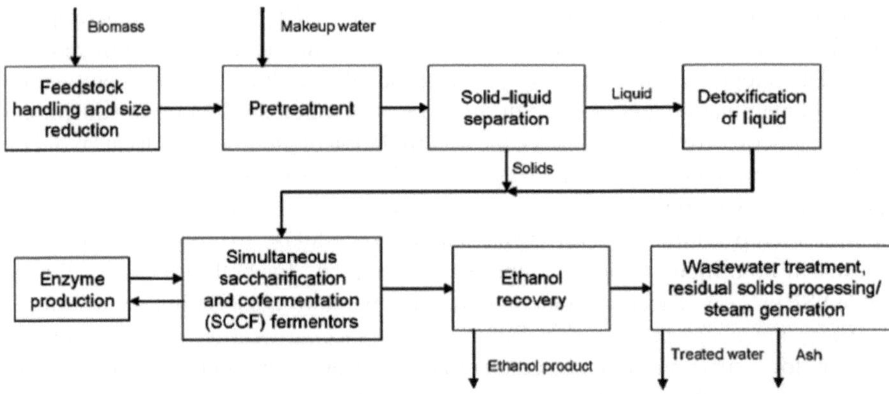

Fig. 5.5 Flow diagram of fuel ethanol production from biomass by the simultaneous saccharification and cofermentation (*SSCF*) process

phase is higher than that in the liquid phase, and this vapor is condensed to obtain ethanol with a higher concentration. The ethanol can be distilled from fermented mash by evaporation-condensation in the distillation column several times, and a high concentration of ethanol can be gained as well as fuel oil and vinasse. After distillation, ethanol can be dehydrated by chemical reaction, azeotropic distillation, extractive distillation, 3A molecular sieve adsorption, membrane separation, vacuum distillation, or the ion exchange resin method (Sun 2010). Finally, fuel ethanol with a water content less than 0.8 % (volume fraction) is obtained after the addition of denaturant (Table 5.3).

I carried out a serial study of the key technologies in straw component fractionation, cellulase solid-state fermentation, and process coupling. Some major key technology breakthroughs for straw ethanol production by enzymolysis and fermentation have been achieved, and an industrialization demonstration project of 3,000 t ethanol annual production was established in 2006 by Shangdong Zesheng Biological Corporation. The project built a 100-m^3 solid-state fermentation reactor and a large-scale unpolluted steam explosion system; proposed the triple-coupling technique of solid-state enzymolysis, simultaneous fermentation, and absorptive separation; and successfully developed a 110-m^3 reaction apparatus for the triple-coupling technique. The project provided amplification parameters for industrial ethanol production by straw enzymatic hydrolysis and fermentation (Chen and Qiu 2007).

Problems of and Prospects for Lignocellulosic Ethanol

There are still some technical issues in the raw material pretreatment, hydrolysis, and fermentation process, which result in high production costs and difficult mass production (Yu and Chen 2010). However, because of the gradual depletion of fossil resources, the development of bioethanol could become one of the main

Table 5.3 Analysis of common unit operation problems in lignocellulose ethanol

Unit operations	Problem	Difficulty	Breakthrough point	Prospect	References
Pretreatment	Pretreatment for a single component; high pretreatment costs; pretreatment study is not enough	Reduce pretreatment costs; full utilization of biomass	Development of new pretreatment technologies for energy saving, low power, nonpolluting; development multicomponent of high value-added products	Combined pretreatment	Yang and Wyman (2008), Chen and Liu (2007)
Hydrolysis	High cost and large amount of cellulase	Low hydrolysis efficiency	Effective pretreatment; synergistic effect of enzyme; research on thermal stability of enzymes	Site production of enzymes	Yu and Chen (2010)
Fermentation	Low fermentation efficiency, high energy consumption, heavy pollution	Improve the conversion efficiency	High-yield strains screened transformation; fermentation process optimization; reactor design enlargement	Improve the tolerance of inhibitor	Chen and Jin (2006)
Product recovery	High separation costs and energy consumption, low product yield	Large-scale bioseparation process and integration	Biological reaction process coupling with bioseparation	Bioseparation coupling with membrane separation	Haelssig et al. (2008)

directions of the future. To achieve the large-scale commercial production of ethanol from lignocellulosic resources, some problems should be solved.

First, macro control and government regulation need to be strengthened to ensure steady development. The scale of ethanol production should be compatible with the availability of raw materials. An appropriate fiscal and taxation system should be adopted, according to the raw materials of ethanol production, to determine the amount of subsidies to ensure that manufacturing enterprises have certain economic benefits. This strengthens the supervision of the whole process for ethanol production and sales and strictly controls the quality of the fuel ethanol to meet the relevant standards.

Second, deep research of key technology is necessary. Lignocellulosic materials provide great potential and technical difficulty at the same time. Pretreatment technology needs to be further optimized and perfected. It is imperative to develop efficient cellulase and a new fermentation reactor. Further technological breakthroughs should be achieved for the rational utilization of lignin and degraded sugars from hemicellulose.

Third, multiproduct biorefineries are necessary. We must learn from the experience of the petrochemical industry and stick to a multiproduct mode of biorefining and ethanol generation to maximize the value of straw feedstock.

5.2.1.3 Biogas Dry Fermentation

Biogas dry fermentation, also known as biogas solid fermentation, is the fermentation process in which straw, manure, and other organic wastes as raw materials (dry matter concentration of more than 20 %) are decomposed by anaerobic bacteria to form CH_4, CO_2, H_2S, and other gases.

Biogas dry fermentation has been widely used in large-scale processing of agricultural solid waste, such as manure, crop stalks, and garbage, to produce biogas and organic fertilizer. It has become the hot spot of anaerobic fermentation technology.

Advantages of Biogas Dry Fermentation

Biogas dry fermentation technology mainly has the following advantages compared with traditional wet fermentation (Weiland 2010):

- Less use of water and low energy consumption. Because of the low solid content in wet fermentation, the energy used to maintain the temperature of the reaction system has been largely governed by water. Especially in the Nothern China, more than 30 % of energy is consumed for thermal insulation of the system, which causes high energy waste and greatly limits the application and promotion of biogas projects. The energy for insulation of dry fermentation requires only

10 % of the total energy generated by fermentation, and the energy consumption is observably reduced.
- Low investment costs. Dry fermentation shortens the digestion cycle and increases biogas production. Meanwhile, the fermentation device is simpler, and the construction costs are greatly reduced. After fermentation, the product without dehydration can be used directly as fertilizer, which simplifies the processing operation and reduces costs.
- Operation and low processing costs. The absence of stirring and a reliable operation system in the fermentation process reduce operating costs. Fermentation finishes with almost no biogas slurry, reducing the cost of sewage treatment.

Principles of Biogas Dry Fermentation

The biogas dry fermentation process is essentially the metabolism of materials and energy metabolism by various groups of microorganisms. In this process, the microorganisms are the core of biogas fermentation. Control of the fermentation conditions is closely related to the growth and reproduction of microorganisms.

Anaerobic dry and wet fermentation are essentially the same for the biochemical reaction. The process is that the obligate and facultative anaerobes degrade the organic compounds to produce biogas in an anaerobic environment, including the hydrolysis, acidification, and methanogenic stages. These stages can be completed cooperatively by three microflora: zymogenous, hydrogen-producing, and methanogenic bacteria, respectively (Chen and He 2012). These microorganisms, according to their nutritional requirements, constitute a food chain from the degradation of complex organic matter to methane.

- Hydrolysis stage. The hydrolysis stage converts the polysaccharides, proteins, lipids, cellulose, and so on in the raw materials into acetic acid, propionic acid, butyric acid, and other long-chain fatty acids and alcohols and a certain amount of hydrogen and carbon dioxide. In this process, the microorganisms are mainly strictly anaerobic bacteria, such as *Clostridium difficile* and *Bifidobacterium* species.
- Acidification stage. The hydrogen-producing acetogenic bacteria, typically *Acetobacterium woodii* and *Clostridium aceticum*, convert volatile fatty acids into acetic acid, hydrogen, and carbon dioxide in the acidification stage. In the fermentation process, the level of partial pressure of hydrogen has a regulatory role in the degradation of organic matter, and hydrogen-producing microorganisms only grow under the existence of hydrogen-consuming microorganisms.
- Methanogenic stage. This stage can be completed by methanogens. Methane-producing bacteria are mainly of two groups, those that use hydrogen and those that use acetic acid as the substrate, such as *Methanosarcina barkeri*, *Methanonococcus mazei*, and *Methanotrix soehngenii*. The methanogens are a special group of microorganisms. They are strictly anaerobic microorganisms

that are very sensitive to oxygen and an oxidizing agent and are suitable for surviving and reproducing in a neutral or slightly alkaline environment. These microorganisms require basic substances for growth because they absorb carbon dioxide and hydrogen and excrete methane to maintain the growth. Because of the difficulty of isolation, culture, and preservation of the methanogens, few pure strains have been obtained so far and cannot be used for production, which has a direct impact on research progress in methane fermentation and has led to the slow improvement of gas production.

Microorganisms in the stage of hydrolysis and acidification and methanogens work together to complete the biogas dry fermentation process. Their relation is one of interdependence and mutual restraint, with the aspects mainly presented in the discussion that follows.

Nonmethanogens Provide Nutrition for Methanogens

Carbohydrates, protein, and fat in raw materials and other complex organic compounds cannot be directly absorbed and utilized by the methanogens; they must be converted into simple soluble compounds by the hydrolysis of nonmethanogens and further decomposed to form the methane-producing bacterial fermentation substrate. So, the microorganisms in the hydrolysis and acidification stage provide a matrix and energy for methanogen cell synthesis. Meanwhile, the methanogens continuously convert the acetic acid, hydrogen, and carbon dioxide produced by nonmethanogens into methane to avoid the accumulation of acid and hydrogen in anaerobic digestion. Because of the synergistic effect of methanogens and nonmethanogens, the biogas fermentation process finally achieves the dynamic balance of acid and methane production and stable operation.

Nonmethanogens Provide an Anaerobic Environment for Methanogens

At the beginning of fermentation, much air enters the reactor when the materials and water are added to it, which is obviously harmful to methanogens. However, because of the gradual decrease of oxidation-reduction potential in the fermentation liquid by the activities of aerobes and facultative anaerobes, an anaerobic environment is created for methanogens to grow and produce methanol.

Nonmethanogens Remove Toxic Substances for Methanogens

The raw materials for biogas fermentation from industrial and agricultural waste always contain some substrates that have toxic effects on methanogens, such as phenols, benzoic acid, cyanide, long-chain fatty acids, and heavy metals. Many nonmethanogens can degrade and utilize these materials and eliminate the substances toxic for methanogens. For example, the H_2S generated by nonmethanogens can react with heavy metal ions to form the precipitates and relieve the toxic effects of heavy metals.

Nonmethanogens Maintain a Suitable pH Environment for Methanogens

In the early stage of biogas fermentation, large amounts of organic acids are produced by the hydrolysis and acidification of starch and sugar in the raw material. Meanwhile, the carbon dioxide produced is also partially soluble in water and decreases the pH value of the fermentation broth. However, NH_3 produced from the ammonification of ammonifiers can partly neutralize the organic acid. At the same time, because of the constant use of acetate, hydrogen, and carbon dioxide by methanogens, the concentration of organic acids and carbon dioxide in the fermentation liquid is gradually decreased. The combined action of two types of bacteria stabilizes the pH at a suitable range. Consequently, in normal fermentation digesters, pH is always able to be maintained in a suitable state without artificial control (Zhu et al. 2009).

The anaerobic dry fermentation process is complex, and its mechanism is still in the exploratory stage. A good and applicable model for anaerobic dry fermentation has not been established, and the design of a fermentation system to a large extent can only be based on experience. So, further studies of the fermentation process are required to provide additional theoretical support for system design and to enhance the operational capabilities of the system.

Process of Biogas Dry Fermentation

Overview of Process Conditions

1. The C/N ratio of the feedstock. The C/N ratio of the fermentation feedstock refers to the ratio between the organic carbon and nitrogen content in the raw materials. In the methane fermentation process, because CH_4 is continuously generated, the C/N ratio of the feedstock declines. At present, the suitable C/N ratio is generally 25–30 for biogas dry fermentation (Isci and Demirer 2007). To ensure the appropriate C/N ratio, some regulation of nutrients in the fermentation process is necessary. Song and Chen (2008) selected wheat bran, yeast extract, urea, and NH_4HCO_3 as different nitrogen sources for methane fermentation and ultimately determined that the best supplemental nitrogen source for biogas fermentation from steam-exploded straw was NH_4HCO_3.
2. Pretreatment of feedstock. The pretreatment of feedstock includes physical, chemical, and biological pretreatment. The current pretreatment could be carried out through use of dilute acid, steam explosion, alkali, hot water, and microwaves. Song and Chen (2008)) studied the applications of steam explosion pretreatment of corn straw for biogas production (Table 5.4).

   Through the analysis of the strength coefficients R_0, the results showed that methane yield increased with the increase of steam explosion strength. The methane yield reached a maximum of 138.2 ml/g when the steam explosion condition was 1.5 MPa for 6 min. However, a continuous increase of steam explosion conditions led to a decrease of methane production. From the perspective of material physical state, when the steam explosion intensity is low, both

Table 5.4 Effects of different steam explosions on anaerobic digestion of corn stover

Steam explosion conditions				
Saturated steam pressure (MPa)	Saturated steam temperature (°C)	High-pressure maintenance time (min)	Strength coefficient R_0	Methane yield (ml/gTS)
–	–	–	0	39.6
1.3	195	3	1,483	45.4
1.3	195	5	2,472	78.4
1.3	195	6	2,967	91.0
1.3	195	8	3,955	116.2
1.3	195	10	4,944	122.2
1.5	201.3	3	2,882	89.8
1.5	201.3	5	3,894	121.8
1.5	201.3	6	4,672	138.2
1.5	201.3	8	6,230	132.4

the degradation degree of hemicellulose and the damage degree of the straw compact structure are low, and the overall availability of straw is not high. So, methane yield increased gradually with the increase of steam explosion intensity. When steam explosion intensity was high, the hard structure of the epidermis was completely destroyed, and a large amount of hemicellulose was degraded into monosaccharides, which were partly converted into furfural and other inhibitors. Therefore, the vast loss of monosaccharides and formation of harmful substances caused the decline of methane yield. Consequently, the optimized steam explosion conditions for biogas production from anaerobic digestion of corn straw were 1.5 MPa and 6 min, for which the methane yield was 3.5 times that of untreated straw.

3. Total solid content. A total solid content concentration that is too high will lead to accumulation of volatile acids, cause acidosis, and result in the termination of the dry fermentation process (Debebe et al. 2011). The study results of Leng Chenbao et al. (2001) showed that by controlling total C/N (to about 30:1), the 25–30 % dry matter concentration was ideal for well-run anaerobic digestion.
4. Inoculum. The high quality and quantity of inoculum are important to ensure a smooth start for biogas dry fermentation. Liu et al. (2010) focused on urban organic waste dry fermentation research and selected the anaerobic digestion sludge as an inoculum. According to the total solid content of test raw materials with the amount of inoculum (10:1) inoculated, the results showed that fermentation and methane production were normal.
5. Fermentation temperature. Anaerobic dry fermentation is divided into normal temperature, mesophilic, and high-temperature fermentation, which depend on the fermentation temperature. In normal temperature fermentation, with a long cycle of biogas production, methane yield is low and is influenced by the ambient temperature. Compared with normal temperature fermentation, mesophilic fermentation, 30–38 °C, is conducive to large-scale production, which speeds decomposition and gives a high yield and good methane quality.

The characteristics of high-temperature, 50–55 °C, fermentation are rapid decomposition, high yield, good environmental effect, and more energy.

6. pH. The suitable pH for anaerobic fermentation is 6. 8–7.4; a pH less than 6.4 or more than 7.6 curbs gas production. At pH values below 5.5, the activities of methanogens are fully exposed to suppression (Li et al. 2011).
   If the hydrolysis stage and acid-producing stage reaction speeds are in excess of the methanogenic stage, the pH value will decrease, affecting the methanogen living environment. This phenomenon often occurs at the start of dry fermentation and is called acidosis. In the fermentation process, the pH value is monitored, and when the pH values are below 6.4, lime or ammonia may be added to adjust the pH, ensuring the process proceeds without a hitch.
7. Stirring. Suitable stirring can be full contact of the fermented feedstock and fermentation microbials to expand the active layer so that the biogas generated is easily separated and escapes and the rate of methane production is improved (Li et al. 2006). The stirring method uses liquid flow stirring and mechanical stirring. In liquid flow stirring, the fermentation broth is withdrawn from the bottom of the reactor from outside and then returns from the top of the reactor at a certain spray angle. The garage type dry fermentation system and leachate storage barrel type dry fermentation system both use liquid flow as mixing measures.

Biogas dry fermentation process conditions mainly include two aspects. The first is the process to meet suitable conditions, anaerobic fermentation of microbial growth and reproduction, to achieve strong fermentation and gas production. This includes the formation of an anaerobic environment, the pretreatment of the raw material, the substrate's C/N ratio, the dry matter content, the fermentation temperature, the pH value, the amount of inoculum, and other parameters of reasonable control. The second is the process to meet the production problems of biogas dry fermentation engineering. Dry fermentation feedstock is solid; there is difficulty in continuous inlet and outlet materials within the reactor under anaerobic conditions, which is particularly important in large-scale production.

Batch Process

The dry fermentation process can be divided into the continuous process and the batch process. The continuous dry fermentation process is complicated and expensive; therefore, it fails to be promoted. The first commercial batch reactor was built in the Netherlands. It produced 260 L methane from 1 kg VS (volatile solids) (Fig. 5.6).

The raw material for batch biogas dry fermentation is a mixture of cow dung and straw. The process is divided into the following steps, as shown in Fig. 5.6 (Li et al. 2010):

1. Cow dung and crushed straw are mixed well. The cow dung and straw ratio is about 3:1.
2. Mixed raw materials after aerobic fermentation are brought into the anaerobic reactor, and then the doors are sealed.

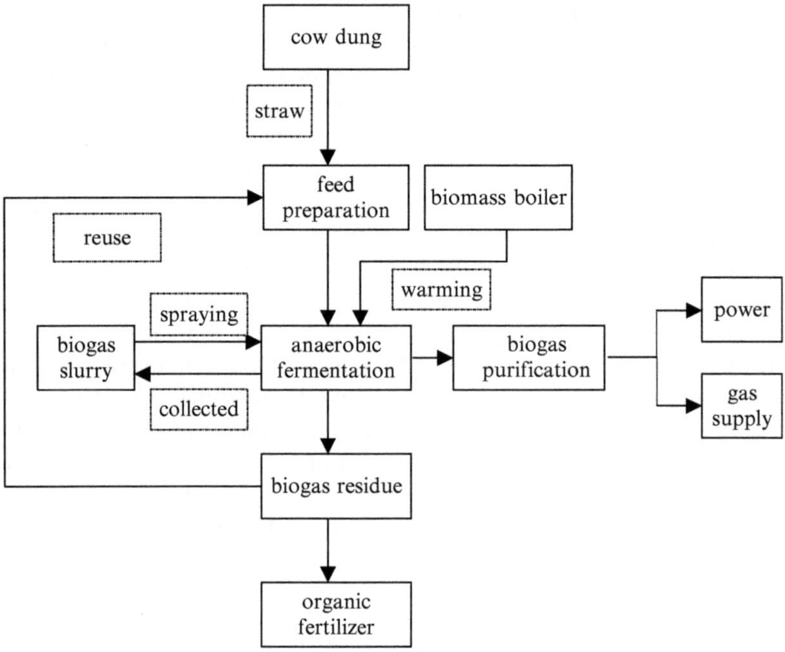

Fig. 5.6 Process flow diagram of batch methane dry fermentation technology

3. Biogas slurry is sprayed to the raw material from the top of the reactor by a spray system. The biogas is collected in the pool.
4. Biomass boilers are used to warm the raw materials through the geothermal pipe at the bottom of the reaction chamber.
5. The dry fermentation system sets up an automatic monitoring and controlling system. This system can monitor the water content of raw materials, pressure, temperature, moisture content, and biogas components and control the whole fermentation process so it runs automatically.
6. After 2 days of fermentation, the raw material can produce biogas normally. The production cycle is 30 days. After purification, the biogas is supplied to residents or used to generate electricity through the transmission and distribution system. After it is dried and screened, the residue is used as fertilizer.

Problems and Prospects

There are some disadvantages for biogas production by dry fermentation: the serious concentration gradient in the raw materials, the difficulties in heat and mass transfer, the difficulties in the control of pH and temperature, and the high technology requirements. All result in difficulty in the dry fermentation technique process control. So far, only a few parameters can be measured online in the fermentation process. There are no simple and suitable parameters because of dry fermentation complexity. Therefore, the process control theory and technology must be studied in depth.

Dry fermentation technology and the equipment for biogas began late in China, and special equipment were lacking. To provide reliable equipment and promote the fermentation effect, many future studies are needed, including those for the following: development of multifunctional straw-grinding technology and equipment; production of dry fermentation equipment and material lifting equipment on a large scale; effective granular fertilizer molding technology and equipment; and improvement of the engineering quality of dry fermentation.

5.2.2 Pure Culture Fermentation

The primary condition of modern solid-state fermentation is pure culture, and the key is cultivation of a pure strain at a larger scale. Considering the problems existing in traditional fermentation, modern solid-state fermentation has received much research attention in limited microbe large-scale cultivation and full utilization of the advantages of solid-state fermentation, which meet the needs of industrial production for modern fermentation.

The pure strain solid-state fermentation process includes material mixing, sterilization, cooling, inoculation, fermentation, drying, crushing, and packaging. There are many problems in pure strain solid-state fermentation technology that need to be solved, such as a complex transfer process that includes gas-solid, gas–liquid, liquid–solid types, and so on; the serious concentration gradient; and the difficulty in heat and mass transfer, which all are the biggest differences between solid-state and liquid fermentation. Chen and Wan (2008)) invented a gas double dynamic solid-state fermentation reactor based on pressure pulsation. This technology achieves airflow exchange among multiple fermentation reactors and forms gas phase double dynamically (pressure pulsation and air circulation), which can effectively improve heat and oxygen transfer, promote cell growth and metabolism, achieve pure culture and large-scale applications, and solve the technical bottleneck of traditional solid-state fermentation.

So far, there are few reports of pure strain anaerobic solid-state fermentation; mainly, these were in the field of food fermentation. Lu et al. (2005) studied pickling radish by culturing *Lactobacillus plantarum* B_2 and natural fermentation. The acidity, pH value, concentration of lactic acid, fermentation cycle, and volatile constituents among the different products were compared by solid-phase microextraction and gas chromatography–mass spectrometry (GC-MS) and sensory characteristics. The results showed that the product fermented by *L. plantarum* B_2 culture was excellent over the natural mixed fermentation product. For industrially producing traditional artificial pickled vegetables and reducing environmental pollution from the pickling liquid, the useful continuous method of pickling liquid was studied in semi-solid-state pure fermentation technology for low-salt pickled cucumber (*Cucumis sativus* L.) by Wu et al. (2009). Predehydrated cucumber with a water content of 65–70 % was fermented continuously with a low-salt concentration in anaerobic fermentation equipment. For reuse of the brine, brine was used as a fermenter for process prefermentation combined with pure culture of lactic acid bacteria. The results showed that the ratio 18:6:2:1 of raw materials, brine, 10^8 cfu/ml starter, and 4 % salt solution was the optimized condition for reusing brine, and 2.24 % low-salt product was obtained. The product texture was not significantly influenced ($P > 0.01$) by stirring under lower intensity (<100 rpm), and textural hardness was kept above 1209.1 N. The yield rate was up to 115.8 %. The vitamin E, total ascorbic acid, protein, fat, and dietary fiber contents were improved via comparison with traditional technology and reached 3.592 %, 0.601×10^{-2} mg/g, 7.843×10^{-2} mg/g, 1.991 %, and 0.375 %, respectively. Du et al. (2011) studied pickle production technology by semisolid fermentation of bacteriocin-producing *Lactobacillus plantarum* P158. The optimal fermentation parameters were determined by orthogonal tests and sensory evaluation using a fuzzy approach. The results indicated that the optimal conditions for pickle production were 4 % salt, 3 % sugar, 0.1 % inoculum of *L. plantarum* P158, and a fermentation temperature of 34 °C. The pickles produced under these conditions were simple and pure in taste with good quality. Therefore, this study provided a new means of pickle production.

5.3 Anaerobic Solid-State Fermentation Reactor

From brewing barrels and cans used in ancient times to the automated production equipment in modern pharmaceutical, food, and environmental protection industries, the development of science and technology enable bioreactors to absorb the latest technological fruits, and various new types of bioreactors spring up continuously. New bioreactors are the dominating equipment for fermentation industries. They provide a suitable environment for microorganism metabolism. Because the microbes are divided into two categories, anaerobic and aerobic, the bioreactors are different. The solid-state fermentation reactor is an important limiting factor for modern biological reaction engineering. The reactor design needs to consider many problems, such as sterilization, inoculation, heat and mass transfer, sampling, gas supply, parameter monitoring and control, and more. So far, there are many types of solid-state fermentation reactors, including those at the laboratory, pilot, and industrial levels. Some bioreactors are used in the fields of edible fungi, enzyme preparations, animal feed, soil remediation, and so on.

The reactors for solid-state fermentation include aerobic fermentors and anaerobic fermentors based on the oxygen supply. Based on structure, the reactors include the stirring, drum, air-lift, pressure pulsation, and peristaltic types. But, the stirring devices cause side effects on microorganism growth. Fermentation equipment also can be divided into static or dynamic reactors according to substrate movement in the fermentor (Chen et al. 2002).

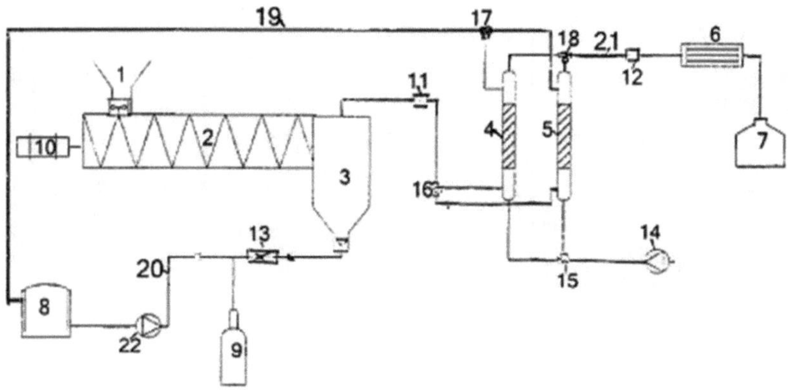

Fig. 5.7 Alcohol preparation device by solid-state fermentation from the sweet sorghum stalk. *1* Air seal machinery. *2* Fermentor. *3* Gas stripping tank. *4* First carbon adsorption column. *5* Second carbon adsorption column. *6* Condensator. *7* Receptacle. *8* Gasholder. *9* CO_2 gas bottle. *10* Stepping motor. *11* First alcohol concentration gauge. *12* Second alcohol concentration gauge. *13* Mass flow controller. *14* Air pump. *15* First three-way valve. *16* First valve. *17* Second three-way valve. *18* Second valve. *19* Pipeline 20 CO_2 recycle gas pipeline. *21* Air pipeline. *22* Compression pump

5.3.1 Ethanol Fermentation Reactor

Chen et al. (2002) invented a novel method and reactor equipment to produce ethanol using sweet sorghum stalks for solid-state fermentation, as shown in Fig. 5.7. In this device, as shown in Fig. 5.7, after sterilization and inoculation, the sweet sorghum stalk is added to the fermentor (2) at a certain rate. While fermenting ethanol, the sweet sorghum stalk goes forward at a predetermined speed. When the substrate reaches the gas-stripping tank (3), ethanol is gas stripped with CO_2. The mixture of ethanol and CO_2 goes through an activated carbon adsorption column (4, 5), and ethanol is absorbed; the rest of the CO_2 is recycled. Ethanol is desorbed by heating the activated carbon adsorption column and recovered in a receptor (7) by a condensator (6).

The method provides continuous solid-state fermentation equipment that acts in five ways: fermentation, air stripping, adsorption, condensation, and recycling. The equipment cooperatively completes the fermentation process. In this device, the fermentation tank and gas-stripping tank are linked vertically and are sealed as a whole. The complete fermented feedstock falls into the gas-stripping tank automatically. From raw material preparation to ethanol production, all the operations are completed in this device, which avoids the influence of the external environment and saves time and labor. This device is suitable for all granular feedstock fermentation for ethanol production. The fermentation process is coupled with ethanol fermentation and ethanol separation. Carbon dioxide as the loop carrier gas can separate ethanol from the fermentation. It can also reduce the impact of the heat generated in the fermentation process and ethanol inhibition. There is no need for

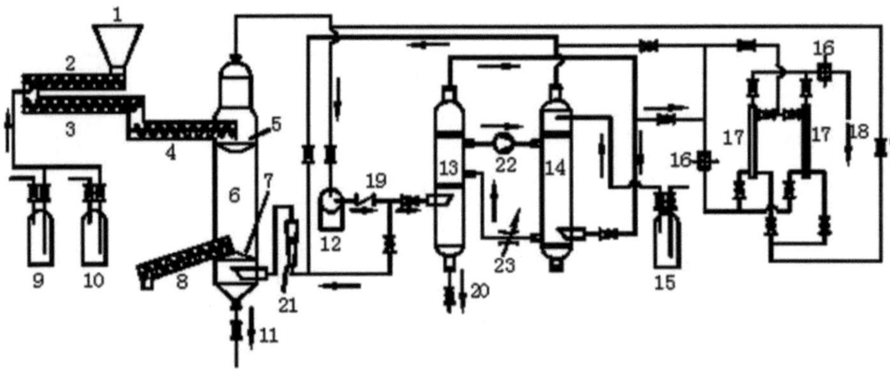

Fig. 5.8 Continuous solid-state fermentation device coupled with separation. *1* The feed port. *2* The first screwed conveyor. *3* The secondary screwed conveyor. *4* The third screwed conveyor. *5* Distributor. *6* Fermentation tower. *7* Unloading conveyor. *8* Screwed conveyor. *9* Seed tank. *10* Nutritive salt tank. *11* Sewage outfall. *12* Roots blower. *13* Heat pump evaporator. *14* Heat pump condenser. *15* Humidifier. *16* Solenoid valve. *17* Activated carbon absorber. *18* Exhaust port. *19* Check valve. *20* Alcohol separation port. *21* Flow meter. *22* Compressor. *23* Throttle valve

squeezing; this reduces the production process and saves on operating costs. The methods are carried out in a sealed device, and the pollution problems of liquid fermentation are solved.

Based on the improvement in material handling, equipment sealing, fermentation, gas stripping, heat and mass transfer, and saving energy, Chen et al. (2008) invented a continuous solid-state fermentation technology that couples with product separation using a heat pump, as shown in Fig. 5.8. Ethanol production by solid-state fermentation using sweet sorghum is an example that illustrates this equipment and process.

Sweet sorghum is added from the feed port as shown in Fig. 5.8 (1). The first screwed conveyor (2) carries the material and has a sealing effect. The secondary screwed conveyor carries the material and simultaneously couples with inoculation. The screwed conveyor (4) feeds the feedstock into the tower. The feedstock is distributed evenly and loosely into the fermentation tower (6) by the distributor (5). The fermentation is completed in the fermentation tower (6). The fermentation residue is discharged outside the fermentation tower by the screwed conveyor (8). To ensure aerobic conditions in the fermentation process, CO_2 is supplied to exchange oxygen and remove the oxygen. The feedstock completes the yeast oxygen-requiring stage in the first and secondary screwed conveyor transfer process. When the temperature rises to 32–35 °C by the circulating water in heating sets and CO_2 cycle gas is heated through the heat pump, fermentation starts. The fermentation tower can be separated into three stages: fermentation prophase, primary fermentation, and postfermentation. The fermented material passes the dry layer and is discharged by the unloading conveyor. The unloading conveyor (7) plays a role in stirring and gas distribution. CO_2 flows circularly and the feedstock in a fixed bed is fluidized. The feedstock moves downward by gravity; at the same time, CO_2 flows upward and gives the feedstock an upward force, which makes the feedstock become inflated or semi-inflated. Ethanol is separated with the CO_2 cycle.

CO_2 carrier gas goes into the fermentation tower from the bottom and moves upward, driven by the pressure difference between the top and bottom. CO_2 carrier gas overcomes the resistance of wet feedstock and makes the feedstock expand in solid-state fermentation. At the same time, CO_2 carrier gas achieves the separation of ethanol and heat and mass transfer, which would maintain few microaerobic conditions to enhance the vitality of yeast.

After CO_2 carrier gas departs the wet feedstock, the ethanol and water content in the gas phase reach a maximum. Ethanol steam, 35–40 °C at the tower top, is carried to the heat pump evaporator (13) by the roots blower (12) under a pressure of about 0.5 kg/cm and temperature of about 5–10 °C. The water and ethanol in the CO_2 vapor are condensed and separated. The crude ethanol is removed from the bottom of the heat pump evaporator. The ethanol and water content reach a minimum in the residual CO_2 carrier gas. The residual CO_2 carrier gas is carried to the heat pump condenser (14) to adjust the humidity and adjust the heat to the set temperature value (above 35 °C) and then is carried into the bottom of the separation tower to complete the process cycle again. The excess CO_2 generated during the process is sent to the activated carbon adsorption apparatus (17) from the top of the heat pump evaporator, and the adsorbed CO_2 is used in the future.

Continuous solid-state fermentation coupled with online separation and enrichment using nonfood raw materials such as sweet sorghum, cassava, kudzu root, straw, and so on is achieved by this device.

The device has many advantages:

1. Four screwed conveyors achieve continuous feeding and unloading of the solid-state feedstock and continuous solid-state fermentation. The equipment sealing is good and is not easily contaminated. It is also suitable for pure strain fermentation.
2. Constant temperature fermentation and coupling with separation using gas stripping is achieved.
3. Heat pump technology is introduced to control the transformation process between the gas and liquid phases. There is recovery of the heat generated from the fermentation for the gas-stripping separation of products. The fermentation, product separation device, and heat pump constitute a heated closed-loop system that achieves full energy use.
4. This process has low energy consumption, little water consumption, and low production cost; is nonpolluting; and is a clean production (Chen 2009b).

5.3.2 *Biogas Dry Fermentation Reactor*

China has the maximum output of agricultural waste in the world. Livestock manure, crop stalks, vegetable waste, township garbage, and human excrement are all ideal raw materials for the production of organic fertilizers or biogas. Biogas is clean energy. It can obtain high-quality organic fertilizer and reduce the environmental pollution of agricultural wastes effectively in the biogas fermentation process. Forced by the dual pressures of energy and environmental protection, large-scale biogas dry fermentation technology causes great concern at home and abroad. The main large-scale biogas dry fermentation equipment in Europe includes garage, biogas slurry storage tank, and vertical storage tank types, which have been put into practice. The structural performance of biogas fermentation equipment and the production status in China are listed in Table 5.5 (Zhu et al. 2011).

Fast feeding/unloading and sealing at a large scale are difficulties in the engineering studies of biogas dry fermentation. The Chinese Academy of Agricultural Engineering has developed a new type of biogas dry fermentation reactor (membrane-covered trough [MCT] bioreactor) (Fig. 5.9). This reactor solves the contradiction between large-scale fast feeding/unloading and reactor anaerobic sealing, the problems of engineering temperature regulation, and the combination of the fermentation process with material pretreatment and postprocessing (Han et al. 2008). The new fermentation devices use loaders to solve the difficulties of feeding/uploading in the biogas dry fermentation process. Based on the autoinsulation properties of the deep heaping layer of solid materials combined with efficient insulation measures, the biogas-producing stage can be maintained in a "warm" status, which increase biogas yield effectively and reduces the energy consumption and operating costs. It can indicate the biogas volume intuitively using a flexible-film-covered biogas reactor tank that is intuitive and avoids safety accidents. Currently, this device has reached the practical engineering level.

The MCT reactor (Han et al. 2008), shown in Fig. 5.9, includes a tank body for accommodating material, a flexible membrane, and a link unit for coupling the flexible film sealing cover to the groove body. Solid organic feedstock is first aerobically fermented to increase material temperature inside the reactor uncovered by the flexible film, then the reactor is covered to produce biogas in mesophilic anaerobic fermentation (35–42 °C), obtained by the aerobic fermentation of bioenergy without an external heat source. Finally, the flexible membrane is removed. The residue goes on to aerobic fermentation to remove excess moisture and to produce organic fertilizers. The dry biogas project with the MCT reactor core is generally composed of multiple MCT reactors using a whole-in and whole-out batch process for feeding/unloading technology to coproduce biogas and organic fertilizer.

The MCT dry fermentation system achieves large-scale transformation of crop residues, animal manure, and other agricultural organic solid waste to biogas and fertilizer. Its technical maturity degree reaches the level of practical engineering, and it provides a new way for resource utilization of agricultural waste. Its main innovations include the following:

1. The flexible membrane and the rigid tank are quickly sealed or unsealed using hose-inflatable pressure sealing technology. An anaerobic or aerobic environment can be built quickly. This solves the problems of feeding/uploading materials using loaders.

Table 5.5 China major straw biogas fermentation device basic information comparison

Fermentation device name	All mixed fermentation reactor	Membrane covered trough bioreactor	Underground exposure covered film fermentation tank	Flexible apical membrane of garage dry fermentation device
Fermentation process type	Wet fermentation	Dry fermentation	Dry fermentation	Dry fermentation
Fermentation device structure	Horizontal tank $\Phi 3 \times \times 8$ m	Aboveground groove type $10 \times \times 6 \times \times 1.5$ m	Underground pool type $10 \times \times 2.5 \times \times 3$ m	Aboveground garage type $10 \times \times 4 \times \times 3$ m
Feed concentration	10 %	30–40 %	20–40 %	20–40 %
Feeding/unloading method	Conveyor feeding, screw conveyor unloading	Loader	Artificial feeding, grab unloading	Loader
Adaptability to multifeedstock	Poor	Good	Better	Good
Mass transfer characteristics	Good (stir)	Worse	Poor	Better
Biogas residue water content	High	Low	Low	Low
Volumetric biogas production	High	Lower	Low	Lower
Stability of gas production	Good	Poor	Worse	Worse
Project investment	Great	More	Less	Less
Running cost	High	Lower	Low	Lower
Management difficulties	Large	Less	Small	Less

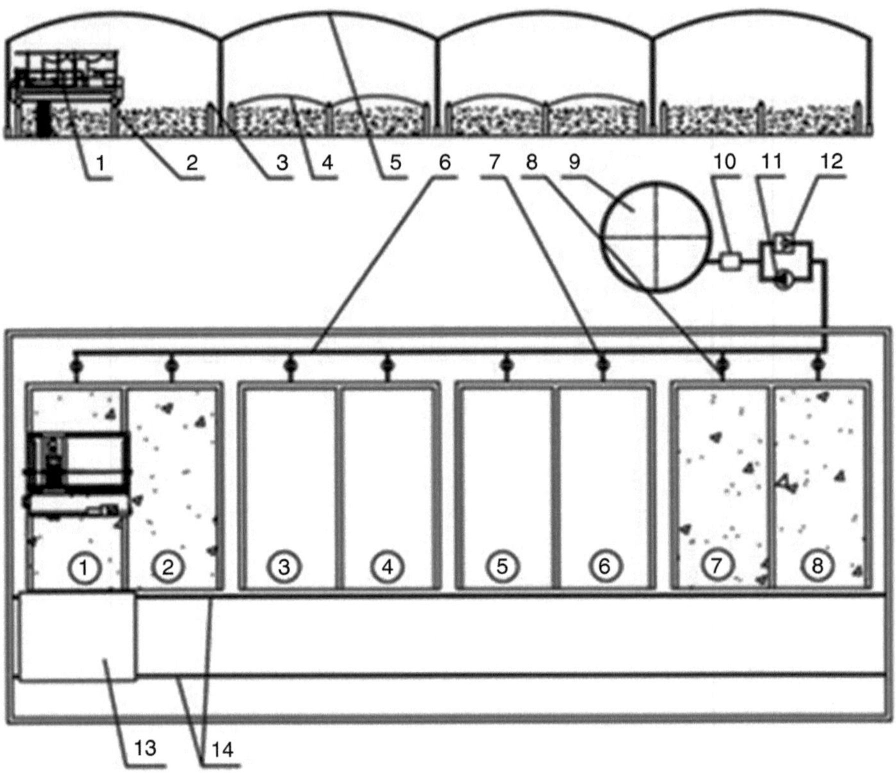

Fig. 5.9 Membrane covered trough dry fermentation system. *1* Special mixing equipment. *2* Reactor trough (add thermal insulation layer). *3* Special mixing equipment orbit. *4* Flexible membrane. *5* Greenhouse. *6* Gas transportation main pipeline. *7* Ball valve. *8* Gas transportation branch pipeline. *9* Gas storage. *10* Biogas purifier. *11* Biogas compressor. *12* Check valve. *13* Shift trough machine of special mixing equipment. *14* Track of shift trough machine

2. Using the heat produced in the aerobic fermentation, based on the autoinsulation properties of the deep heaping layer of solid materials and without an external heat source, the materials are fermented at a middle temperature (35–42 °C) to produce biogas, which improves biogas yield and reduces energy consumption and operating costs.
3. Biogas volume is indicated intuitively using a flexible-film-covered biogas reactor tank, and accidents are avoided. The tedious operation using carbon dioxide for replacement of residual biogas is avoided, reducing the safe operation difficulty and the investment costs and promoting economy and practicality.
4. The unit design can meet the needs of different biogas consumption by adjusting the number of anaerobic reactors.
5. It is suitable for a wide range of raw materials. Crop residues, animal manure, and other organic solid waste can be used for conversion to biogas and fertilizer.

The bioreactor is the key of the fermentation process. An ideal anaerobic solid-state fermentation bioreactor has the following characteristics:

1. The materials used for the reactor must be strong, corrosion resisting, as well as nontoxic for microbial fermentation.
2. The sealing design can prevent pollutants and release of the fermentation products into the environment.
3. It is effective for controlling the operating parameters (temperature, water activity, oxygen concentration, and so on) by effective ventilation adjustment and removal of heat.
4. It is also a critical factor for minimizing the thermal gradient by maintaining the uniformity of the substrate.
5. Operation is convenient for the total solid-state fermentation process.

There are many limiting factors for the anaerobic solid-state fermentation reactor used in industrial production, such as the solid state of the reaction substrate; complexity of the reaction system (heat transfer, mass transfer, momentum transfer); high demand of anaerobic fermentation sealing; and lack of automatic control and monitoring (Liao and Zheng 2005).

5.4 Application of Anaerobic Solid-State Fermentation

Anaerobic solid-state fermentation can be used in such production fields as those for bioenergy, chemicals, traditional food, agricultural feed, environmental protection, and other aspects. In the future, solid-state fermentation will have broad application prospects.

5.4.1 Application in Silage

5.4.1.1 Silage

Straw resources as a feed source are extremely abundant, which can save much food and indirectly provide animal protein products for human beings. Therefore, it is the key point of the scale development and utilization of straw as feed resources by scientific pretreatment technology, such as anaerobic solid-state fermentation, to improve the nutritional value and palatability of straw feed as raw material. Solid-state fermentation using straw to produce feed is a process that uses the acid produced by lactic acid bacteria to reduce the pH value of feed and inhibit the growth and reproduction of harmful microorganisms. At the same time, the straw can be transformed into good quality feed with higher nutritional value and palatability, such as silage and microstorage of feed (Table 5.6) (Gao et al. 2008).

Table 5.6 Comparison of silage and microbial silage

Fermentation	Silage	Microbial silage
Feedstock requirements	Feedstock moisture content 65–70 %	Feedstock moisture content 55–65 %
Processing principles	Lactic acid fermentation under anaerobic environment	Cellulose anaerobic decomposition to product acid
Processing method	Using lactic acid bacteria in the feedstock, cannot add any substance	Add high activity of bacteria and nutrients
Treatment time	Only in mowing season	Put to use year round
Processing results	Loss of a few nutrients, especially vitamins	Loss of a few nutrients
Disposal characteristics	Natural fermentation	Directional fermentation

Green stalks of crops such as cornstalks, sorghum, and millet crops with a moisture content of 65–75 % are chopped and fermented under sealed anaerobic conditions to produce silage. The key factor of the silage is the degree of lactic acid fermentation.

5.4.1.2 Ensiling Process

Three interacting factors need to be controlled in the ensiling process: the chemical composition of raw materials, the amount of air in the silage feedstock, and the bacterial activity. According to changes of environment, microbial populations, and materials, the general silage fermentation process can be divided into three stages: aerobic respiration, anaerobic lactic acid fermentation, and silage stable stages.

Aerobic Respiration Phase

Fresh silage material is put into a sealed silage container; plant cells do not die immediately, but in 1–3 days, and the organic matter is decomposed into acid through their respiration. When the oxygen is depleted, the anaerobic solid-state fermentation starts.

At the beginning of silage, aerobic microorganisms, such as yeasts, molds, and lactic acid bacteria, can attach to the raw materials and rapidly reproduce using soluble carbohydrate. The oxygen is quickly depleted by the activity of aerobic microbes and respiration of plant cells, which results in the formation of an anaerobic environment. In addition, heat is generated by the respiration of plant cells, enzymatic oxidation, and microbial activities. The anaerobic and warm environment creates suitable conditions for lactic acid fermentation. The stage is a necessary process for the formation of an anaerobic environment from the aerobic environment in silage.

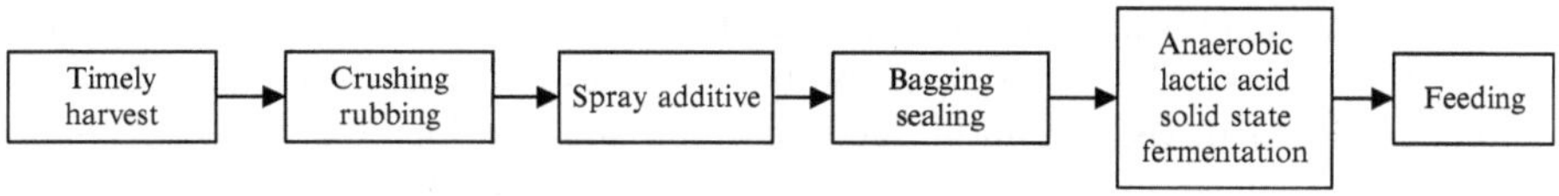

Fig. 5.10 Silage process flow diagram

The aerobic respiration stage should be as short as possible. If the oxygen content in the silage is too high, the plant respiration time is too long, and aerobic microbial activity is too vigorous, the temperature of the raw materials will rise. Sometimes, the temperature can reach up to 60 °C, thus weakening the competitiveness of the lactic acid bacteria and other microorganisms. The silage nutrients are lost, and quality declines excessively.

Anaerobic Lactic Acid Fermentation Stage

With the completion of the period of anaerobic fermentation, small amounts of *Lactobacillus* begin to reproduce rapidly in the plant material. The carbohydrates can be converted into lactic acid by *Lactobacillus*, and the pH value of the crop straw can be reduced by the formation of a large amount of lactic acid, which results in the inhibition of bacterial growth and enzyme activity and the promotion of aerobic microorganism activities. When the pH is decreased to below 4.2, a variety of harmful microorganisms cannot survive, and even the *Lactococcus lactis* activities are also suppressed. When the pH is decreased to 3 continuously, the activity of *Lactobacillus* is inhibited, which is basically the end of the lactic acid fermentation process.

Silage Stable Stage

When the pH is decreased below 4.2, all kinds of microbes sign off; only a small amount of lactobacilli exist, and silage no longer loses nutrients. Silage in the anaerobic and acidic environmental condition can be stored properly. This period continues for about 2–3 weeks. If the sealing method is effective, silage can be stored continuously.

5.4.1.3 Silage Process

Silage is produced when the chopped straw is put into a sealed anaerobic environment and fermented by lactic acid bacteria to produce high-quality feed that can be stored for a long time without mildewing. The major domestic silage technology includes the pit, stacking, and bagged methods (Fig. 5.10).

There are many benefits of silage; for example, it has less nutrient loss, is rich in digestible nutrients, has high digestibility, increases appetite, promotes digestion, and is available all year. With in-depth research and development of new technology for straw silage, the yield and quality of silage can be constantly improved, and the potential commercialization of the silage can be continuously discovered.

The processing route for corn stalk silage is discussed next.

Timely Harvest

Corn stover harvest must not affect corn production; the stover should be harvested in the ripening stage (stage with four green leaves), with a dry matter content of 25–35 %, to ensure the nutritional content of the feed and the moisture content of straw. Harvested straw should not have too long stubble or roots.

Straw Rubbing

Stalks should be transported to a location immediately after harvest, cut short, and rubbed with a rubbing machine to destroy the cuticle in the straw surface and internode. Compared to cutting straw, rubbing straw is 6–8 cm long, soft, and easily compacted. As a result, the probability of punctured plastic bags is greatly reduced, and feed palatability is substantially increased. Rubbing straw needs to consume much energy, but the quality of forage is improved, which is still very cost effective.

Spraying Additives

Silage needs additives to reduce the loss of nutrients in the straw ensiling process, to prevent corruption, and to ensure and improve the quality of the silage. Straw should be compacted for air removal after spraying additives to shorten the straw silage aerobic stage and to ensure feed quality.

Bagging and Sealing

Stalks will be loaded into a special bag. In the bagging process, air should be excluded as much as possible, and then the bag is tied. The aims of bagging and sealing are the exclusion of oxygen and achieving lactic acid fermentation. In addition, the small package method is beneficial because it avoids the loss of nutrients and facilitates transport.

Access

Silage can be layered for access after fermentation for 45 days. It needs to be sealed after access to prevent secondary fermentation. The accessed silage should be run out the same day, not left overnight, to avoid deterioration. The principle is that at the start of feeding silage process should be gradually increased from less to more; at the end of feeding, it should be gradually decreased from more to less.

The procedure of rubbing, compacting, and bagging should be carried out continuously. Otherwise, the procedure not only will lead to the loss of nutrients but also will affect silage quality. The high-quality silage color is green or yellow-green with a rich fruit acid flavor or wine acid flavor; the silage is soft and slightly moist and loose and has a pH of 4–4.2.

5.4.1.4 Current State of Straw Silage Technology

Conversion of straw to silage is achieved by anaerobic solid-state fermentation. This is an important measure for the development of animal husbandry and the direction of straw utilization. The improvements of the silage process and equipment and the study of the mechanisms of silage fermentation are essential for the development of animal husbandry in China.

Utilization of Silage Additives

Currently, the addition of silage additives is the most effective control method to promote the silage fermentation process (Filya et al. 2007). Domestic and foreign researchers have done much work on silage additives and have confirmed that a lactic acid fermentation accelerator has natural, nontoxic, and efficient advantages and great potential in the development of roughage resources (Wang et al. 2010; Filya et al. 2006). Therefore, the enhancement of the silage process by a lactic acid fermentation accelerator will be the direction of future silage development. Research by Dönmez et al. (2003) showed the concentration of lactic acid increased with the addition of additives to the silage; the contents of water-soluble carbohydrate and organic acid were increased by the addition of the cellulase to the silage. At the same time, the silage time was shortened. Zhao et al. (2010) obtained the best combination of lactic acid bacteria, and the silage process could be promoted by the addition of cellulase and sodium nitrite as fermentation accelerators.

Mechanization Production of Straw Silage

To achieve the large-scale population and industrial production of straw silage technology, mechanized production of straw silage has become a key research trend. Chen et al. (2006) designed new bag silage equipment, 9SYG-1 (silage compression filling machine), to improve the compaction degree and quality of silage; this was helpful for commercialization of silage based on the characteristics and principle of silage.

Cao (2012) tested the performance and production of the self-developed straw silage equipment, the YJR-4 corn straw kneading machine and YD-85 electrohydraulic horizontal-type balers. The results showed that the equipment could meet the silage quality requirements.

Silage corn harvesting time is only 30 days. Because of lack of systematization and miniaturization research for silage machinery, the pulverizer and the baler remain to be applied, so independent research and advanced straw silage mechanization equipment are necessary.

5.4.2 Solid-State Anaerobic Treatment of Organic Solid Waste

Organic solid waste refers to material that has a high content of organic matter and low moisture content and is easily degraded by microorganisms. These materials contain a large amount of biomass energy; the effective utilization of these resources has a great impact on environmental protection and sustainable economic development.

5.4.2.1 Overview of Anaerobic Composting

Currently, there are three types of disposal methods for organic solid waste: sanitary landfill, composting, and incineration. With the development of the concept of a recycling economy in the treatment of urban organic solid waste, composting has strong vitality and becomes important in the achievement of urban organic solid waste reduction, recycling minimization, and use of a harmless resource.

Composting is the mineralization and humification process of organic matter, and the organic matter is transformed into fertilizer by anaerobic microorganisms. During the process, many effective nitrogen, phosphorus, and potassium compounds are generated that can be absorbed and used by plants. The new synthesis of polymer organic matter-humus constitutes creation of important active substances for soil fertility (Farrell and Jones 2009). In the composting process, along with the decomposition of organic matter and the formation of humic substances, the compost changes significantly in volume and weight. The conversion of organic matter into volatile components such as carbon and nitrogen is coupled with the reduction of water, so the weight and volume of compost will be reduced by about half.

The aerobic composting process needs to consume large amounts of oxygen and energy. Yet, anaerobic composting is an anaerobic solid-state fermentation process that not only does not need a lot of energy, but also produces biogas, which is

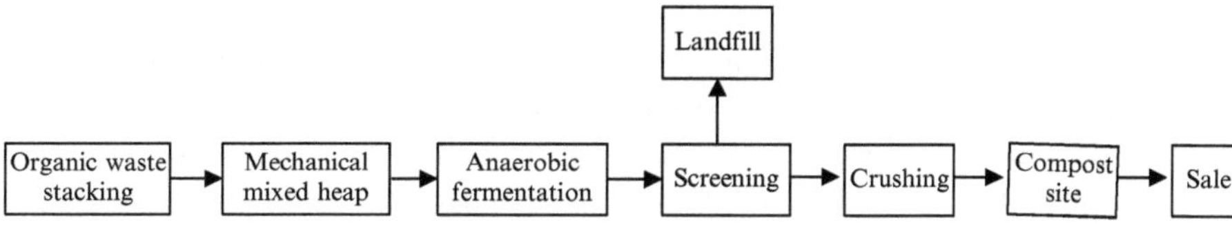

Fig. 5.11 High-temperature composting technology in Jixi City

deemed clean energy. Therefore, anaerobic composting is an ideal approach from the perspective of a circular economy.

5.4.2.2 Principles of Anaerobic Composting

Anaerobic composting is decomposition of organic matter by an anaerobe; it mainly has two stages (Jiao 2007): acidic and alkaline fermentation. In the acidic fermentation stage (including hydrolysis and acidification), the produced small-molecule substance is converted to fatty acids, alcohols, ammonia, carbon dioxide, sulfides, hydrogen, and energy. In the initial decomposition stage, organic acids would greatly accumulate, which results in the decrease in pH value. In the alkaline fermentation stage, the methanogens start to decompose organic matter and alcohol to produce methane and carbon dioxide. With the reproduction of methanogenic microbes and the decomposition of organic acids, the pH value rises rapidly.

The microbial reaction in the composting process is similar to that in liquid fermentation, which is also a biological oxidation-reduction reaction. The composting complexity is caused by various metabolisms of microorganisms, which play a leading role in the different stages of the reaction.

5.4.2.3 Anaerobic Composting

Anaerobic Composting Process

According to the northeastern China winter and summer temperature characteristics, Jixi environmental health research institutes in Jixi City initiated the first two sets of a fermentation process. In the hot summer, garbage and feces could easily lead to pollution in the urban environment, so a quick and harmless solution was key to garbage disposal. The innocuous treatment of garbage includes sorting, crushing, mixing with feces, and aerobic fermentation. In the winter, the weather is cold; the minimum temperature can reach −38 °C. The pollution caused by garbage and feces seems less serious than in the summer; consequently, the garbage and feces should be anaerobically fermented for 6 months (November to April) and then separated and crushed (Fig. 5.11).

After many years of research and practical experience, the city used a 7:3 mixing ratio of garbage and feces in the composting fermentation for 10 days. Some of the fermentation residues were removed and used as a heat source; the other parts were covered by new garbage and feces (7:3 ratio of garbage to feces) to make a new heap. The fermentation proceeded from November of 1 year to April of the second year, and then fermentation residues used as mature fertilizer were screened, crushed, and directly sold. The temperature of anaerobic composting should be tracked and monitored to prove that the composting temperature rose to 55–60 °C and harmless process requirements were reached.

Elements of the Composting Process

There are many factors that affect the composting process, such as moisture, temperature, and pH values (Li et al. 2011). It is said that the regulation of pH is the base, the regulation of humidity is the key, and the regulation of temperature is guaranteed.

Water

Because of the different physicochemical properties of various materials, suitable water contents for fermentation are different. A low or high material water content will result in the sharp rise in temperature of the compost body. In the composting process, the water will gradually evaporate, particularly in high-moisture materials, which will be helpful for reducing the drying cost in the later period. The moisture control of the material should use the following general principles:

1. The water content should be properly lowered in the southern region and increased in the northern region. Because the water content of the air in the southern region is high, water evaporation in the composting process is weak.
2. The water content should be properly lowered in the rainy season because of the influence of humid air on natural evaporation.
3. The water content should be properly lowered in the cold seasons because the rise in temperature is low. The fermentation temperature rises relatively slowly because of increased heat loss, less volatile water.
4. The water content of stale material should be properly lowered. Because the organic matter has been decomposed by environmental microbiology in a way that weakened the intensity of the biochemical reaction, the moisture demand is relatively small.
5. If the carbon-to-nitrogen ratio is properly low, the water content should be lower because a low carbon-to-nitrogen ratio means fewer biodegradable carbohydrates and less demand for water for biochemical reactions.

In short, the regulation of water content in composting should be based on the characteristics of geography, climate, materials, and formulations. Researchers should carefully observe moisture change and its impact on composting materials and make timely adjustment measures.

Temperature

In the anaerobic composting process, the heat will be generated by the metabolism of microorganisms. The temperature change of the compost is the most direct and sensitive indicator to reflect fermentation status. Temperature has an extremely close relationship to moisture, permeability, and other compost-controlling factors, so it is also the most complex factor. The temperature change trend for compost can be summarized as follows: The temperature rises stably in the early stage, the temperature is moderate in the middle stage, and the temperature declines slowly in the end stage. The perfect temperature for rapid composting is around 50–60 °C and not above 70 °C. The time length for high temperature reflects the formula adjustment and the qualities of pretreated raw material in the early stage. The ideal time length for high temperature is generally 5–10 days.

pH

The pH is also important to the composting process. Ammonia can be produced from nitrogen-containing organic compounds in the stacking decomposition, which will cause a rise in pH values. However, the organic acid generated will lead to a drop of pH values. Both high and low pH values will cause composting process difficulties. Usually, the pH values should be maintained between 7.5 and 8.5 to obtain the most efficient composting.

5.4.2.4 Prospects for Anaerobic Composting Technology

Aerobic composting, especially a high-temperature aerobic composting process, has advantages such as complete substance decomposition, short fermentation period, and appropriate mechanization operations. So, current composting research mainly is focused on high-temperature aerobic composting (Wei et al. 2007). In view of the long time needed for anaerobic composting, much effort should be made to investigate it, such as the pretreatment of substrate and the optimization of the digestion process (Zhang et al. 2002).

1. Add inoculants to improve the efficiency of composting. Strengthening research into compost microbial agents is needed to increase the types and quantities of microbes in compost raw materials, to shorten the period of high-temperature aerobic composting, to enhance the decomposition activity of the microorganisms, and to accelerate the organic matter decomposition speed. The decomposition of organic waste involves complex microbial metabolic processes. A specific and stable decomposition microbial system should be built under certain suitable environments to promote the decomposition rate. The specific and stable decomposition microbial system should play an important role in the early composting process, which is an extremely important factor that affects composting fermentation progress.

2. Composted materials quickly decompose. Another key of the composting process is the rapid decomposition of material that is difficult to break down. The largest amount of organic waste is lignocellulosic material, which consists of cellulose, lignin, and other biodegradable ingredients. Therefore, the decomposition degree of organic solid waste is the important factor that influences the whole organic solid waste composting process.

Steam explosion technology is expected to accelerate biodegradation. The waste materials are maintained in the steam conditions at high temperature and high pressure and are rapidly expanded instantaneously. The cellulose crystalline structure of steam-exploded straw is changed, and the hemicellulose and lignin are degraded. This pretreatment can accelerate the degradation of cellulose, speed up the composting process, and improve the efficiency of anaerobic digestion, thereby increasing the volume reduction rate of the garbage.

In solid-state anaerobic digestion, an external stimulus can strengthen the digestive process and mass transfer, heat transfer, momentum transfer, and information transfer effects (Zhang et al. 2005). The pressure pulsation is external stimulation technology that can accelerate the rate of digestion and biogas yield in the solid-state anaerobic digestion process. Pressure pulsation is the process by which the gas pressure gradually increases to a certain extent and then rapidly decreases during fermentation. Here, the pressure pulsation has a stirring effect to overcome the shortcoming caused by a stirring method under loading with a substrate with high solids (Li and Chen 2005).

In addition, the combination of anaerobic fermentation and high-temperature aerobic composting to produce organic fertilizer and biogas and biogas residue not only can greatly reduce investment but also can produce higher fertilizer efficiency and better utilization of municipal solid waste resources.

Overall, anaerobic solid-state fermentation can be widely used in the traditional food industry for a unique flavor, such as for soy sauce, vinegar, tempeh, fermented bean curd, cheese, yogurt, pickles, and sausages. With the sharp increase in the pressure for environmental protection and resource conservation, anaerobic solid-state fermentation with its unique advantages of energy saving and environmental protection again attracts the attention of people around the world. Anaerobic solid-state fermentation has shown tremendous application potential; for example, it is widely used in the conversion of biomass resources to biofuels such as fuel ethanol, biohydrogen, and biogas and the conversion of biomass to energy, food, feed, and fertilizer.

References

Alvira P, Tomás-Pejó E, Ballesteros M, Negro M. Pretreatment technologies for an efficient bioethanol production process based on enzymatic hydrolysis: a review. Bioresour Technol. 2010;101:4851–61.

Cao SW. Significance and benefits analysis of mechanized production of storage ensilage. Agric Sci Technol Equip. 2012;206:65–7.

Chen HZ. Process engineering in plant-based products. New York: Nova Science; 2009a.
Chen HZ. Principles and applications of anaerobic solid state fermentation. Beijing: China Agricultural Science and Technology Press; 2009b.
Chen HZ, He Q. Value-added bioconversion of biomass by solid-state fermentation. J Chem Technol Biotechnol. 2012;87(12):1619–25. doi:10.1002/jctb.3901.
Chen HZ, Jin SY. Effect of ethanol and yeast on cellulase activity and hydrolysis of crystalline cellulose. Enzyme Microb Technol. 2006;39:1430–2.
Chen HZ, Liu LY. Unpolluted fractionation of wheat straw by steam explosion and ethanol extraction. Bioresour Technol. 2007;98:666–76.
Chen HZ, Qiu WH. The crucial problems and recent advance on producing fuel alcohol by fermentation of straw. Prog Chem. 2007;19:1116–21.
Chen HZ, Qiu WH. Key technologies for bioethanol production from lignocellulose. Biotechnol Adv. 2010;28:556–62.
Chen HZ, Wan L. Research progress on key process and integrated eco-industrial chains of biobased products-proposal of biobased product process engineering. Chin J Proc Eng. 2008;8:676–81.
Chen HZ, Xu FJ, Tian ZH, Li ZH. A novel industrial-level reactor with two dynamic changes of air for solid-state fermentation. J Biosci Bioeng. 2002;93:211–4.
Chen HX, Zhang CZ, Dong DJ. Research on the technology and devices of the straw ensiling. J Agric Mech Res. 2006;1:34–6.
Chen HZ, Ding WY, Dai SH et al. Methods and equipment of continuous solid-state fermentation and separation by product gas stripping coupling heat pump. China Patent 200810134423.7; 2008.
Debebe YD, Stefan T, Majid HM, et al. Improved bio-energy yields via sequential ethanol fermentation and biogas digestion of steam exploded oat straw. Bioresour Technol. 2011;102:4449–55.
Dong YS, Liu TJ. Study on fuel ethanol production by solid-state fermentation of sorgo straw. China Brewing. 2008;1:44–6.
Dönmez N, Karslı M, Çınar A, Aksu T, Baytok E. The effects of different silage additives on rumen protozoan number and volatile fatty acid concentration in sheep fed corn silage. Small Ruminant Res. 2003;48:227–31.
Du XH, Liu SL, Pu B, et al. Production technology of pickle by semi-solid fermentation of *Lactobacillus plantarum*. China Brewing. 2011;1:63–6.
Farrell M, Jones DL. Critical evaluation of municipal solid waste composting and potential compost markets. Bioresour Technol. 2009;100:4301–10.
Filya I, Sucu E, Karabulut A. The effects of *Propionibacterium acidipropionici* and *Lactobacillus plantarum*, applied at ensiling, on the fermentation and aerobic stability of low dry matter corn and sorghum silages. J Ind Microbiol Biotechnol. 2006;33:353–8.
Filya I, Muck RE, Contreras-Govea FE. Inoculant effects on alfalfa silage: fermentation products and nutritive value. J Dairy Sci. 2007;90:5108–14.
Gao LJ, Yang HY, Cui ZJ, et al. Rice straw fermentation using lactic acid bacteria. Bioresour Technol. 2008;99:2742–8.
Haelssig JB, Tremblay AY, Thibault J. Technical and economic considerations for various recovery schemes in ethanol production by fermentation. Ind Eng Chem Res. 2008;47:6185–91.
Han J, Xiang X, Li X. Technology and equipment of large scale biogas dry fermentation in membrane covered trough. Trans CSAE. 2008;24:100–4.
Han B, Wang L, Li SZ, et al. Ethanol production from sweet sorghum stalks by advanced solid state fermentation (ASSF) technology. Chin J Biotechnol. 2010;26:966–73.
Isci A, Demirer GN. Biogas production potential from cotton wastes. Renew Energy. 2007;32:750–7.
Jiao DF. Research on anaerobic compost of municipal dewatering sludge. J Changchun Norm Univ. 2007;26:35–8.

Leng CB, Xiao B, Yang JK, et al. Study of house refuse by dry anaerobic digestion in "underground river". Environ Eng. 2001;19:45–7.

Li HQ, Chen HZ. The periodic change of environment factors in solid state fermentation and effect on microorganism fermentation. Chin J Biotechnol. 2005;21:440–5.

Li HQ, Chen HZ. Studies of process engineering of cellulosic bioconversion. Biotechnol Biobus. 2010;1:40–6.

Li X, Zhao LX, Han J, Xiang X. The new direction of agricultural residues utilization in China: biogas dry fermentation technology. China Biogas. 2006;24:23–7.

Li CB, Luo GH, Guo YZ. Research of intermittent type dry process fermentation of methane. J Agric Mech Res. 2010;3:225–7.

Li YB, Stephen YP, Zhu JY. Solid-state anaerobic digestion for methane production from organic waste. Renew Sustain Energy Rev. 2011;15:821–6.

Liao CY, Zheng YG. Solid state fermentation bioreactor. Microbiol China. 2005;32:99–103.

Liu XF, Yuan YX, Yan ZY. Progress on biogas technology and engineering. Chin J Biotechnol. 2010;26:924–30.

Lu LX, Wang XF, Xiong XH, et al. Study on pickling radish by *Lactobacillus plantarum* B_2 culture and natural fermentation. Sci Technol Food Ind. 2005;26:59–60.

Lv CF, Li JH, Yu Y. Three changes in the solid-state fermentation process of liquor. Liquor Mak. 2003;30:61–2.

Mustafa B, Havva B. Recent trends in global production and utilization of bio-ethanol fuel. Appl Energy. 2009;86:2273–82.

Salimi F, Zhuang K, Mahadevan R. Genome-scale metabolic modeling of a clostridial co-culture for consolidated bioprocessing. Biotechnol J. 2010;5:726–38.

Song YM, Chen HZ. Study on biogas production by thermophilic solid-state fermentation from steam exploded corn stalk. Chin J Environ Eng. 2008;2:1564–70.

Sun Q. Technology lectures of fuel ethanol (IV) ethanol distillation and dehydration. Renew Energy Resour. 2010;28:154–6.

Vane LM. A review of pervaporation for product recovery from biomass fermentation processes. J Chem Technol Biotechnol. 2005;80:603–29.

Wang KK, Yu Z, Shao T, Liu PD. Effects of lactobacillus on fermentation quality of mixed silage of *Medicago sativa* and *Elymus dahuricus*. Acta Partacult Sin. 2010;19:94–100.

Wei ZM, Xi BD, Zhao Y, et al. Effect of inoculating microbes in municipal solid waste composting on characteristics of humic acid. Chemosphere. 2007;68:368–74.

Weiland P. Biogas production: current state and perspectives. Appl Microbiol Biotechnol. 2010;85:849–60.

Wu JH, Wang CL, Wang YR, et al. Pilot test on semi-solid-state pure-fermentation technology of low-salt pickled cucumber. Trans CSAE. 2009;25:265–70.

Yang B, Wyman CE. Pretreatment: the key to unlocking low-cost cellulosic ethanol. Biofuels Bioprod Biorefining. 2008;2:26–40.

Yu B, Chen HZ. Effect of the ash on enzymatic hydrolysis of steam-exploded rice straw. Bioresour Technol. 2010;101:9114–9.

Zhang AJ, Chen HZ, Li ZH. The present situation and progress of study of solid-state anaerobic digestion of organic solid wastes. Res Environ Sci. 2002;15:52–4.

Zhang AJ, Chen HZ, Li ZH. Air pressure pulsation solid state production of alkaline protease by *Bacillus pumilus*. Proc Biochem. 2005;40:1547–51.

Zhao H, Yi MH, Wang HJ, et al. Study on mixed lactic acid fermentation production of corn stover silage. Cereal Feed Ind. 2010;10:52–4.

Zhou DQ. Microbiology. Beijing: High Education Press; 2002.

Zhu SQ, Zhang YL, Zhang WQ, Zhang JF. The progress of dry anaerobic fermentation technology. Renew Energy Resour. 2009;27:46–51.

Zhu DW, Cao CM, Chen YS, et al. Key technologies and discussion of the straw dry-anaerobic fermentation. Chin Agric Mech. 2011;4:56–9.

Zhuang XS, Yuan ZH, Ma LL. Fuel ethanol application and production efficiency evaluation in China. Acta Energy Solaris Sin. 2009;30:526–31.

Chapter 6
Principles and Application of Solid-State Fermentation Carried Out on Inert Support Materials (Adsorbed Carrier Solid-State Fermentation)

Abstract Despite the potential of solid-state fermentation (SSF) systems, some physical aspects related to heterogeneity and nutrition of the media are serious constraints. To overcome these problems, it is convenient to use supports that have well-characterized properties and a minimum of interaction with the biological process. SSF carried out on inert support materials (adsorbed carrier solid-state fermentation, ACSSF) is proposed based on the principle of SSF and is regarded as a future development for SSF systems. The ACSSF system is composed of porous, heterogeneous, biologically inactive material, called an inert support, in which defined culture media and inoculum are absorbed. Cell growth takes place under controlled aerated conditions within suitable reactors kept at a constant temperature. In this chapter, the properties of inert carriers, the techniques, and fermentors of ACSSF and its process optimization are expatiated. Some typical examples are used to project the future of ACSSF. In addition, the economy and the problems of ACSSF are discussed.

Keywords Adsorbed carrier solid-state fermentation • Inert support pretreatment • Application

6.1 Introduction to Solid-State Fermentation Carried Out on Inert Support Materials

6.1.1 Concept

Traditional solid-state fermentation (SSF) is only applicable for products with low-purity requirements or products easily separated and purified. Examples of the former are enzyme preparation, seasoning, biological feed, biopesticides; the latter include ethanol as well as antibiotics and organic acids. Generally, the supports utilized are of agricultural origin, such as maize, broomcorn, wheat, cereal, soybean, lignocelluloses, and so on, and have a heterogeneous structure and composition.

H. Chen, *Modern Solid State Fermentation: Theory and Practice*,
DOI 10.1007/978-94-007-6043-1_6, © Springer Science+Business Media Dordrecht 2013

Furthermore, the support and the substrate are usually associated with bringing about changes in the structure because of microbial growth.

Despite the potential of SSF systems, some physical aspects related to heterogeneity and nutrition of the media are a serious constraint. The disadvantages of traditional SSF are as follows: (1) SSF substrate is a source of nutrition for the growth of microorganisms; at the same time, it provides a microenvironment for microbial growth and a gathering place for the fermentation products. In the fermentation process, the media are gradually decomposed, the substrate is agglomerated, and the porosity decreases, which results in mass and heat transfer limitations. (2) Impurities in the media are easily integrated into the fermentation extracts, which does not benefit the separation and purification of fermentation products. (3) Some problems derived from heterogeneity are the difficulty in accurate and reproducible measurement of key variables and unavailability of general strategies for bioreactor control and scale-up procedures. Aside from classical key variables such as cell growth, substrate consumption, and product formation, in SSF there are other physical variables, such as bulk and inlet air temperatures, water availability, water activity, and void space availability, that have a strong effect on the physiological and biochemical activities of the organism used and, consequently, on the global effectiveness of the system. Mathematical modeling for SSF involving these variables together is scarce in the literature for SSF systems (Chen and Xu 2004; Pandey 2003).

To overcome these problems, it is convenient to use supports that have well-characterized properties and that have minimum interaction with the biological process. SSF carried out on inert support materials (adsorbed carrier solid-state fermentation, ACSSF), regarded as one of the future developments of SSF systems, is proposed based on the principle of SSF. The ACSSF system is composed of a porous, heterogeneous, biologically inactive material, called an inert support, in which defined culture media and inoculum are absorbed. Cell growth takes place under controlled aerated conditions within suitable reactors kept at a constant temperature. The inert support does not interact with the microorganism and does not change its characteristics throughout the fermentation.

In 1935, Cahn (1935) achieved citric acid production by *Aspergillus niger* cultured on a sucrose solution absorbed on sugarcane bagasse or sugar beet pulp. In 1965, Meyrath (1965) described a system in which *Aspergillus oryzae* was grown on vermiculite imbibed with a starchy liquid substrate. High yields of dextrinogenic amylases with *Aspergillus oryzae* on vermiculite impregnated with starch solution were reported.

In 1988, Oriol et al. (1988) used a synthetic liquid medium absorbed on a solid support to demonstrate that this culture method was suitable for the growth of filamentous fungi, and it allowed the utilization of high-concentration substrate solutions. The water activity of the liquid phase, the support particle size, and the amount of spore inoculum were found to be critical factors for the growth of molds, with the growth accounting for different spore germination patterns between solid and submerged cultures. This kind of cultivation method might broaden the use of solid-state cultures for producing fungal metabolites with low-cost technology.

In 1990, Richard et al. (1990) grew *Aspergillus niger* on ion exchange resin (Amberlite IRA 900) impregnated with a defined medium. The ion exchange resin studied can be used as a model for SSF. The growth pattern was similar to that obtained on other agricultural supports. It has several interesting characteristics: It is inert; it remains unchanged during fermentation; growth is only superficial; the mold is easily removed from the substrate; and it has good media retention without drainage. It has a spherical shape, well-characterized properties, and high and homogeneous packing density in the fermentors. The facts that the size of the support does not change during fermentation and that growth is only superficial allow relating the pressure drop with the biomass produced.

In 1996, Nampoothiri and Pandey (1996) studied the cultivation of *Brevibacterium* sp. on sugarcane bagasse impregnated with a medium containing glucose, urea, mineral salts, and vitamins for producing L-glutamic acid. Maximum yields (80 mg glutamic acid per gram dry bagasse with biomass and substrate, mg/gds [milligrams/g dry substrate]) were obtained when mixed particle size bagasse was moistened at an 85–90 % moisture level with the medium containing 10 % glucose. This is the first report of the cultivation of *Brevibacterium* sp. in solid cultures for production of glutamic acid.

In 2010, Weng and Chen (2010) grew *Acetobacter xylinum* to produce bacterial cellulose on polyurethane foam (PUF), which served as an inert support. ACSSF overcame the disadvantages of static and dynamic fermentation. Compared with the volumetric productivity (0.39 g/L · per day) of submerged fermentation, it reached 1.62 g/L · per day in SSF using PUF as an inert support, and the fermentation period was reduced from 10 to 3 days.

6.1.2 Advantages

It is not necessary to combine the role of support and substrate, but rather the conditions of low water activity and high oxygen transference are reproduced using a nutritionally inert material soaked with a nutrient solution (Pandey 2003). Solid-state fermentation is carried out on inert support materials, which differs from the process of microbial growth on or in solid particles floating in a liquid medium. The use of solid inert material impregnated with suitable liquid medium would provide homogeneous aerobic conditions throughout the fermentor, and the purity of the product would also be comparatively high (Nampoothiri and Pandey 1996).

6.1.2.1 Advantages of ACSSF Compared with Traditional SSF

Solid-state fermentation using inert supports impregnated with chemically defined liquid media has several potential applications in both scientific studies and the industrial production of high-value products, such as metabolites, biological control agents, and enzymes. As a result of its more defined system, SSF on inert supports offers numerous advantages, such as improved process control and monitoring and enhanced process consistency, compared with cultivation on natural solid substrates. Except for the advantages of traditional SSF, it can overcome many problems of traditional SSF (Chen and Xu 2004).

1. The structure of the inert support is not modified by microorganisms or metabolites in microbial growth; therefore, it can overcome agglomeration and the heat and mass transfer limitation.
2. The culture medium is synthetic, which can be the same as the liquid medium to optimize the medium to meet the fermentation process. The medium component is a key factor for fermentation. The nutritional component types and the proportion of the components result in differences of microbial growth and metabolism. Traditional SSF uses a natural substrate as medium; therefore, it is not easy to change the composition of the medium and to evaluate which components affect fermentation. The ACSSF medium is synthetic, and the components are defined. Each component can be evaluated for microbial growth and fermentation, and the fermentation process is repeatable, which overcomes the heterogeneity among different batches of conventional SSF.
3. Suitable inert carriers can avoid mixing of carrier impurities into the fermentation extract and can simplify the purification process. The extracellular products of interest can be easily extracted from the inert support; therefore, products can be obtained with fewer impurities compared with the natural substrates, and the support can be reused. For example, spores of *Penicillium roquefortii* formed inside buckwheat grains can only be recovered following disruption of the substrate. By contrast, spores produced on pozolano particles (volcanic material) can be readily extracted without destroying the particles; therefore, the recovery process is simplified, and the particles can be reused (Larroche and Gros 1989).
4. There is high target product yield and fewer by-products. The SSF processes using conventional, nutritionally rich substrates such as wheat bran have certain inherent problems that could largely be overcome using nutritionally inert supports. In 1995, Nagendra Prabhu and Chandrasekaran evaluated the use of polystyrene as an inert solid support for the production of L-glutaminase using *Vibrio costicola*. L-Glutaminase with much higher specific activity could be recovered using polystyrene as a support than when using wheat bran (Table 6.1). Enzyme synthesis was induced by L-glutamine; wheat bran without glutamine failed to support enzyme production. The extract obtained from the polystyrene support was also much less viscous (mean viscosity 0.966 Ns/m^2) than that from wheat bran (2.072 Ns/m^2) and was free of undesired proteins. The wheat bran extract contained amylase (200–400 U/gds) and cellulase (filter paper activity 1.0–1.6 U/gds) besides glutaminase. Although wheat bran is cheaper than polystyrene, the latter could be more economical for the production of enzymes because of higher specific activity and purity and cheaper downstream processing, as seen in glutaminase production. Maximal recovery of the leachate (92–96 %) and reusability of beads are added advantages of using polystyrene for SSF.
5. ACSSF can maintain the uniformity and consistency of the environment in the SSF process so that ACSSF is repetitive.

Table 6.1 L-Glutaminase production by *Vibrio costicola* during solid-state fermentation on polystyrene or wheat bran

	Polystyrene		Wheat bran	
Time (h)	Yield (U/gds)	Sp. act (U/mg protein)	Yield (U/gds)	Specific activity (U/mg protein)
0	Nil	Nil	Nil	Nil
12	35	5.5	56	0.4
24	69	4.2	111	0.8
36	88	4.4	114	0.8
48	56	3.9	92	0.7
60	43	3.8	70	0.5
72	39	3.7	67	0.5

gds grams dry substrate

6. The physiological metabolic and fermentation kinetics of microorganisms in the ACSSF process can be studied in depth, which will help control the fermentation process as well as the construction of a dynamics model and provide a theoretical basis for process improvement, process control, and reactor design and scale.

6.1.2.2 Advantages of ACSSF Compared with Submerged Fermentation

Compared with submerged fermentation (SmF), there are certain advantages of ACSSF.

Improve the Oxygen Transfer Rate with the Low Energy Costs of the Aeration Agitation Process

For ACSSF, inert carriers provide a large surface for the growth of microorganisms, which can be maintained at a high level throughout ACSSF. With little ventilation or no ventilation, the microorganisms can obtain the oxygen needed for growth from the outside. Therefore, the microorganism do not need pass into the large amount of sterile oxygen or air; it does not need strong stirring, saving energy. For liquid fermentation, oxygen diffuses into the fermentation liquid from the surface of the liquid film; the speed of this depends on the concentration difference of oxygen between the internal liquid film and external liquid film, viscosity, temperature, and the specific surface area of fermentation broth. Therefore, SmF needs vigorous stirring and continuous ventilation to maintain satisfactory oxygen.

In 1994, Soccol et al. (1994) studied the production of L(+)-lactic acid by *Rhizopus oryzae* in solid medium on sugarcane bagasse impregnated with a nutrient solution containing glucose and $CaCO_3$. A comparative study was undertaken in submerged and solid-state cultures. As shown in Table 6.2, the optimal concentrations in glucose were 120 g/L in liquid culture and 180 g/L in SSF, corresponding to production of L(+)-lactic acid of 93.8 and 137.0 g/L, respectively. The productivity was 1.38 g/L per hour in liquid medium and 1.43 g/L per hour in

Table 6.2 Comparison of physical (aeration and agitation) and chemical (pH) conditions in relation to fungal biomass production by *Rhizopus oryzae* NRRL 395 in different fermentations

Fermentation type	Time (h)	Aeration (v/v/h)	Agitation (rpm)	Glucose (g/L) Initial	Glucose (g/L) Final	L(+)-Lactic acid (g/L)	Yield (g/g)	Productivity (g/lh)
Flask	72	ND	250	120	0	89	0.74	1.24
Fermentor (2 L)	68	75	400	120	0	93.8	0.78	1.38
Solid-state fermentation column	96	4.8	0	180	2.5	137	0.77	1.43

ND not determined lh:liter per hour

solid medium. The highest fermentation yields (78 %) were obtained in liquid fermentation in 2-L fermentors and in SSF on an impregnated support (77 %). The lowest yield (74 %) was obtained in Erlenmeyer flasks. The high level in solid medium was explained by the use of higher substrate concentrations (mainly glucose), which is hardly possible in SmF. The yield was comparable to 15 % higher productivity. In addition, SSF needed little ventilation without stirring; therefore, little energy was consumed

Reduced Wastewater, Shorter Fermentation Period, and Increased Yield

Compared with some SmF, ACSSF can reduce wastewater, shorten the fermentation period, and increase product yield (Weng and Chen 2010).

Xanthan gum-containing solutions, there is stagnant fluid in the fermentor when the gum concentration is high. The oxygen transfer resistance caused by these stagnant zones is a primary restriction in xanthan production. Xanthan gum is an extracellular heteropolysaccharide produced by *Xanthomonas campestris*, which is an aerobic microorganism. Sufficient supply of oxygen is necessary for microbial growth and product formation; thus, strengthening oxygen transfer is necessary to improve the xanthan yield. Zhang and Chen (2010) evaluated the feasibility of SSF on PUF for xanthan production. As shown in Table 6.3, the differences between xanthan final concentrations in static SSF and SmF (shake flask and stirred tank) were statistically insignificant when the initial glucose concentration was low (such as 20 and 40 g/L). When the glucose concentration was increased to 60 and 80 g/L, the post hoc test showed that the final xanthan concentrations obtained by static SSF were significantly higher than those obtained by SMF. In addition, there was a positive correlation between the final gum concentration and the initial glucose concentration in static SSF. These results indicated that SSF on PUF is suitable for xanthan gum production, especially when media with high glucose concentrations (above 40 g/L) are used. The enormous surface area of PUFs provides a stable interface for oxygen transfer; therefore, SSF improves the aeration conditions and promotes the production of xanthan.

Table 6.3 Comparison of xanthan production for submerged fermentation (SmF) and static solid-state fermentation (SSF)

		Initial glucose concentration ($g\,L^{-1}$)	Xanthan production ($g\,L^{-1}$)	Biomass concentration ($g\,L^{-1}$)	Residual glucose ($g\,L^{-1}$)	Xanthan yield on glucose ($g\,g^{-1}$)
SmF	In shake flask	20	11.21 ± 0.71	1.01 ± 0.21	1.45 ± 0.20	0.56
		40	21.85 ± 0.36	1.49 ± 0.29	5.39 ± 0.11	0.55
		60	22.81 ± 0.89	2.59 ± 0.08	21.30 ± 0.05	0.38
		80	21.39 ± 0.74	3.94 ± 0.19	38.10 ± 0.23	0.27
	In stirred bioreactor	40	22.4	1.18	4.23	0.56
		80	23.24	3.66	35.35	0.27
SSF	Static SSF	20	11.79 ± 0.53	0.98 ± 0.24	0	0.59
		40	22.19 ± 0.44	1.81 ± 0.23	0.37 ± 0.12	0.55
		60	30.73 ± 0.98	2.88 ± 0.33	0.46 ± 0.11	0.51
		80	38.65 ± 1.01	3.84 ± 0.37	1.65 ± 0.3	0.48

When the glucose concentrations were 20 and 40 g/L, the fermentation time was 48 h; when glucose concentrations were 60 and 80 g/L, the fermentation time was 96 h respectively

Table 6.4 Average performance of five penicillin solid and liquid fermentations with *Penicillium chrysogenum*

System	P_{max} (U/ml)	Time (h)	Yield (U/mg glucose)	Productivity (U/ml)
SSF	686	49	10.77	2.01
SmF	38.5	166	1.5	0.23

SmF submerged fermentation, *SSF* solid-state fermentation

In 1988, Barrios-Gonzalez et al. (1988) studied the production of penicillin by nonsterile SSF on bagasse impregnated with culture medium. As shown in Table 6.4, it can be observed that average production by SSF was 17 times higher than obtained by SmF, and this was achieved in one-third of the time. An approximate calculation revealed that the efficiency of sugar utilization to produce the antibiotic (yield) was seven times higher in SSF. Finally, SSF showed 8.7 times greater volumetric productivity.

Improved Product Quality

Spores produced by aerial mycelia of *Trichoderma harzianum* PI, a potential biocontrol agent, showed both higher ultraviolet (UV) resistance and longer viability after storage than those produced within liquid media ("submerged" spores) (Munoz et al. 1995). Aerial spores were produced in clusters, had a thick outer wall, and had few organelles. Trehalose content was significantly lower than in submerged spores. Conversely, submerged spores were mostly collapsed, not clustered, and larger than aerial spores. They had many cytoplasmic organelles and a thinner outer wall. These spores were hydrophilic; aerial ones were highly hydrophobic. On analysis, the latter was related to the presence of a single major low molecular mass protein (<14 kDa). This protein was nearly absent in extracts from walls of submerged spores but was found in the extracellular medium. Pectinolytic enzymes produced by SSF were more thermostable than those obtained by SmF. Pectinases produced by SSF had more stable properties in relation to extreme pH values than those produced by SmF because the pectinases produced by SSF had broader pH profiles of enzyme activities and were denatured more slowly at pH values other than optimal as compared to pectinases produced by SmF (Acuña-Argüelles et al. 1995).

6.2 Material Properties of ACSSF

6.2.1 Material Characteristics of ACSSF

The inert carrier not only provides an interface to support the growth of microorganisms but also, more importantly, provides a suitable stable and uniform environment for the growth and metabolism of microorganisms. This environment is conducive to the retention of water and nutrients, to the enhancement of mass and heat transfer, and to ventilation. The substrate used in an SSF process must have properties such as the following:

1. The substrate needs structure homogeneity, regular particle form, and uniform size.
2. There is a need for high porosity and large specific surface area.
3. The carrier is highly water absorbent and has a high water retention capacity. Dehydration or hygroscopicity has no effect on the geometry of carriers.
4. Carriers have low density and good mechanical resistance and a certain strength to maintain their shape against the shearing force in the mixing process without destroying the support and to protect the microorganisms from shear force impact.
5. The structure is not modified by microorganisms or metabolites in microbial growth.
6. There are nontoxic side effects on microbial growth and product production.
7. There are many sources, and they are easy to utilized.

6.2.2 Types of Inert Carrier Materials

At present, the inert carriers used in solid-state fermentation can be divided into two types. One type is natural materials, such as lignocelluloses (sugarcane bagasse (Pandey et al. 2000), cassava residue, cocoa coconut, and agricultural residues) and rock (vermiculite (Meyrath 1965), perlite). Another type is synthetic carriers, such as ion exchange resins (Richard et al. 1990), polystyrene (Nagendra Prabhu and

Table 6.5 Inert carriers used in solid-state fermentation

Inert carriers	Microorganism	Products	Authors (year)
Bagasse	*Aspergillus niger*	Citric acid	Cahn (1935)
	Aspergillus niger	–	Oriol et al. (1988)
	Candida utilis	–	Christen et al. (1993)
	Claviceps purpurea	Ergot alkaloids	Hernandez et al. (1993)
	Penicillium chrysogenum	Penicillin	Barrios-Gonzalez et al. (1993)
	Rhizopus oryzae	L(+)-Lactic acid	Soccol et al. (1994)
	Aspergillus niger	Pectinase	Huerta et al. (1994)
	Brevibacterium sp.	L-Glutamic acid	Nampoothiri and Pandey (1996)
	Gibberella fujikuroi	Gibberellic acid	Tomasini et al. (1997)
Hemp	*Coniothyrium minitans*	Spore	Weber et al. (1999))
Vermiculite	*Aspergillus niger*	Amylase	Meyrath (1965)
Ion exchange resin	*Aspergillus niger*	–	Richard et al. (1990)
	Candida utilis	–	Christen et al. (1993)
Polystyrene	*Vibrio costicola*	L-Glutaminase	Nagendra Prabhu and Chandrasekaran (1995, 1997)
	Aspergillus ficuum, Rhizopus oligosporus	Phytase	Gautam et al. (2002)
Polyurethane foam	*Penicillium citrinum*	Nuclease P1	Zhu et al. (1994)
	Aspergillus oryzae	Alkaline protease	Ozawa et al. (1996)
	Mucor bacilliformis	Spore	Lareo et al. (2006)
	Bacillus pumilus	Alkaline protease	Chen et al. (2006)
	Trichoderma veride	Chitinases	Marin-Cervantes et al. (2008)
	Aspergillus terreus	Lovastatin	Baños et al. (2009)
	Xanthomonas campestris	Xanthan	Zhang and Chen (2010)
	Acetobacter xylinum	Bacterial cellulose	Weng and Chen (2010)
	Trichoderma veride	Cellulase	Hu et al. (2011)
SiO_2 xerogel	*Aspergillus niger Phanerochaete chrysosporium*	–	Peralta-Perez et al. (2001)

Chandrasekaran 1995), PUF (Zhang and Chen 2010), and silica (SiO_2) xerogel (Peralta-Perez et al. 2001). Lignocellulosic materials are usually low in nutritional value and cannot be easily used by microbes compared with the culture medium. Synthetic materials are stable chemically and cannot be degraded by microorganisms. The selection of carriers depends on the types of microorganism and the fermentors. Tomasini et al. (1997) compared the virtues and limitations of gibberellin production by different SSF systems: cassava flour, sugarcane bagasse, and low-density polyurethane. SSF on bagasse showed excellent growth but presented gibberellin extraction problems. Very low production and growth were observed in SSF with low-density polyurethane as an inert support. SSF on cassava flour showed high production (250 mg/kg of dry solid medium) in a very short time (36 h). Table 6.5 summarizes the inert carriers used in SSF.

6.2.2.1 Natural Carriers

Agricultural by-products are abundant and easy to handle; therefore, when they are used as inert carriers, the cost is low. But, compared with synthetic carriers, the substrates in SSF, which are usually by-products of agroindustry, have some disadvantages, such as excessive substrate layer thickness, low porosity, inadequate internal structures that disturb aeration and heat removal, and inefficient nutrient uptake. In most of these systems, it is impossible to separate residual solid substrate from biomass. Agricultural by-products are nutrition-based carriers relative to other microorganisms, and the system is easy to contaminate.

Solid-state fermentation produces a highly concentrated enzymatic extract with very active chitinases. Readily available sugarcane bagasse was used as a support for fungal growth in fermentation process; however, such SSF showed problems of substrate sterilization, temperature and pH control, as well contamination of the system, which was mainly detected at long process times. Also, the enzymatic extracts carried impurities from the organic support, which involved a more difficult purification process (Marin-Cervantes et al. 2008).

When sugarcane bagasse was used as an inert support for SSF, *Gibberella fujikuroi* displayed excellent growth. However, only traces of gibberellin were extracted and quantified. To see if gibberellin was being efficiently extracted from the solid medium, 300 mg/kg of exogenous gibberellin were absorbed in fermented and nonfermented solid media. Extraction efficiencies of 15–18 % were obtained using water or ethyl acetate and centrifugation for 20 min (Tomasini et al. 1997).

6.2.2.2 Synthetic Carriers

Polyurethane foam has been used as a suitable support for SSF because it presents high porosity, low density, and relatively high water absorption. PUFs have an adequate pore size, which provides a satisfactory environment for fungal growth. Moreover, the biomass and support are easily separated into an enzymatic extract with few impurities, which facilitates further purification. The absorbent capacity can be up to 20 ml/g. There are many small pores (approximately 700 μm in diameter) supported by the polyurethane skeleton in the PUF. If the pores were modeled as spheres, the specific surface area of the PUF was about $4.5 \times 10^3\ m^2/m^3$. Accordingly, the thickness of the broth distributed on the inner surface of the PUF was about 0.1 mm, given that the liquid-to-solid ratio was 10 ml/g PUF cube. In SmF production of xanthan, the primary restriction was oxygen transfer resistance caused by stagnant regions; when SSF was performed, the broth was dispersed over many stable surfaces of the PUF cubes. The large inner surface and the thin broth film were helpful for oxygen transfer from the air to the liquid. Moreover, the gas phase above the inner surface of the PUF was continuous, which contributed to oxygen transfer from the outside to the inside. Therefore,

PUF is an ideal carrier material. The aperture of PUF cannot be too large, which can weaken capillary action, decrease water-holding capacity, and make an uneven distribution of water, which all result in the fermentation not being carried out smoothly (Weber et al. 1999).

6.2.3 *Inert Support Pretreatment*

6.2.3.1 Plant-Based Carriers

For sugarcane bagasse (Oriol et al. 1988; Nampoothiri and Pandey 1996; Soccol et al. 1994), milled sugarcane bagasse was sieved to obtain particles 0.45–3.00 mm in size. After washing twice with deionized water (using ten times water volume each time) to remove soluble sugars and other soluble constituents adhering to the particles, the bagasse was dried in a hot air oven at 60 °C for 30 h and stored at room temperature in sealed polythene bags. For each set of the experiment, an adequate amount of the bagasse was autoclaved at 121 °C for 30 min.

6.2.3.2 Ion Exchange Resin

Ion exchange resin (Amberlite IRA 900) is an anionic ion exchange resin with a macroreticular structure. Its main characteristics are as follows: spheric form with a mean particle diameter of 530 μm; true density of 1.07 g/cm^3; an apparent density of 0.672 g/cm^3; maximum water absorption of 0.59 g H_2O/g wet resin. The support was received with a water content of 59 %. It was suspended in water in a proportion of 1 g to 4 ml, and the pH was adjusted to 5.5 with 1 mol/L HCl. It was then filtered, and a 0.25 mol/L phosphate buffer solution was added in a proportion of 1 g to 2 ml. The support was further dried under the same conditions.

6.2.3.3 Polyurethane Foam

Marin-Cervantes et al. (2008) studied the effect of the PUF shape as supports on the growth of *Lecanicillium lecanii* and the production of chitinases. Finely minced PUF (MPUF) and roughly cut PUF (CPUF) were used as inert support for the growth and chitinase production of *Lecanicillium lecanii* by solid substrate fermentation.

The MPUF and CPUF (approximate density of 0.018 g/cm^3) were evaluated on their properties to retain the culture media because of their capillarity and porosity. The water storage capacity of MPUF and CPUF were determined to be 19.8 and 21.4 ml/g, respectively (Table 6.6). CPUF presented a honeycomb of polygonal cells of the polymer; MPUF, for which cells of the polymer were destroyed during the mincing process, was observed as irregular forms. The averages of the interstitial

Table 6.6 Characteristics of polyurethane foams (PUFs) used in this study

PUF type	Porosity (φ)	Maximum water absorption (ml/g)	Interstitial spacing (μm^2)
Minced PUF	0.912 ± 0.01	19.8	99.48 ± 6.83
Cut PUF	0.955 ± 0.05	21.4	73.77 ± 4.59

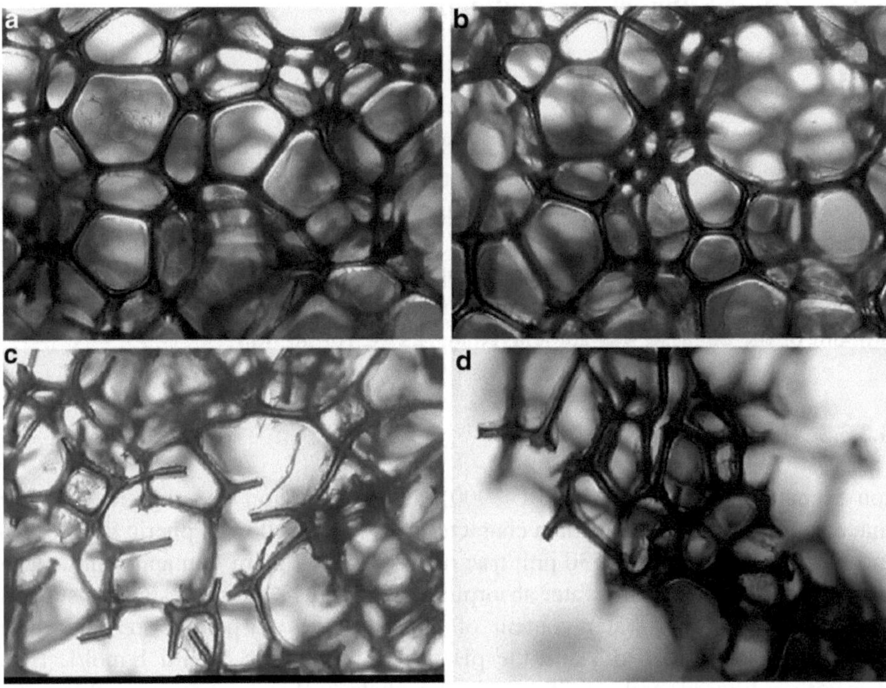

Fig. 6.1 Stereomicroscopic micrographs of polyurethane foams (*PUF*s) (×4): (**a**) and (**b**) cut PUF; (**c**) and (**d**) minced PUF

areas were different for CPUF and MPUF; the former was 73.77 mm^2, and the latter was 99.48 mm^2; the porosity of the CPUF was 0.955 and for the MPUF was 0.912. According to these data, MPUF presented a decrease in capillarity because of the decreased porosity, which had a significant effect on the absorption of the culture media compared with CPUF (Table 6.6). The CPUF was subjected to less mechanical force, causing minor destruction of pores, than MPUF; therefore, it presented higher porosity than the MPUF (Fig. 6.1). The effect of surface tension (capillarity) is to draw liquid back into the borders. Because of the existence of highly curved borders in CPUF, the capillarity pressure was higher than in the MPUF thin films (Fig. 6.1).

The distribution of the fungal cells in MPUF with NG media (Czapeck media with added colloidal chitin) was loose and disperse, whereas in CPUF the growth was in loosely packed hyphae, as shown in Fig. 6.2a, c. The fungal growths in

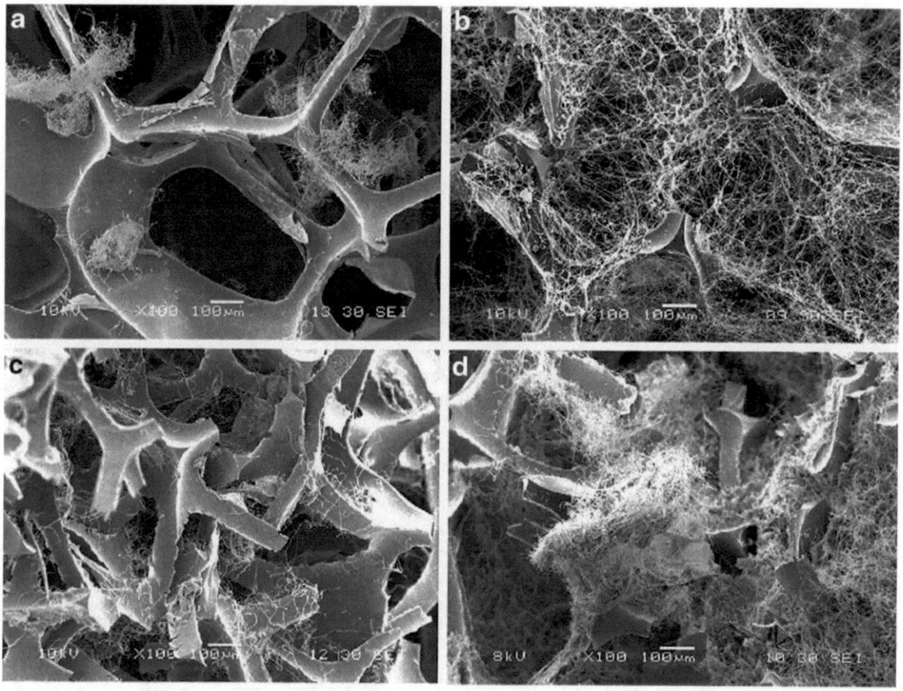

Fig. 6.2 Scanning electron micrographs of *Lecanicillium lecanii* ATCC 26854 growth with colloidal chitin at 85 % moisture content on cut polyurethane foam (*PUF*): (**a**) without addition of glucose, (**b**) with addition of glucose; and on minced PUF: (**c**) without addition of glucose, (**d**) with addition of glucose

CPUF covered the surface by formation of a net of intertwined hyphae in PG (Czapeck media with added colloidal chitin and glucose) media; in MPUF, growing *L. lecanii* formed aggregates in the ends of the branches made from the sectioned cells of the PUF (Fig. 6.2b, d). The formation of filamentous mycelia in PG media might be caused by media composition, which contained glucose, which favors this fungal morphology. However, it is remarkable that mycelial aggregation was observed in MPUF and may be because of loss of water during SSF.

6.2.4 Growth Characteristics of Microorganisms on Inert Carriers

In 2001, Peralta-Perez et al. (2001) prepared SiO_2 xerogel by the hydrolysis-condensation reactions of the sol–gel method. Silica xerogel was used as a support for the growth of two filamentous fungi: *Aspergillus niger* ATCC 9642 and *Phanerochaete chrysosporium* A594. In both cases, apparent abundant mycelia growth was observed.

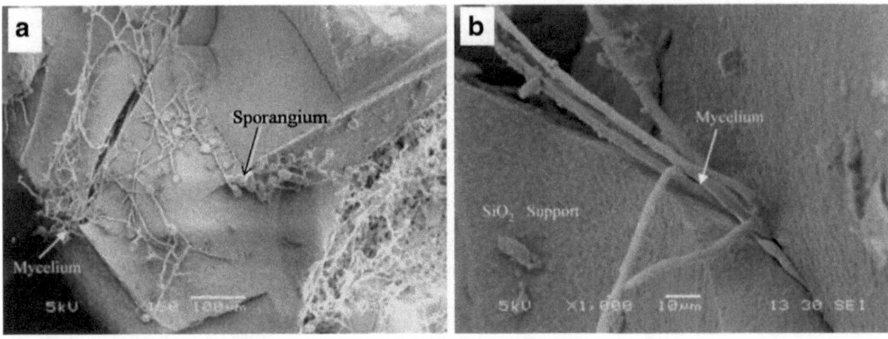

Fig. 6.3 Scanning electron microscopy of *Aspergillus niger* ATCC9642 mycelia growing on xerogel at different fermentation times: (**a**) 146 h (original magnification ×160); (**b**) 146 h (original magnification ×1,000)

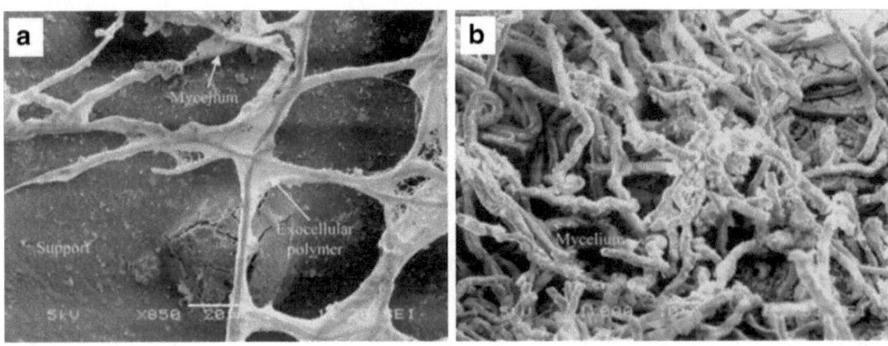

Fig. 6.4 Scanning electron microscopy of *Penicillium chrysosporium* A594 mycelia growing on xerogel at different fermentation times: (**a**) 146 h (original magnification ×850); (**b**) 146 h (original magnification ×1,000)

The xerogel obtained had a specific surface area of 759 m^2/g and an average pore size of 3.0 nm. The internal surface was wrinkled, and although no pores larger than 3.0 nm were seen, a large crack was detected between two separate fragments of the gel, where the fungus could grow using the internal walls preferentially. The water retention capacity was normally about 40–50 %.

Scanning electron microscopy was used to study the colonization of the xerogel by both fungi. Figure 6.3 shows the growth of *A. niger* at 146 h. The growth took place on the surface of the xerogel and on the large crevices or cracks that originated by fracture of the xerogel, whose internal walls were clearly suitable for colonization. The typical width of the hyphae was approximately 3 μm, and that of the sporangium observed in Fig. 6.3a was 16 μm.

Figure 6.4 shows the presence of an extracellular polymer surrounding the mycelia. In the natural environments of fungi, the function of the polymer is to

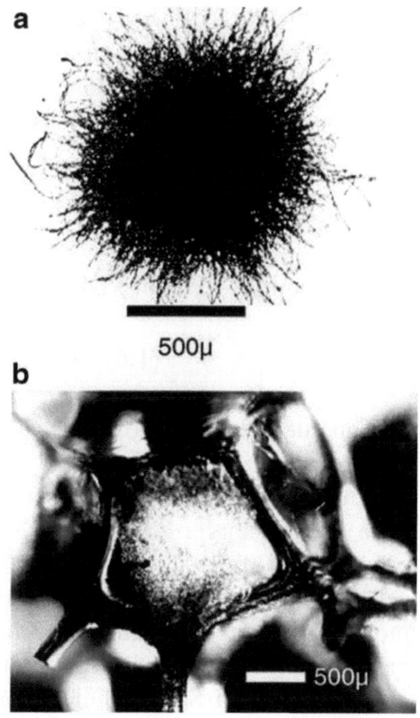

Fig. 6.5 (**a**) Micrograph of *Aspergillus niger* C28B25 growing under submerged culture conditions with 100 g of sucrose per liter after 24 h of incubation (×50). (**b**) Micrograph of *A. niger* C28B25 growing in polyurethane foam under solid-state fermentation conditions after 6 h of culture (×100)

wrap the wood covalently to start its depolymerization. This extracellular polymer apparently has an important role in the transport of wood depolymerizing enzymes. The occurrence of the structures previously observed in this fungus on a natural substrate confirmed that using silica xerogel as a support allows the study of this kind of microorganism, with the great advantage of using a well-defined medium that enables study and analysis without interference. Thus, SiO_2 xerogels show great potential as supports for microbial growth. Furthermore, their use provides an improved understanding of the phenomena involved during SSF.

In 2003, Viniegra-Gonzalez et al. (2003) compared microscopic and geometric characteristics of SSF and SmF. Cultures of *A. niger* developed by the SmF technique usually produce quasi-spherical aggregates called pellets, as indicated in Fig. 6.5a. Such aggregates have a diameter in the millimeter range (i.e., 3,000 μm) with two well-defined zones: an outer layer made of loose mycelia, which dyes lightly with methylene blue and has an approximate thickness in the range of 3–10 cm (i.e., $h\text{P} = 200$ μm), and a central core made of tightly woven mycelia, which dyes strongly with methylene blue.

The fine structure of PUF is made of a honeycomb of impermeable polymer pillars approximately 300 μm thick and forming polyhedral cells with a diameter close to 1,000 μm (Fig. 6.5b). Added liquid broth (about 20 cm^3/g PUF) is distributed evenly by capillarity in a thin layer having a thickness $h\text{W} = 60 \pm 5$ μm. When inoculated with *A. niger* spores, a population of mycelial aggregates was formed that grew inside the water layer and spread out as mycelial slabs having an average thickness of $hX = 300 \pm 50\ \mu m$. Average polymer density of light PUF is close to 15 g/L. Using a simple geometrical model in which volume is equal to the surface divided by the thickness of the layer, it is possible to estimate an area-to-volume ratio of a completely dispersed liquid broth as follows:

$$\frac{A}{V} \approx \frac{1}{h_W} \approx 1.67 \times 10^2\ \text{cm}^{-1}$$

For example, 2.5 g of PUF loaded with 50 cm^3 of liquid broth will have an area

$$A = \left(1.67\ 10^2\ \text{cm}^{-1}\right)\left(50\ \text{cm}^3\right) = 8.35\ 10^3\ \text{cm}^2$$

This surface area exposed to passive gas exchanges is much larger than the usual surface area of the air-to-liquid interphase in a 250-cm^3 shake flask filled with 50 cm^3 of broth (~20 cm^2). This geometrical consideration shows that SSF cultures have a much larger air-to-liquid interphase than conventional SmF cultures, for example, 400 times higher. Thus, SSF cultures are grown initially in thin slabs of cell aggregates having a very large surface area for gas exchange, whereas SmF mold cultures are grown in pellets having large diameters and small surface area for gas exchange. For single-cell cultures, such as yeast suspensions, it has been observed that an SSF technique on PUF produces large cell aggregates (nearly 10^2 cells) spread out in the thin layers of liquid broth having a large *A*/*V* ratio. SmF cultures (shake flasks), however, produce small aggregates (~10 cells) with a much smaller *A*/*V* ratio, as indicated previously here (unpublished results). Such geometrical differences between SSF and SmF cultures may account for important physiological differences encountered experimentally and discussed in the following material in terms of processes limited by diffusion of gases, substrates, and products.

As for SmF mold cultures in the pellet form, it is well known that the active aerobic layer has a thickness $h \approx 10^2\ \mu m$, which is in the same order of magnitude as the measured thickness of the experimental loose layer in Fig. 6.5a. Hence, whenever the diameter of the pellet becomes on the order of $10^3\ \mu m$, most of the pellet will be anoxic with a rather low level of sugar inside the pellet core. When considering a mold culture made of disperse mycelia, most of those cells would be in full contact with a similar concentration of sugar and subjected to strong catabolite repression. For SSF cultures, geometrical and physical restrictions are quite different from those observed in SmF cultures. Oxygen diffusion is perpendicular to large cell aggregates, which are dispersed in a thin layer of broth. In such a system, sugar diffusion is horizontal along the thin layer (Fig. 6.5b). It is therefore possible to create concentration gradients of sugars within large cell aggregates because there is no mixing within the static layer of liquid broth, itself finely dispersed on the solid substrate.

6.3 ACSSF Techniques and Fermentors

6.3.1 *Packed Bed Bioreactor*

In the SSF process, problems, such as aeration and heat accumulation, have been partially solved by using thin layers or agitated bioreactors. However, not enough effort has been devoted to the design of SSF bioreactors and their control systems.

A semipilot, plant-scale, packed bed bioreactor (130 L, 10–25 kg dry matter) implemented with temperature and air humidity control was used for pectinase production with *Aspergillus niger* using absorbed substrate fermentation techniques (Huerta et al. 1994). Pectinolytic enzyme activity and relative CO_2 production were used as indicators of metabolic activity. Absorbed substrate fermentation is an efficient process for pectinase production and is also an interesting model because the culture medium, water, nutrients, and specific inducers can be designed at the desired concentrations.

The reactor consists of a rectangular stainless steel box ($50 \times 40 \times 65$ cm) with ten heat exchanger plates spaced at 5-cm intervals. A distribution device for aeration was placed at the bottom of the box (Fig. 6.6).

The controlled variables were temperature of the medium and temperature and humidity of the inlet air. To control the medium temperature, a water line with a system of solenoid valves and heat exchanger plates was used. A thermocouple was located in the medium, operating the valves through the heating or cooling systems. The temperature and humidity of the air were controlled using a humidifying packed tower. CO_2 in the gas phase was collected into an alkaline solution (pH 11) by bubbling the exit gases and automatically titrated with 4 M NaOH solution. A qualitative determination of CO_2 production was obtained from the consumption rate of NaOH solution divided by the total NaOH solution consumed during the fermentation process.

Temperature of the medium was controlled at 35 °C, and moisture in the medium was observed to be constant at 62 %. This low water content, the low pH values (2–3) during the culture, and the heavy initial inoculum kept the culture from becoming contaminated. Heat and mass transfer problems were eliminated by using a material such as sugarcane bagasse with a high porosity.

6.3.2 *Repeated Batch Bioreactor*

The process performance of SSF is less than that of SmF because of the following problems: process monitoring, scaling up, quality control, and regulation of culture condition. Culture condition is hard to control because it depends on medium components, such as wheat bran, rice, and so on. Furthermore, separation of product is complicated, and it is hard to reuse the fungi for enzyme production. To

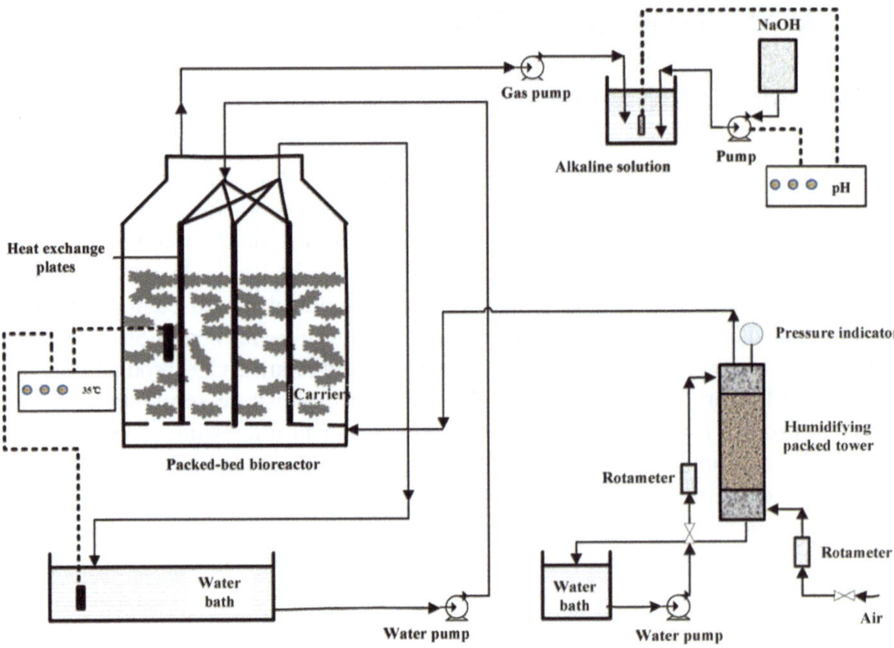

Fig. 6.6 Diagram of the equipment used for the adsorbed carrier solid-state fermentation (*ACSSF*) process (Huerta et al. 1994)

overcome these problems, modified SSF methods using asbestos, ion exchange resin, or urethane foam as carriers of liquid medium have been developed.

In 1996, Ozawa et al. (1996) immobilized *A. oryzae* on a PUF carrier; a novel SSF system was developed to repeatedly produce alkaline protease (Fig. 6.7). After 36- or 48-h cultivation, repeated batch production of alkaline protease was carried out by exchanging part of the culture broth with fresh medium every 12 h. These results indicated that the repeated batch production of alkaline protease could be done by maintaining alkaline protease activity at a high value and recovering a part of the enzyme produced periodically.

The novel fermentation method was compared with wheat bran fermentation (Table 6.7). SmF was also tested, but alkaline protease activity had not been detected in the culture broth; in the batch production of the novel method, alkaline protease productivity was 3.9 times higher than that of wheat bran fermentation. Applying the repeated batch process, by feeding soluble starch solution and soluble starch and polypeptone solution, 3,100 and 2,900 U/L bulk volume/h of alkaline protease were obtained at 168 and 300 h of cultivation time, respectively. These values were smaller than that obtained in the batch production. These productivities of repeated batch productions can be valued higher than that of batch production by including preculture time and fermentor cleaning time to the fermentation process.

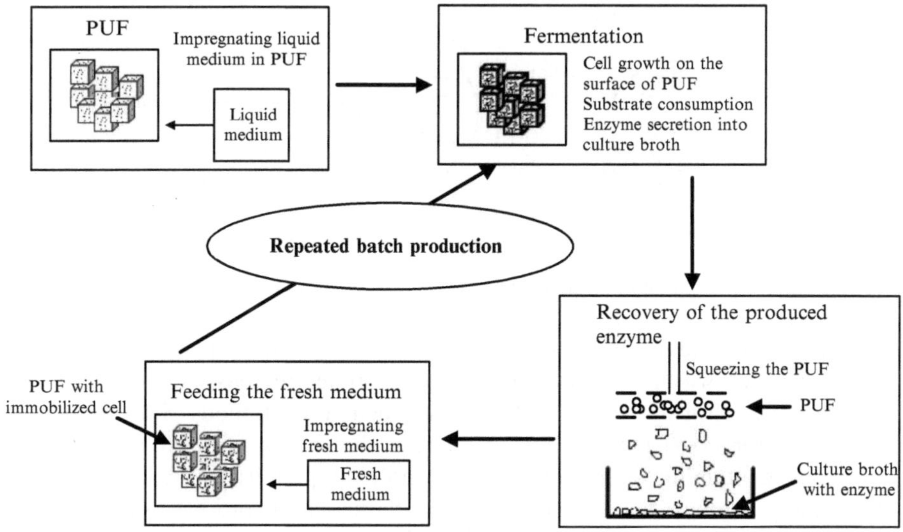

Fig. 6.7 Schematic diagram of repeated batch production of enzyme by solid-state fermentation using polyurethane foam (*PUF*) (Ozawa et al. 1996)

Table 6.7 Comparison of alkaline protease productivity in several fermentation methods

Fermentation type	Time (d)	P_{max} (U/L bulk volume)	Productivity (U/L bulk volume/h)
SSF using wheat bran	120	110,000	900
SSF using PUF (batch production)	72	250,000	3,500
Repeated batch production (1)	168	520,000	3,400
Repeated batch production (2)	300	870,000	3,100

Note: Repeated batch production: (1) Solids starch 100 g/L solution was used as feeding medium. (2) Soluble starch 100 g/L and polypeptone 30 g/L solution were used as feeding medium
PUF polyurethane foam, *SSF* solid-state fermentation

The production time of alkaline protease was prolonged by the addition of polypeptone in the feeding medium. The maximum activity of alkaline protease achieved was 870,000 U/L bulk volume at 300 h by adding 30 g/L polypeptone to the feeding medium.

6.3.3 *Continuous Bioreactor*

When crop products are used as substrate for SSF, because of poor mobility, it is difficult to input and discharge substrate, and contamination is easy in the fermentation process; therefore, it is difficult to achieve continuous SSF. Currently, almost all the SSF processes involve batch fermentation. The batch fermentation procedure

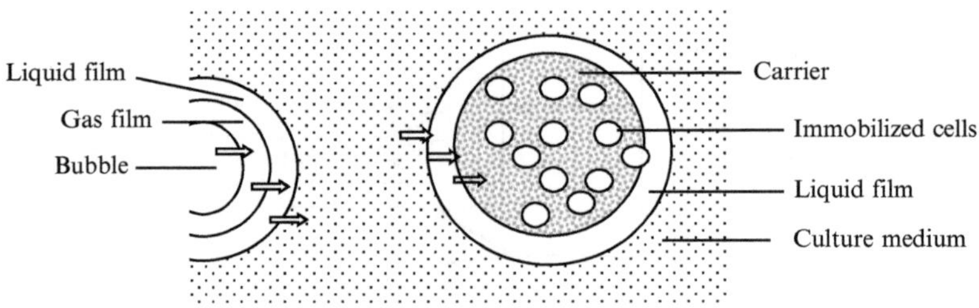

Fig. 6.8 Schematic diagram of gas passing from the gas phase to the immobilized cells

is cumbersome and labor intensive, which leads to lower operating efficiency. A conveyor belt was usually used to convey substrate previously in the continuous SSF process. Hrubant et al. (1989) developed a continuous SSF fermentor. The fermentor consisted of three chambers aligned axially and separated by bulkheads containing a centrally located hole to permit passage of substrate sequentially through the chambers. In the fermentation process, substrate was input at one end of the fermentor, and product was discharged at the other end. This method reduced the breaking off, promoted the continuity of fermentation, and had a certain value. But, in the fermentation process, seed must be inoculated into substrate continuously; therefore, this method cannot be regarded as true continuous SSF.

A true continuous SSF system needs only one inoculation at the beginning of fermentation and can be a continuous feed and outfeed in the fermentation process. The fermentation system remains stable for the whole fermentation process, and input and output maintain a balance. In the conventional SSF process, to meet the fermentation process and not require repeated inoculation, the substrates must be fully mixed with stirring to provide microbial cells or microbial spores to the fresh materials that are continuously added. But, shear force caused by agitation often reduces product yield and spore formation.

When inert carrier-absorbing culture medium is used in SSF, the inert carrier is solid phase; it cannot be utilized by microorganisms. Microorganisms grow on the surface of an inert carrier using absorbed culture medium. This fermentation method provides a chance for continuous SSF. Using inert carrier as the solid phase, the fermentation broth can be replenished continually; after fermentation, the fermentation broth can be discharged continuously, which achieves the continuous SSF process.

There are differences and similarities between ACSSF and immobilized cell fermentation. Cell immobilization refers to positioning cells in a certain space to maintain its catalytic activity; an immobilized cell can be repeated and continuous used. In aerobic immobilized cell fermentation, oxygen is provided by ventilation. As shown in Fig. 6.8, gas passing from the gas phase to the immobilized cells is subjected to a series of rate-limiting steps. For example, after oxygen enters the liquid body, it needs to pass the immobilization carrier first before reaching the immobilized cells. Currently, because it is difficult to provide sufficient oxygen to

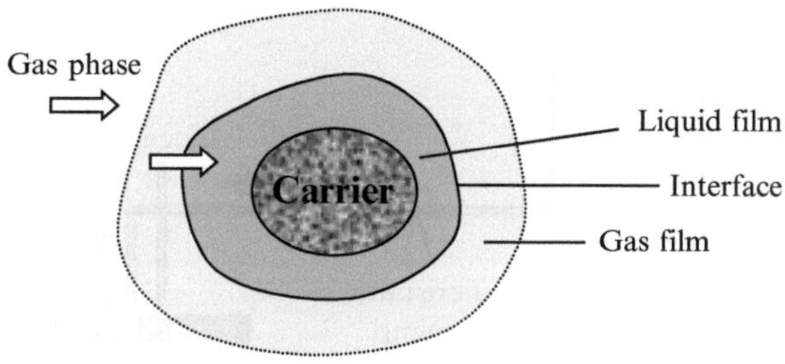

Fig. 6.9 Schematic diagram of adsorbed carrier solid-state fermentation (*ACSSF*) oxygen transfer

the SmF of immobilized cells, immobilized cells are used for the anaerobic fermentation process, such as ethanol and lactic acid fermentation. For ACSSF, air is the continuous phase. Figure 6.9 shows the oxygen transfer steps in ACSSF. By comparison, gas transfer rate-limiting steps in immobilized cell fermentation are more than in ACSSF. In ACSSF, the inert carrier not only has no limitation in the oxygen transfer but also provides a huge and stable specific surface area for the growth of microorganisms, which greatly facilitates the effective delivery of the gas.

Adsorbed carrier solid-state fermentation also has similarities with immobilized cell fermentation. (1) Both are continuous processes. Immobilized cell fermentation uses the metabolic function of the microorganism cells repeatedly and improves the efficiency of the fermentation. In ACSSF, the fermentation broth is input continuously and passes the carrier to complete the fermentation process. (2) In the ACSSF continuous fermentation process, a portion of the microorganisms will be adsorbed on the surface of the carrier for inoculation for fresh fermentation broth. It can be said that absorbing the microorganisms on the inert carrier is also a cell immobilization process. (3) In the ACSSF continuous fermentation process, the adsorption process can be regarded as an inert carrier immobilizing fermentation liquid; microorganisms grow in the adsorbed liquid film. Therefore, to a certain extent, ACSSF can be regarded as immobilized cell fermentation.

Currently, because it is not able to achieve an efficient continuous fermentation process and it is less efficient compared with continuous liquid fermentation, ACSSF has no practical applications. Chen (2004) designed a continuous bioreactor. It can use the advantages of ACSSF and promote its efficiency. A schematic diagram of the continuous bioreactor is shown in Fig. 6.10. The SSF fermentor was made of stainless steel materials and filled with PUF as the support. PUF has a porous honeycomb structure with excellent softness, elasticity, and water-absorbing and water retention capacities.

Prior to fermentation, a certain amount of seed broth was added to the bioreactor first. Seed broth was dispersed uniformly on top of the carrier and then penetrated into the carrier and adsorbed on the surface of the carrier. The inoculation was completed. The fermentation medium was added to the bioreactor uniformly through the peristaltic pump. A filtration plate with holes was placed on top of

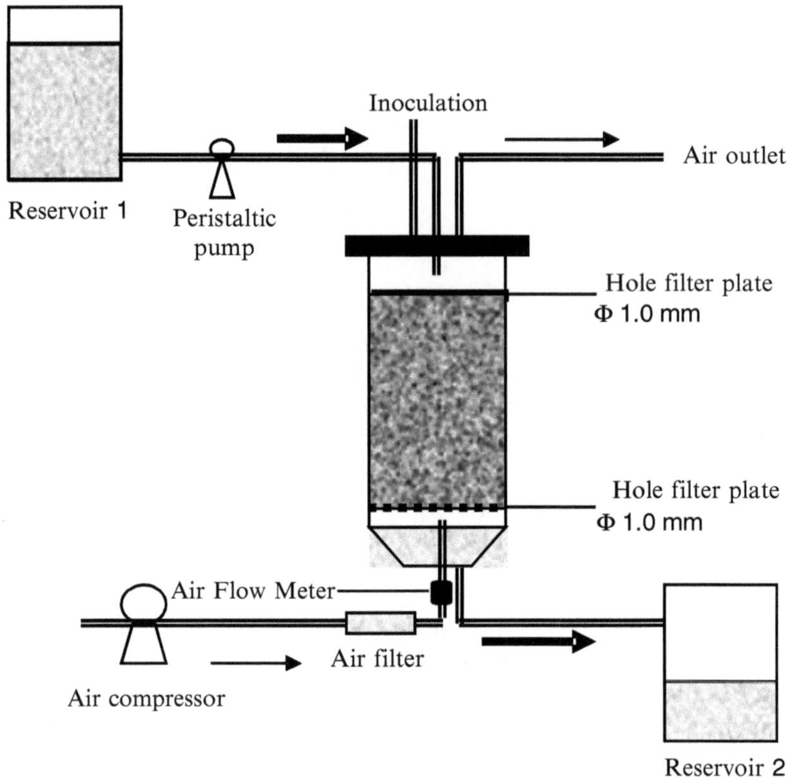

Fig. 6.10 Schematic diagram of continuous bioreactor

the carrier. Fresh medium from the top of the bioreactor spread on the surface of the carrier uniformly through the filtration plate. Under the action of permeation as well as the gravity, the fermentation medium entered the inert carrier and was adsorbed on its surface. With the gradual increase of the added medium, the fermentation medium seepage was top-down. Meanwhile, sterile air passed into the bioreactor from the bottom using the ventilation apparatus. Air flowed upward through the pores of the carrier and finally was discharged from the outlet. In the fermentation process, the fermentation broth gradually flowed to the bottom of the bioreactor, flowed out of the bioreactor, and entered the reservoir. In the fermentation process, fermentation broth was continuously added to the carrier; microbes grew on the surface of the carrier with absorbed medium. The flowing air provided oxygen for the growth of microorganisms, and the whole process was continuous.

In continuous ACSSF, the relationship of the carrier with the fermentation broth and air is shown in Fig. 6.11. The fermentation broth flowed downstream along with carrier, and the air flowed upstream through the pores of the carrier. The inert carrier was the solid phase, and fermentation broth and air were the mobile phase.

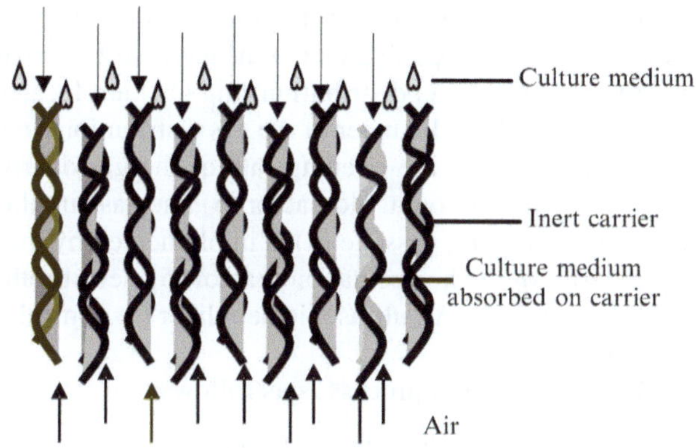

Fig. 6.11 Microscopic structure schematic diagram of adsorbed carrier solid-state fermentation (*ACSSF*) process

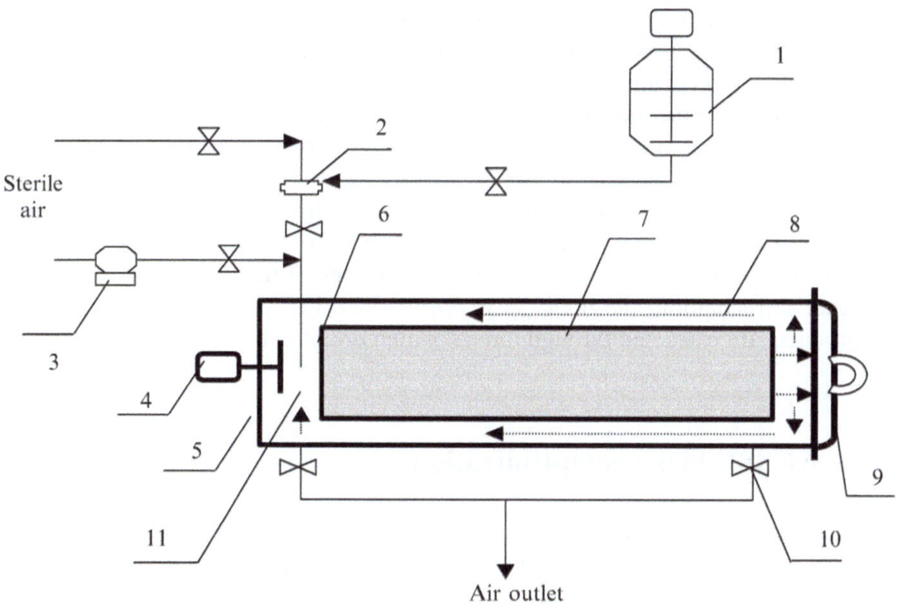

Fig. 6.12 Schematic diagram of horizontal bioreactor. *1* Seed fermentor. *2* Venturi. *3* Ultrasonic nebulizer. *4* Circulating ventilator. *5* Horizontal bioreactor. *6* Gas distributor. *7* Cylinder with porous carrier. *8* Gas circulating passage. *9* Door. *10* Valve. *11* Gas distribution ring

6.3.4 Horizontal Bioreactor

Currently, SSF is difficult to apply to products with high purity. To overcome this disadvantage of SSF, Chen et al. designed a bioreactor with porous supports. As shown in Fig. 6.12, this fermentation system consists of a seed fermentor (1), venturi (2), ultrasonic nebulizer (3), circulating ventilator (4), horizontal bioreactor (5), gas distributor (6), cylinder with porous carrier (7), gas-circulating passage (8), door (9), valve (10), and gas distribution ring (11).

The diameter of the gas distributor (6) is eight-tenths that of the horizontal bioreactor (5); the opening rate of the gas distributor (6) is 50–70 %. The circulating ventilator (4) and gas distribution ring (11) are located at an end of the horizontal bioreactor (5). The diameter of the gas distribution ring (11) is two-thirds that of the gas distributor (6); the opening rate of the gas distributor (6) is 20–50 %. In the interior of the horizontal bioreactor (5), the gas circulation between the cylinder (7) and gas-circulating passage (8) is implemented by the circulating ventilator (4). In the exterior of the horizontal bioreactor (5), seed broth is inoculated into the carrier by the venturi (2). An ultrasonic nebulizer (3) controls the humidity of the horizontal bioreactor (5).

The fermentation process is as follows:

1. Porous carriers that absorb fermentation medium, such as PUF, pulp, zeolite, vermiculite, and the like, are located in the cylinder (7) of the horizontal bioreactor (5). The carrier-fermentation medium ratio is 1:2; sterilization occurs in situ at 120 °C for 20–30 min. The circulating ventilator (4) is turned on. The horizontal bioreactor (5) is cooled to 40 °C. Seed broth is inoculated by the venturi (2), and the inoculation volume is 5 %. The temperature of the horizontal bioreactor (5) is controlled at 30 °C; the humidity is controlled at 85–90 %. Fermentation is started.
2. In the fermentation process, the humidity is controlled by the ultrasonic nebulizer. The flow rate of sterilized air is 0.8–1.2 vvm. The air volume entering the bioreactor per minute is 0.8–1.2 times the volume of the bioreactor.
3. After fermentation, the door (9) is opened and the cylinder is taken out. The products are extracted through three countercurrent extractions.

6.4 ACSSF Process Optimization

The factors affecting the kinetics of ACSSF include the carrier particle size, initial moisture, initial water activity, pH, temperature, stirring, ventilation, strain ages, amount of inoculation, and so on.

6.4.1 Effect of Moisture Content

The three main ingredients of the medium are the support, the substrate, and water. The initial moisture content could be set by varying one ingredient and water independently of the third.

Oriol et al. (1988) studied the growth kinetics of *A. niger* growing on the bagasse absorbing glucose and inorganic salt. Two series of experiments were run; in the first series, the support/water ratio was changed, and in the second, the substrate/water ratio was changed. Cultures with a constant substrate/water ratio (corresponding to a glucose concentration in the liquid phase of 17 % w/v) and increasing support/ water ratios (corresponding to initial moisture contents ranging from 40 to 75 %) showed little variation of growth kinetics, resulting in similar specific growth rates and fermentation times (Table 6.8). The amount of moisture content had no effect on the mass and heat transfer; therefore, it had no effect on the fermentation of *A. niger*.

For the production of glutamic acid using ACSSF by *Brevibacterium* sp. (Nampoothiri and Pandey 1996), five different initial moisture levels, as shown in Table 6.9, were set in the substrate. The total substrate in each flask was always 20 g (wet weight). Although bacterial growth, as observed by glutamic acid production, occurred even at the lowest experimental moisture content (i.e., 65–70 %), maximum production was observed with 85–90 % moisture substrate in 96 h. By this time, 78 % of the glucose was consumed by the bacterial culture, and the conversion rate was 29.88 % (based on glucose consumed and that 81.74 % is the theoretical conversion of glucose to L-glutamic acid). For the homogeneous distribution of the nutrients, the moistening agent (production medium) was required in high amounts, so the initial moisture level also was required in high amounts. The amount of moistening agent was shown to influence the physiochemical properties of the solids. At higher moisture contents (90 % or more), the support may form clumps, which affect bacterial activity adversely.

Weng and Chen (2010) added different amounts of fermentation broth to PUF; the solid/liquid ratio was in the range 1:10–1:18. The effect of the solid/liquid ratio on the yield of bacterial cellulose was studied. The result is shown in Fig. 6.13. When the solid–liquid ratio was 1:16, the bacterial cellulose yield was the highest. Fermentation broth could be absorbed on the surface of PUF uniformly, and there was no significant gradient distribution of fermentation medium. With the decrease in solid/liquid ratio, the yield decreased; this was because there was too little medium to satisfy microorganism growth. When the solid/liquid ratio further increased, free medium appeared because of the limited water-absorbing capacity. The liquid film thickness on the surface of PUF increased, and the porosity decreased; the oxygen transfer was limited, which inhibited the growth of microorganisms.

6.4.2 *Effect of Water Activity*

Water activity α_w is an indicator of water availability and is a critical factor in the SSF of different kinds of substrates. Water not only provides a nutritionally adequate water environment for the growth of microorganisms in SSF and has an impact on nutrient transfer and oxygen utilization, but also affects microorganism growth, physiology, and metabolism (Acuña-Argüelles et al. 1994).

Oriol et al. (1988) carried out trials in which they maintained a constant support/ water ratio and increased the substrate concentration in the liquid phase; this showed differences in the fermentation profiles (Table 6.8). The germination time increased drastically, and the average specific growth rate decreased when the α_w of

Table 6.8 Effect of moisture content and water activity on the growth parameters of *Aspergillus niger* cultured on support with an absorbed liquid glucose medium (Oriol et al. 1988)

Bagasse/water ratio (w/w)	Glucose concentration (g/L)	Initial moisture content (%)	Initial A_w	Time (h)	Specific growth rate (h^{-1})	Final cell mass	Glucose consumption	Final cell mass/glucose consumption
1.30	168	40	0.975	18	0.453	–	–	–
0.83	168	50	0.981	18	0.388	–	2.31	–
0.50	168	60	0.977	16	0.429	–	2.30	–
0.36	168	65	0.977	15	0.473	–	2.72	–
0.25	168	70	0.974	17	0.464	–	2.42	–
0.19	168	75	0.973	18	0.429	–	–	–
0.36	57	72	0.986	12	0.541	0.58	1.10	0.527
0.36	168	65	0.962	17	0.433	1.4	2.81	0.498
0.36	260	59	0.939	26	0.354	2.39	5.08	0.470
0.36	330	55	0.92	31	0.317	2.95	5.99	0.492
0.36	417	50	0.899	41	0.198	3.82	7.53	0.507

Table 6.9 Influence of initial moisture on glutamic acid production in solid-state fermentation by *Brevibacterium* sp. (Nampoothiri and Pandey 1996)

Initial moisture level (%)	Solid–liquid ratio (ml/g)	Glucose consumption (%)	Glutamic acid (mg/gds)	Protein (mg/gds)
65–70	14:6	66.70	33.30	133.21
70–75	15:5	66.65	42.14	200.87
75–80	16:4	69.72	48.30	242.73
80–85	17:3	77.20	62.69	306.82
85–90	18:2	78.32	75.39	354.32

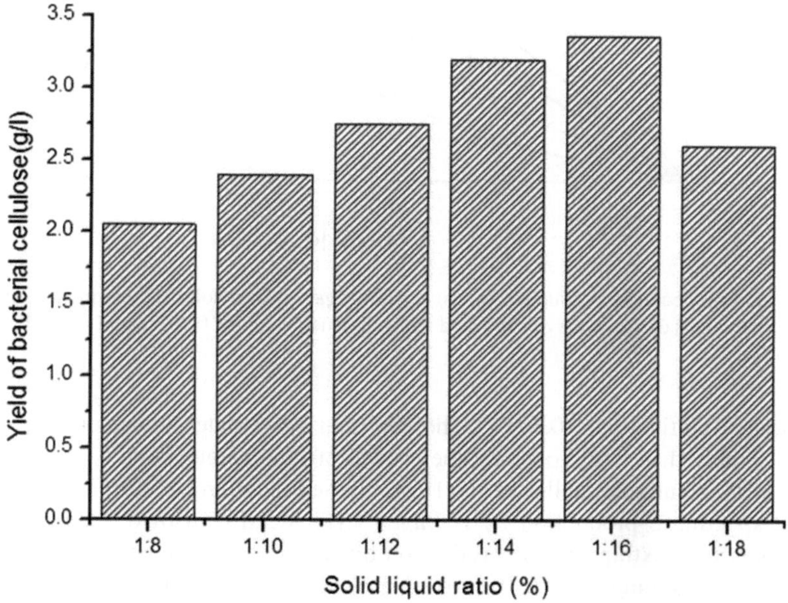

Fig. 6.13 Effect of solid–liquid ratio on the production of bacterial cellulose using adsorbed carrier solid-state fermentation (*ACSSF*)

the medium was lowered. It may be hypothesized that *A. niger* had the ability to grow at low α_w. These results demonstrated that the water activity and thus the osmotic pressure of the medium were more important than the water content for the germination and the growth pattern of *A. niger* on the solid phase with an absorbed nutritive liquid. SSF provided unique conditions for the microorganism/air/water interface and allowed growth at a high glucose concentration (42 % w/v in the liquid phase) and thus high-tonicity media (α_w 0.88) with satisfactory specific growth rates (0.20 h^{-1}). When using greater glucose concentrations (more than 450 g/L) in the liquid media, condensation of water at the lower and upper parts of the column resulted in locally higher α_w, and gradients of growth were observed. Nevertheless, some average specific growth rates calculated here for low-tonicity media (Table 6.8) were higher than those normally reported with fungi in

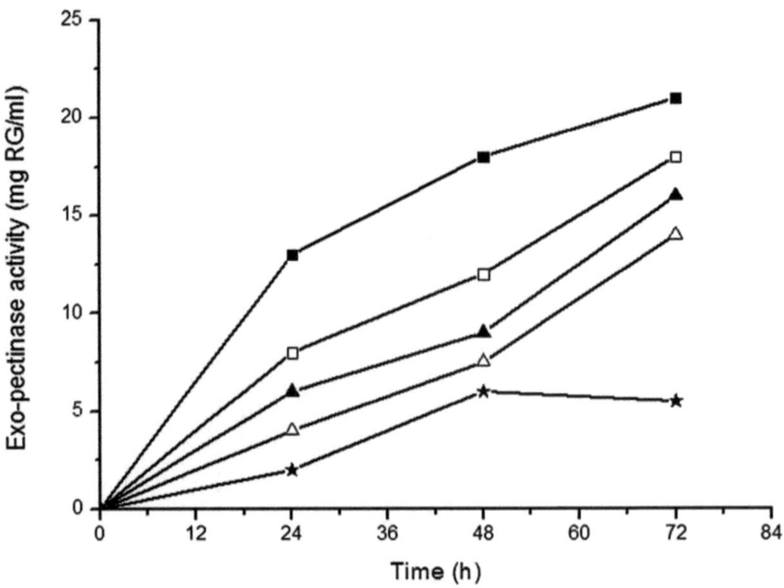

Fig. 6.14 Exo-pectinase production by *Aspergillus niger* CH4 in solid-state fermentation (SSF) at different values of water activity adjusted with ethylene glycol. ★ 0.90, △ 0.92, ▲ 0.94, □ 0.97, ■ 0.9837

submerged cultivations (0.3 h^{-1}), showing that solid-state culture conditions are more suitable than liquid ones for the growth of filamentous fungi.

In 1994, Acuña-Argüelles et al. (1994) studied the effect of different α_w values in extrapectinase production by *A. niger* CH4, dealing with different types of α_w depressors. Extrapectinase production by *A. niger* CH4 in SSF was evaluated at α_w values ranging from 0.90 to 0.98. Results are shown in Fig. 6.14. Without ethylene glycol, enzyme production was maximal at 72 h, but with ethylene glycol ($\alpha_w < 0.98$), extrapectinase production decreased. It is noteworthy that even at α_w values as low as 0.90, nearly 27 % of the pectinolytic activity was produced. It has been reported that low α_w values generally diminished both growth and enzyme production. In this work, cultures of *Aspergillus niger* CH4 at α_w values of 0.94 produced 80 % of the enzymatic activity produced without the depressor. This capacity to produce the enzymes even at α_w values of 0.90 may be correlated with the nature of the α_w depressor.

The extracellular protein concentration was also lower at low α_w values, and it was more reduced than extrapectinase production. These decreases in the extracellular protein concentration might be caused mainly not only by a decrease in growth but also by the selective production of extrapectinases in the extracellular fraction. This effect is clearly shown in Fig. 6.15; specific extrapectinase activity increased by decreasing water activity. The fact that specific extrapectinase activity obtained at $\alpha_w = 0.90$ was nearly 4.5 times that of the activity produced without ethylene

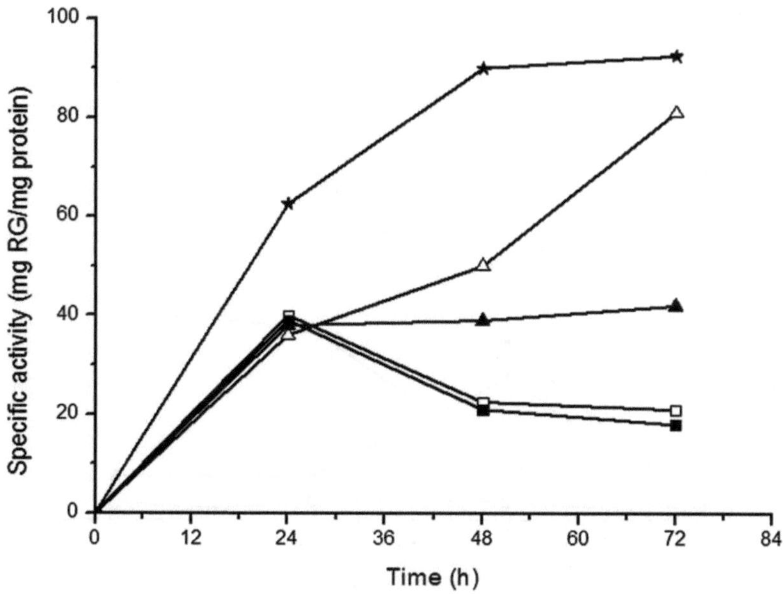

Fig. 6.15 Specific activity of extrapectinase production at different values of water activity by *Aspergillus niger* CH4 in solid-state fermentation (*SSF*). ★ 0.90, △ 0.92, ▲ 0.94, □ 0.97, ■ 0.9837

glycol ($\alpha_w = 0.98$) suggested a certain selectivity for the kind of enzymes produced by *A. niger* CH4 in SSF.

It has been reported that a decrease in α_w of growing cultures of fungi led to modifications in the phospholipid fatty acid saturation and, consequently, to reduction in membrane fluidity and permeability. To evaluate this hypothesis, reducing groups in file culture medium at different α_w values were measured. As seen in Fig. 6.16, there was accumulation of the reducing groups at low α_w values (0.92 and 0.90). This accumulation suggested that although the nonreducing polymers used as a carbon source (pectin and sucrose) were hydrolyzed, the monomers could accumulate in the culture medium, probably because of membrane transport limitations.

6.4.3 Effect of Support Particle Size

The carrier particle size and specific surface area are important factors affecting microbial activity and the oxygen transfer rate in SSF. The smaller the particle size is, the larger the specific surface area is, which is conducive to the adsorption and growth of microorganisms. But, the gap between the carriers and the porosity will decrease, which affects oxygen transfer. Larger particle size will obtain greater porosity, but the specific surface area will be reduced. Therefore, choosing a

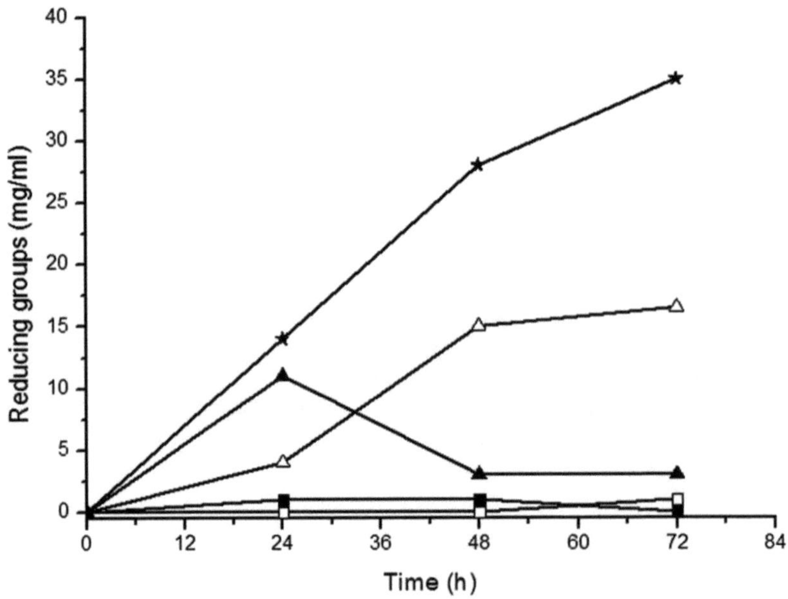

Fig. 6.16 Reducing groups accumulated during culture of *Aspergillus niger* CH4 in solid-state fermentation (*SSF*) at different water activity values. ★ 0.90, △ 0.92, ▲ 0.94, □ 0.97, ■ 0.9837

suitable particle size will meet the growth of microorganisms and the oxygen transfer rate (Nampoothiri and Pandey 1996).

Nampoothiri and Pandey (1996) studied the effect of sugarcane bagasse particle size on glutamic acid production in ACSSF. The particle size, and therefore specific surface area, of the substrate is of great importance in SSF. As shown in Table 6.10, substrates with particles of four different individual sizes and three of mixed sizes (see table for composition of substrates with mixed mesh) were used. Maximum glutamic acid production (82.22 mg/gds) was obtained with the substrate containing particles of four different sizes in equal amounts. Among the substrates with a single particle size, the best yields (43.67 mg/gds) were received with the substrate with 1-mm particles. Thus, evidently substrates with a mixed particle size are the better choice. With smaller particles, the available surface area for microbial growth is larger, but the interparticle space and hence porosity decrease. With a substrate with bigger particles, the situation is reversed. These two opposing factors probably interact and thus determine the growth and activity of microorganisms. The role of oxygen transfer into a void space affects the microbial growth, and a compromise always has to be made regarding the composition of substrates with mixed particle sizes for optimal activity and mass transfer effects.

Oriol et al. (1988) ran trials varying the average support particle size from less than 10 to more than 150 mesh (approximately 2.5–0.1 mm). An increase of the average particle size produced a decrease of the growth rate in the deceleration

Table 6.10 Effect of sugar cane bagasse particle size on glutamic acid production in solid-state fermentation by *Brevibacterium* sp. (72 h)

Particle size (mm)	Distribution (%)	Glucose consumption (%)	Glutamic acid (mg/gds)	Proteins (mg/gds)
0.45	100 %	69.20	29.49	72.00
1.00	100 %	73.50	43.67	114.12
2.00	100 %	70.58	30.08	72.82
3.00	100 %	65.32	27.36	81.40
1 and 0.45	50 % each	72.00	46.42	241.13
1 and 2.00	50 % each	68.73	50.10	310.58
0.45, 1, 2, and 3.00	25 % each	70.10	82.22	337.38

Table 6.11 Effect of the support particle size on the parameters of growth of *Aspergillus niger* on support with an absorbed liquid medium containing 280 g/L (w/v) of glucose

Support particle size (mm)	Initial density (kg/L)	Final cell mass[a] (g/column)	Final cell mass[b] (g/column)	Substrate consumption (g/column)	Specific growth rate (h^{-1})	Time (h)
2.50	0.230	2.16	2.05	4.75	0.323	27
1.43	0.295	2.66	0.60	4.95	0.330	26
0.79	0.295	2.52	2.40	5.42	0.348	26
0.51	0.295	2.76	2.60	4.76	0.400	25
0.10	0.485	2.53	2.33	4.58	0.375	25

[a]Calculated from the amino acid content
[b]Calculated from the oxygen utilization rate (OUR)

phase of growth and, for the largest fibers, led to lower final mycelial contents (Table 6.11). As mold growth observed with a microscope appeared to occur only on the support surface, the depressive effect observed with a large bagasse particle size might be caused by a limitation of growth caused by intraparticular diffusion of nutrients. Small average support particle size (0.1 mm) produced high-density packed material in the column (0.5 kg/L against about 0.3 kg/L with larger particles), but no variation in the kinetics of growth or in the final mycelial content was detected (Table 6.11). In addition, high packing density tended to make the carrier compact and resulted in limitation of heat and mass transfer.

6.4.4 Effect of Inert Carrier Depth

Substrate depth is one of the most important factors that affect heat and mass transfer in SSF. With rice husk and wheat bran as solid substrates, five different depths of substrate for cellulase production with PUF were studied and compared with the control without PUF (Fig. 6.17) (Hu et al. 2011). The results showed that the peak cellulase yield of Filter paper activity (FPA) (9.3 IU/gds) and

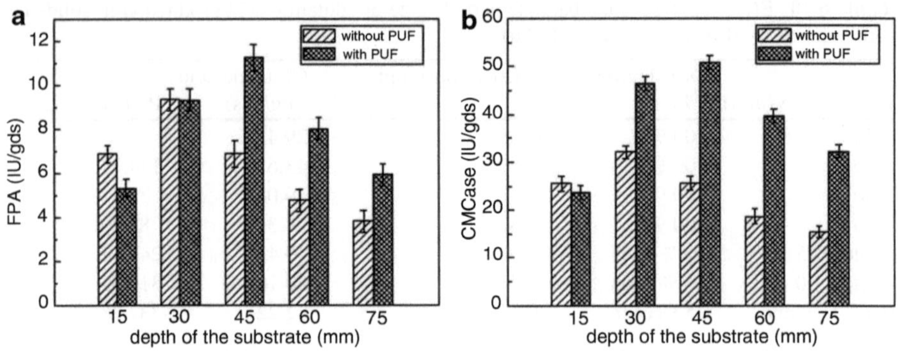

Fig. 6.17 Effect of substrate depth on Filter paper activity (FPA) (**a**) and Carboxyl methyl cellulose enzyme activity (CMC) (**b**) in solid-state fermentation (*SSF*) (Hu et al. 2011). *PUF* polyurethane foam

Carboxyl methyl cellulose enzyme activity (CMC) (32.0 IU/gds) was obtained when the depth was 30 mm for the fermentation without PUF. However, in the SSF with PUF, the highest cellulase yield of FPA (11.2 IU/gds) and CMCase (50.7 IU/gds) was obtained at a depth of 45 mm. Compared with the SSF system without PUF, the highest cellulase yield in SSF with PUF increased by 20.4 % (FPA) and 62.3 % (CMCase). The results implied that the addition of PUF could increase cellulase productivity. After taking the contribution of PUF to the volume of the substrates into account, the highest cellulase yield based on the whole volume of the substrate in SSF with PUF was still 12.8 % (FPA) and 48.5 % (CMCase) higher than the system without PUF, which may have resulted from improving oxygen transfer in solid matrix and benefiting the growth of microorganisms after the addition of PUF.

6.5 ACSSF Applications

To overcome the disadvantages of traditional SSF (limitation in mass transfer and heat transfer, difficulty in biomass determination, fermentation process control and purification), ACSSF is proposed based on the principle of SSF. ACSSF has been studied extensively from the 1930s to the present, as shown in Table 6.5. ACSSF can be used in the production of enzymes (such as alkaline protease, lipase, nuclease, L-glutaminase chitin enzyme, cellulase, pectinase); organic acids (such as clavulanic acid, citric acid, lactic acid); polysaccharides (such as xanthan gum, bacteria cellulose); antibiotics (such as penicillin, lovastatin); spores; biological alkali; and so on. It has been confirmed to have advantages not available in traditional SSF, mainly as follows: The inert carrier is stable chemically and easy to recover and reuse. It can overcome the agglomeration and mass and heat transfer problems. It can void the impurities being mixed into the fermentation extracts. The

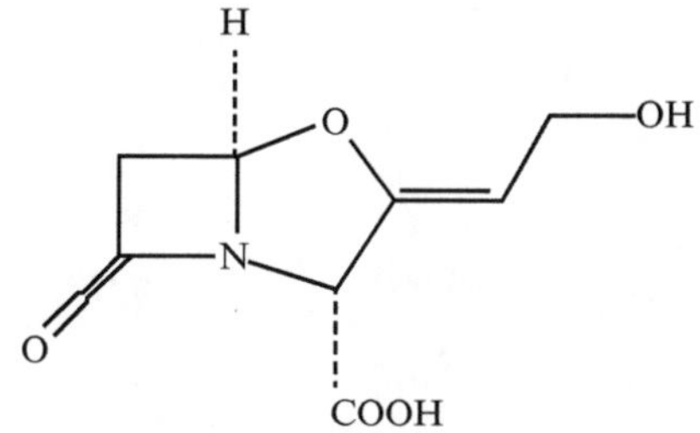

Fig. 6.18 Structure of clavulanic acid

medium is defined. Therefore, ACSSF is a potential fermentation method and will bring a new revolution to SSF.

Even though this fermentation method is still at the laboratory level, it provides a solid foundation and theory for its further industrial scale-up. ACSSF has been studied from carrier selection to process optimization, reactor design, and fermentation kinetics. The depth and breadth of the studies have been constantly expanding. With the rich research and mature research tools, ACSSF will have broader application prospects.

6.5.1 Clavulanic Acid Production by SSF on PUF

It is well recognized that the β-lactamases of both gram-positive and gram-negative bacteria contribute significantly to the resistance of bacteria to β-lactam antibiotics. Clavulanic acid is a potent inhibitor of β-lactamase produced by a wide range of gram-positive and gram-negative bacteria and is produced naturally by *Streptomyces clavuligerus*. The structure is shown in Fig. 6.18. It can protect β-lactam antibiotics against hydrolysis by binding irreversibly to the active center of β-lactamase.

The traditional production method of clavulanic acid was SmF, and the producer was *Streptomyces clavuligerus*. The seed broth and fermentation medium were synthetic medium containing soybean powder and glycerin. In SmF, continuous stirring was required to ensure an adequate supply of oxygen, which not only increased energy consumption but also damaged the mycelia by shear force and affected growth and metabolism. Further, clavulanic acid production was low and easily decomposed, which resulted in high cost for product separation and purification.

According to the characteristics of SSF on an inert carrier, (Wang 2006) prepared PUF using straw polyhydroxy alcohol liquefied products and used PUF as a carrier to produce clavulanic acid in SSF. Compared with SmF, it improved the heat and mass transfer and saved the cost of separation. Porous inert carrier can absorb bacteria and fermentation medium. Its unique interface environment is conducive to heat and mass transfer. Bacteria can obtain an adequate supply of oxygen; do not require mechanical agitation; save energy consumption and avoid the mycelial damage caused by stirring; increase the concentration of the clavulanic

Fig. 6.19 Fermentation of clavulanic acid in flask

acid in the fermentation broth; save the cost of product separation; and can realize continuous production.

6.5.1.1 Fermentation and Analysis of Clavulanic Acid

Fermentation of Clavulanic Acid in a Flask

Polyurethane foam was cut into 5 × 5 × 5 mm cubes and dried in the oven until the weight kept constant. Then, 3 g of PUF pieces were placed in a 250-ml conical flask that had been cleaned and dried. The flask with the PUF was sterilized at 121 °C for 20 min and cooled to room temperature. About 5–35 ml medium and inoculum were added to each flask under strict aseptic conditions and pressed softly by a glass stick to allow the PUF to fully and evenly absorb them. The flasks were then incubated in an autonomous incubator with constant temperature and humidity for a desired length of time. Fermentation of clavulanic acid in the flasks is shown in Fig. 6.19.

Extraction of Clavulanic Acid

From the fermented solid substrate, fermentation broth extraction was carried out using 93 ml distilled water so that the final extraction volume was 100 ml (93 ml distilled water with 7 ml moistening medium). First, the fermented substrates were properly mixed with distilled water, and the flasks were kept on a rotary shaker at 150 rpm for 30 min. After this, the solids were separated from the solution by filtering through a nylon cloth sieve. The solution was centrifuged at 12,000 rpm for 20 min at 4 °C in a refrigerated centrifuge. The supernatant was collected to determine the concentration of clavulanic acid.

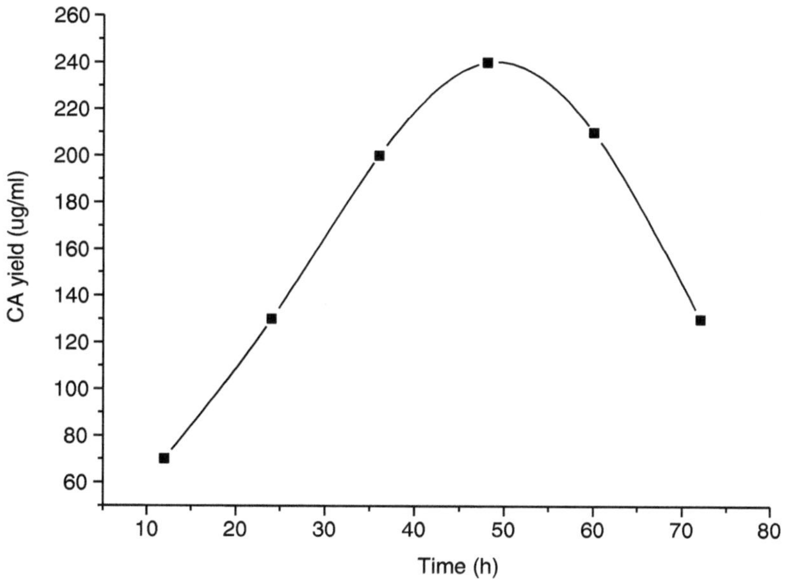

Fig. 6.20 Effect of fermentation time on clavulanic acid production by *Streptomyces clavuligerus*

Determination of Cell Concentration

Seed broth or fermentation broth containing the bacteria was diluted 40-fold, and cells were counted with an optical microscope and blood count plate.

6.5.1.2 Results and Discussion

Effect of Fermentation Time

As shown in Fig. 6.20, with the fermentation time prolonged, the yield of clavulanic acid increased. The yield reached the maximum (240 μg/ml) at 48 h and then decreased. At the same time of clavulanic acid synthesis, clavulanic acid also was degraded. The long fermentation time will lead to excessive consumption of clavulanic acid. Considering the actual production, the fermentation time was controlled in about 48 h.

Effect of Moistening Medium Volume

As shown in Fig. 6.21, by adding different amounts of culture medium to 3 g PUF, the yield of clavulanic acid was maximum with addition of 15 ml of medium.

By adding 5 and 10 ml of culture medium to the fermentation process, the growth of microorganisms and clavulanic acid production were inhibited because of low moisture and water evaporation. As the adsorption amount of PUF reached a

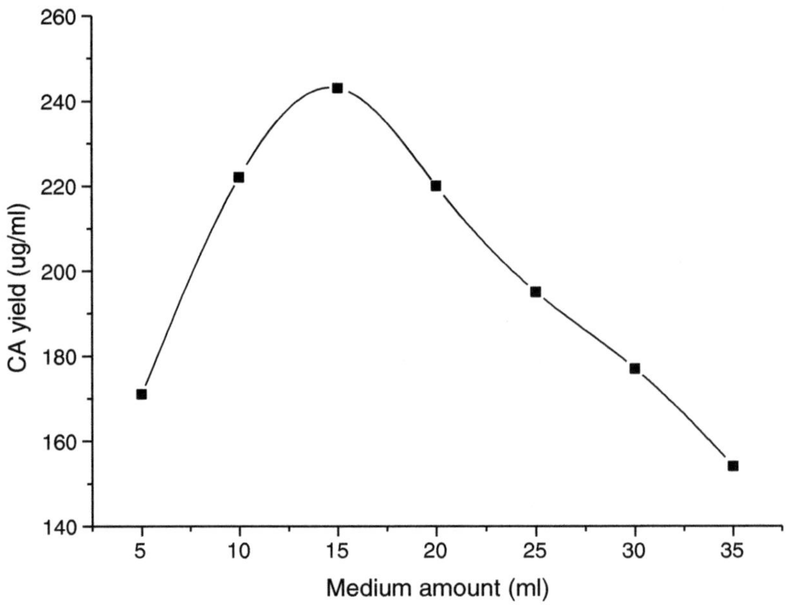

Fig. 6.21 Effect of moistening medium volume on clavulanic acid production by *Streptomyces clavuligerus*

certain extent, the liquid film adsorbed on PUF thickened. The distribution of medium absorbed on the surface of PUF appeared significantly uneven. In the bottom of the flask, there was free medium. A concentration gradient from the bottom to the top appeared. In the top of the flask, the medium was distributed on the surface of PUF uniformly, there was no free medium, and the microorganisms grew well. At the bottom of the flask, there was free medium, and the PUF pores filled with free medium, which affected oxygen transfer and limited the growth of microorganisms.

Effect of Inoculum Amount

As for the ACSSF process without stirring, a certain amount of the seed broth was added to the fermentation broth so the strains were distributed uniformly on the carrier. As shown in Fig. 6.22, the clavulanic acid production increased gradually as the inoculation amount increased from 2 to 10 %. The yield of clavulanic acid increased slowly when the inoculum amount was above 10 %.

The PUF is a porous carrier and can absorb culture medium to form a liquid film. *Streptomyces* attaches on the liquid film, where the nutrient growth is metabolic. The porosity of the foam is limited and can accommodate only a certain number of *Streptomyces clavuligerus* growing on the liquid film. An excessive inoculation amount will cause overcrowding in the internal space, resulting in lack of nutrition.

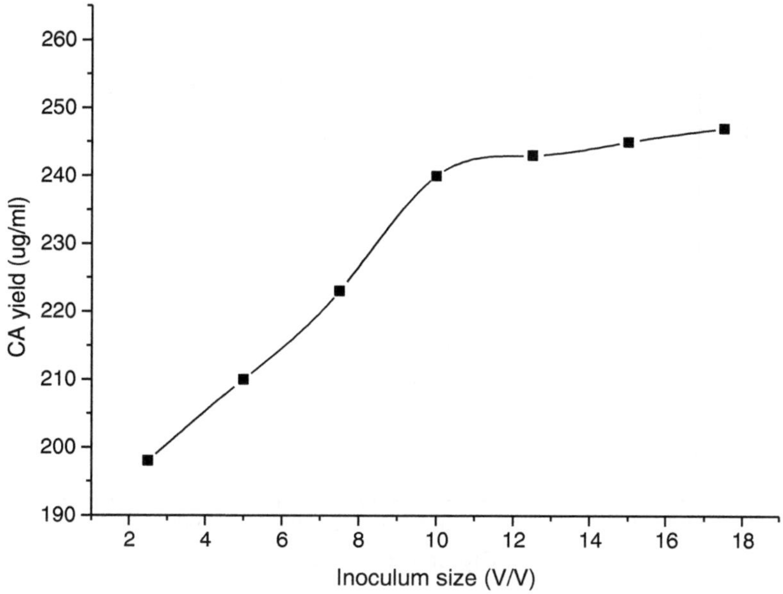

Fig. 6.22 Effect of inoculum amount on clavulanic acid production by *Streptomyces clavuligerus*

Effect of Glycerol

The carbon source can markedly influence the production of clavulanic acid by *S. clavuligerus*. Glycerol is essential for the biosynthesis of clavulanic acid. The carbon skeleton of the β-lactam ring of clavulanic acid was derived from glycerol without any intermediate rearrangement of the three carbons (C5, C6, C7). In the absence of glycerol, no clavulanic acid was formed, but cephamycin C, another metabolite of *S. clavuligerus*, was produced.

As shown in Fig. 6.23, no clavulanic acid was formed by *S. clavuligerus* in the absence of glycerol. With the increase in the amount of glycerol, the clavulanic acid yield increased. When the amount of glycerol was 15 g/L, the yield of clavulanic acid reached up to 245 μg/ml. The biosynthesis of clavulanic acid was inhibited when glycerol was above 15 g/L.

Effect of Lithium Chloride

As shown in Fig. 6.24, when lithium chloride was added to fermentation broth, the yield of clavulanic acid increased. Clavulanic acid combined with lithium chloride can promote chemical stability, and degradation is not easy. In actual production, lithium chloride was added to the fermentation broth; the clavulanic acid was readily obtained in good yield and purity as a salt in a form suited for formulation as pharmaceuticals.

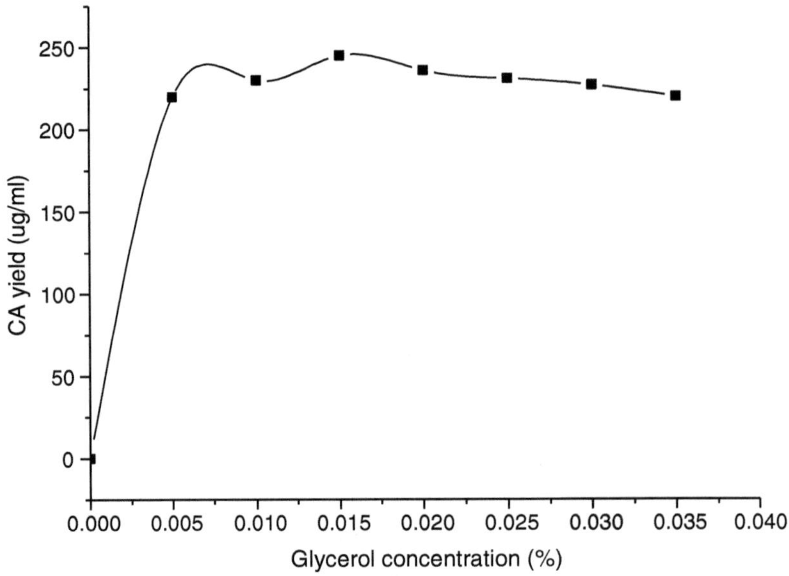

Fig. 6.23 Effect of glycerol concentration on clavulanic acid production by *Streptomyces clavuligerus*

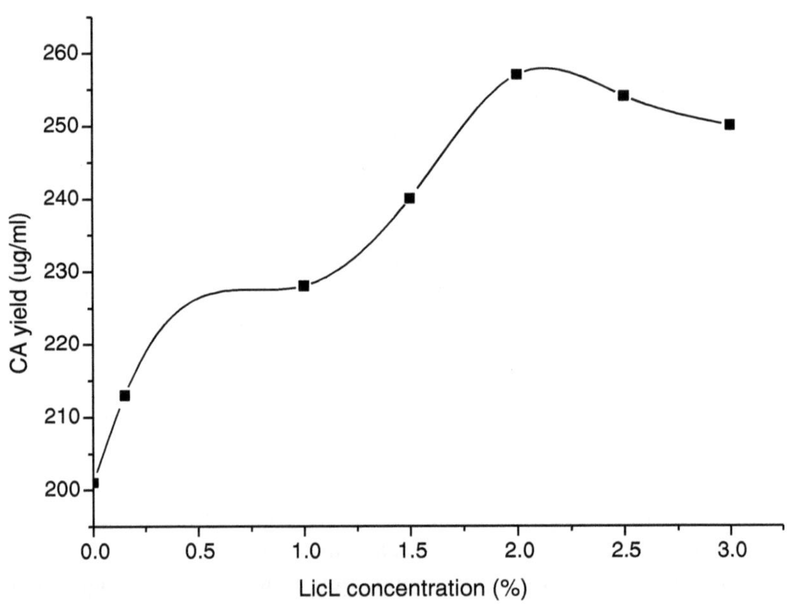

Fig. 6.24 Effect of lithium chloride on clavulanic acid production by *Streptomyces clavuligerus*

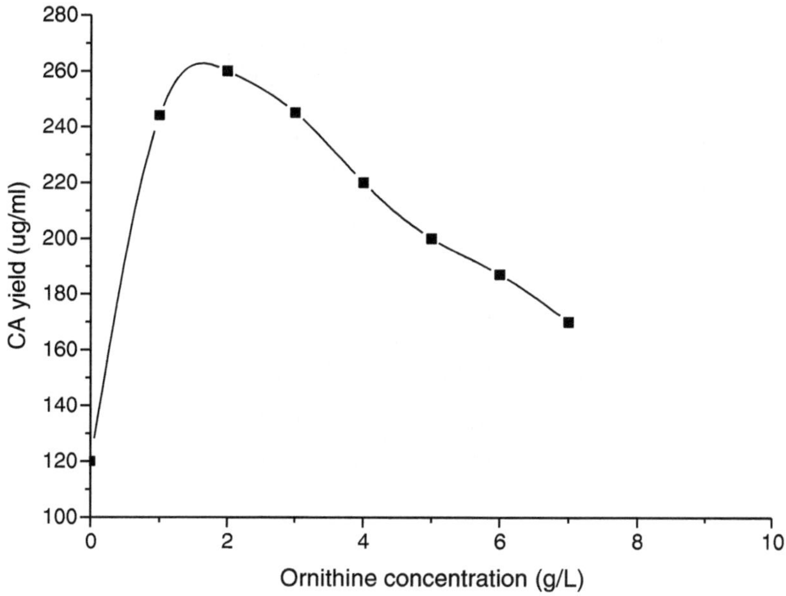

Fig. 6.25 Effect of ornithine on clavulanic acid production by *Streptomyces clavuligerus*

Effect of Ornithine

When fermentation is involved in the formation of products, precursors may be added to the medium to improve the yield or the quality of products. Generally, the precursor molecule or its closely related derivative is incorporated into the product molecules of the fermentation. The primary metabolic precursor of C5 intermediates is well known to be derived from ornithine. Clavulanic acid production could be enhanced significantly by adding ornithine. The feeding of ornithine could allow more C3 precursor for the biosynthesis of clavulanic acid.

As shown in Fig. 6.25, with an addition of 2 g/L ornithine to the culture, the clavulanic acid production (260 mg/L) was about 2.2-fold of that without any addition of amino acids (125 mg/L).

Effect of PUF on Fermentation Product Separation

As shown in Fig. 6.26, after SSF on PUF, the concentration of *Streptomyces clavuligerus* was very low. It indicated that most of the microbes were absorbed on the PUF. This can reduce the impurities in the fermentation broth and simplify the purification process. The strains can be reused, which makes the fermentation continue without inoculation.

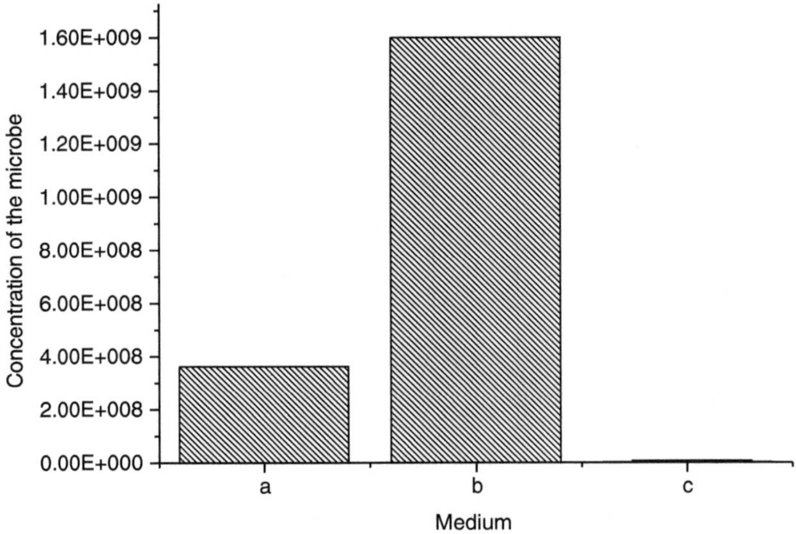

Fig. 6.26 Concentration of *Streptomyces clavuligerus* in fermentation broth. *a* fermentation broth before fermentation; *b* fermentation broth of submerged fermentation (*SmF*); *c* fermentation broth extracted in solid-state fermentation (*SSF*)

6.5.1.3 Conclusion

1. The adsorption capacity of inert carrier is limited. More or less culture medium can influence the fermentation results significantly.
2. PUF is a porous material; the porosity is limited. Excessive inoculum makes the pore space crowded and decreases nutrition.
3. Glycerol is the necessary carbon source. It is the precursor of the C3 intermediate of clavulanic acid.
4. Lithium chloride combined with clavulanic acid protects clavulanic acid from degradation.
5. Ornithine is the precursor of the C5 intermediate of clavulanic acid and influences the yield of clavulanic acid significantly.
6. In SSF on PUF, *Streptomyces clavuligerus* was absorbed on the surface of PUF, which made the fermentation broth have few microorganisms and simplified the purification process.

6.5.2 Alkaline Protease Production by SSF on PUF

Proteases are by far the most important group of enzymes produced commercially and are used in the detergent, protein, brewing, meat, photographic, leather, and dairy industries. These enzymes offer advantages over the use of conventional chemical catalysts for numerous reasons; for example, they exhibit high catalytic activity and a high degree of substrate specificity, can be produced in large amounts, and are economically viable. The detergent industry has now emerged as the single major consumer of several hydrolytic enzymes acting at highly alkaline pH. The major use of detergent-compatible proteases is in laundry detergent formulations.

Until now, alkaline protease has been produced mostly by SmF, a process plagued with problems, including serious pollution, low product concentration, and high production cost. Compared to SmF, SSF has received more attention recently as it uses simpler fermentation medium, requires a smaller space, is easier to aerate, and has higher productivity, lower waste water output, lower energy requirement, and less bacterial contamination. Chen and Wang evaluated the potential of *Bacillus pumilus* AS 1.1625 for producing alkaline protease by SSF on PUF (Wang 2006).

6.5.2.1 Fermentation and Analysis of Alkaline Proteases

Fermentation

The PUF was cut into 5 × 5 × 5 mm cubes and dried in the oven until the weight remained constant. Four grams of PUF pieces were then placed in a 250-ml conical flask that had been cleaned and dried. The flask with the PUF was sterilized at 121 °C for 20 min and cooled to room temperature. About 5–35 ml medium and inoculum were added to each flask under strict aseptic conditions and pressed softly by a glass stick to allow the PUF to fully and evenly absorb them. Then, the flasks were incubated in an autonomous incubator at a constant temperature and humidity for a desired length of time.

Extraction

From the fermented solid substrate, enzyme extraction was carried out using 93 ml distilled water with 0.1 % Tween® 80 so that the final extraction volume was 100 ml (93 ml distilled water with 7 ml moistening medium). First, the fermented substrates were properly mixed with distilled water, and the flasks were kept on a rotary shaker at 150 rpm for 30 min. After this, the solids were separated from the solution by filtering through a nylon cloth sieve. The solution was centrifuged at 12,000 rpm for 20 min at 4 °C in a refrigerated centrifuge. The supernatant was collected and used for enzyme assay.

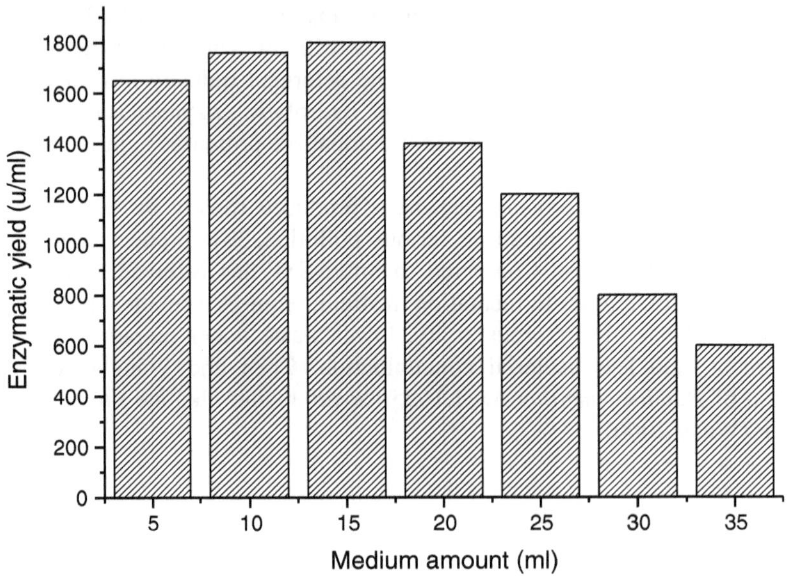

Fig. 6.27 Effects of the volume of medium added on enzymatic productivity. Fermentation temperature, 30 °C; initial pH, 8.0; culture time, 24 h

6.5.2.2 Results and Discussion

Effect of Moistening Medium Volume on Enzyme Activity

The results presented in Fig. 6.27 indicate that alkaline protease production increased as the volume of moistening medium increased up to 15 ml; at that point, the maximal enzyme activity (1,800 U/ml) was obtained. When the volume of moistening medium was less than 15 ml, the microorganism was not able to obtain enough nutrients, resulting in low enzyme activity. On the other hand, when the volume of the moistening medium was more than 15 ml, which exceeded the absorption capacity of the PUF, liquid accumulated on the surface of the PUF, limiting the transfer of oxygen in the pores of the PUF and hindering normal metabolism.

Effect of Fermentation Temperature on Enzyme Activity

As shown in Fig. 6.28, fermentation temperature influenced the yield of alkaline protease for maximal enzyme activity of 1,950 U/L at 32 °C. Fermentation temperature has a complex effect on the growth of *Bacillus pumilus* AS 1.1625. At low temperatures, an increase in temperature enhances the growth of the microorganism and therefore improve enzyme activity. At high temperatures, an increase in temperature negatively influences the microorganism by making some important

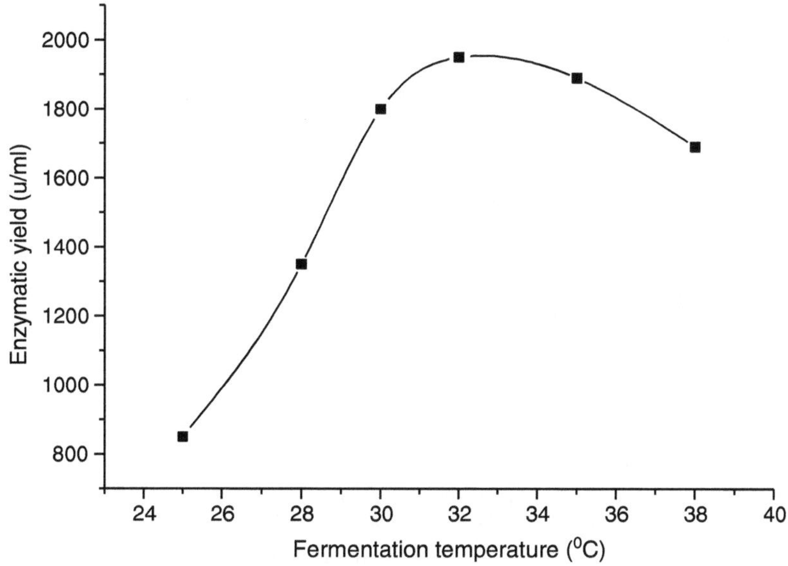

Fig. 6.28 Effects of fermentation temperature on enzymatic productivity. Culture time, 24 h; initial pH, 8.0; volume of medium, 15 ml

protein denaturalize, thus limiting microbial growth. In addition, a high temperature accelerates the volatilization of water in the medium, thus restraining the growth and enzymatic production of the microbe.

Effect of Inoculum Amount on Enzyme Activity

In this study, the inoculum concentration was 8.2×10^8. Figure 6.29 indicates that there was a gradual increase in the synthesis of enzyme when the amount of inoculum was increased up to 1 %. Further increase of inoculum amount reduced the amount of nutrient solution that the microbe obtained.

Effect of Initial pH of Moistening Medium on Enzyme Activity

The optimum pH required for maximal enzyme activity by SSF on PUF was evaluated using various initial pH levels (7–10) of the moistening medium. Figure 6.30 shows that maximal enzyme activity was 1,950 U/ml at pH 9.0, implying that higher or lower pH generally leads to poor growth and results in a decline of enzymatic activity.

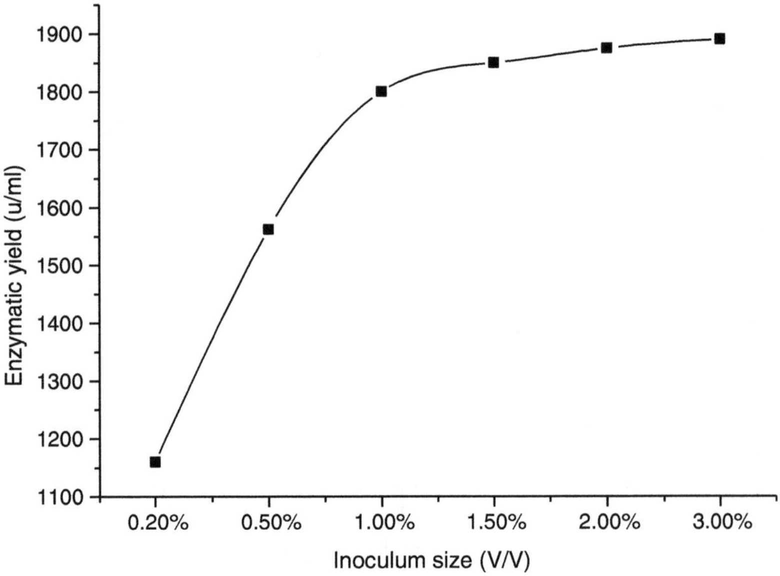

Fig. 6.29 Effects of the inoculum size on enzymatic productivity. Fermentation temperature, 32 °C; culture time, 24 h; initial pH, 8.0

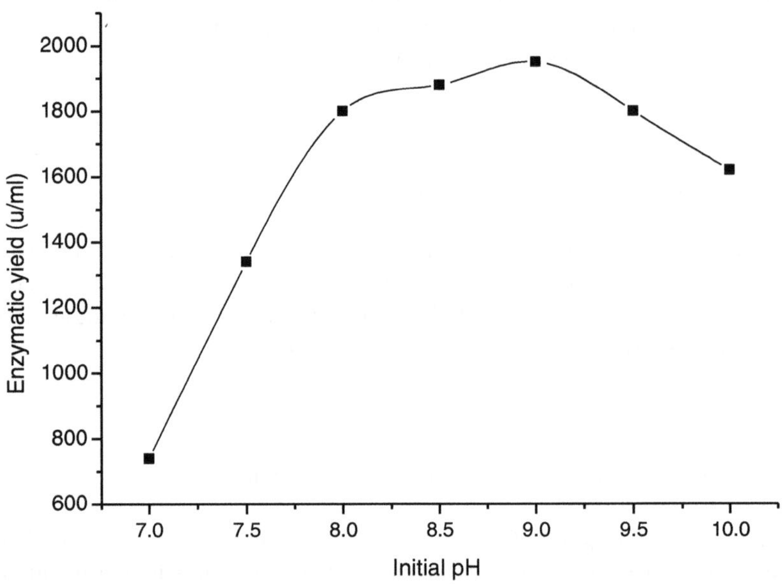

Fig. 6.30 Effect of initial pH on enzymatic productivity. Fermentation temperature, 32 °C; culture time, 24 h; inoculum size, 1 %

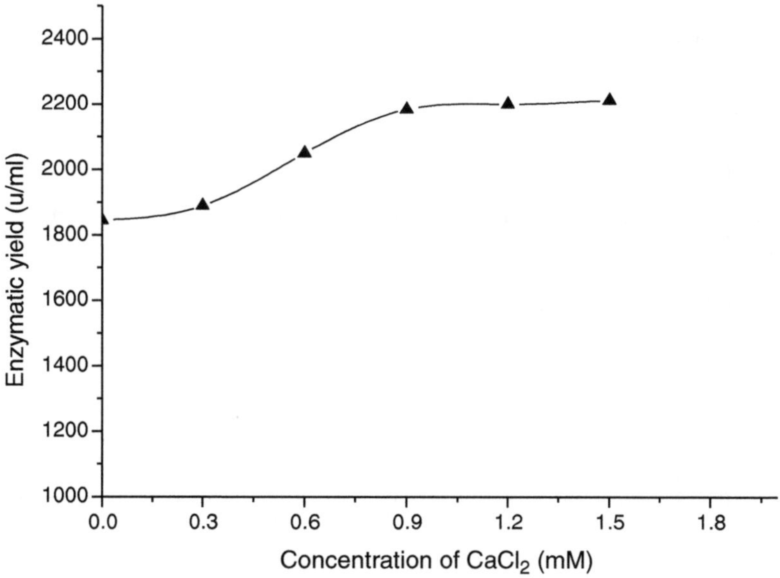

Fig. 6.31 Effect of $CaCl_2$ on enzymatic production. Fermentation temperature, 32 °C; culture time, 24 h

Influence of Added Ca^{2+} and Casein on Enzyme Activity

According to some reports, Ca^{2+} could improve both the activity and the stability of alkaline protease. Different concentrations of casein were also added to the fermentation medium to evaluate its influence on enzyme activity. Figure 6.31 indicates that when the Ca^{2+} supplement was 15 mmol/L, an appropriate amount of Ca^{2+} enhanced alkaline protease production up to the maximal enzyme activity of 2,185 U m/L.

Different concentrations of casein were also added to the fermentation medium to evaluate the induction of alkaline protease by *Bacillus pumilus* AS 1.1625. Figure 6.32 indicates that there was no increase in enzyme activity by casein supplementation at the tested concentrations.

Effect of PUF on the Separation of Fermentation Products

As shown in Fig. 6.33, after SSF on PUF, the concentration of *Bacillus pumilus* AS 1.1625 was very low, indicating that most of the microbes were absorbed on the PUF. This can reduce the impurities in fermentation broth and simplify the purification process. The strains can be reused, which makes the fermentation continue without inoculation.

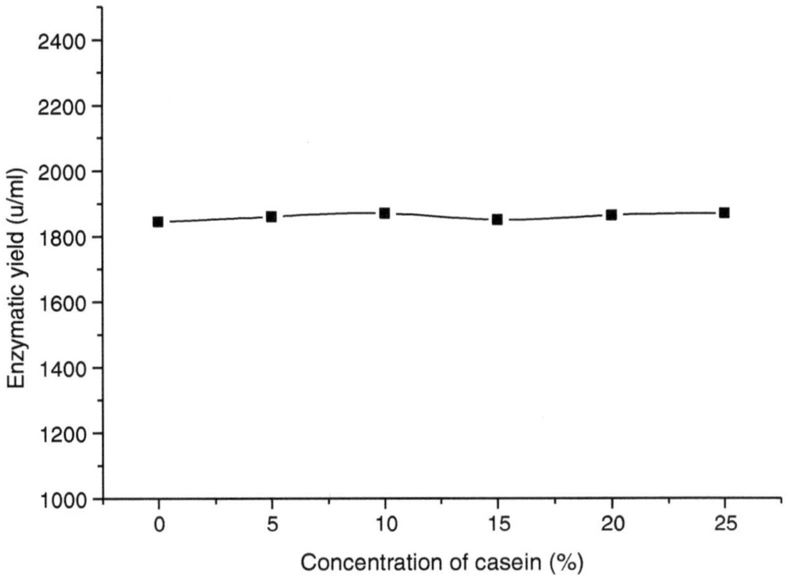

Fig. 6.32 Effect of casein on enzymatic production. Fermentation temperature, 32 °C; culture time, 24 h

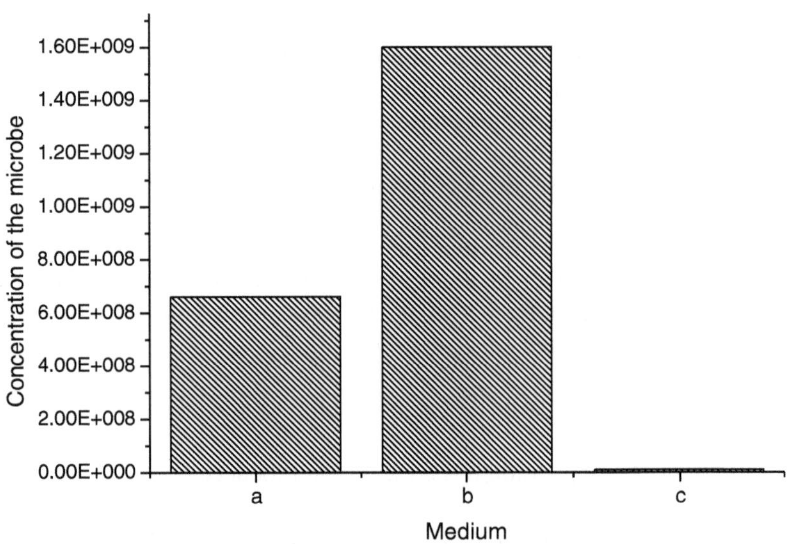

Fig. 6.33 Concentration of *Bacillus pumilus* AS 1.1625 in fermentation broth. *a* fermentation broth before fermentation; *b* fermentation broth of submerged fermentation (*SmF*); *c* fermentation broth extracted in solid-state fermentation (*SSF*)

6.5.2.3 Conclusion

1. The volume of moistening medium influenced the yield of alkaline proteases. When the volume of moistening medium was low, the microorganism was not able to obtain enough nutrients, resulting in low enzyme activity. On the other hand, when the volume of the moistening medium was high, it limited the transfer of oxygen in the pores of the PUF and hindered normal metabolism.
2. Temperature was a key factor for the yield of alkaline proteases. A temperature of 32 °C was optimal. There was no significant effect on the yield of alkaline protease at temperatures between 30 and 35 °C. A lower or higher temperature decreased the yield.
3. Inoculation amount affected the yield of alkaline protease significantly.
4. *Bacillus pumilus* AS 1.1625 grew well in alkaline conditions.
5. Casein had an insignificant effect on the alkaline proteases, and a suitable amount of Ca^{2+} could promote the yield of alkaline proteases significantly.
6. Most of *Bacillus pumilus* AS 1.1625 was absorbed on the surface of the PUF, which could be beneficial for the separation of products.

6.5.3 Clavulanic Acid Production by Continuous SSF on PUF

The effect of PUF synthesized using liquefying steam-exploded wheat straw in SSF was positive from the results discussed in Sects. 6.5.1 and 6.5.2. But, there were great differences between the actual production conditions and shake flask fermentation. There might be unexpected problems in the actual production conditions. Wang (2006) designed a set of continuous fermentation equipment and techniques to produce clavulanic acid on PUF (Fig. 6.10), which could evaluate the feasibility of continuous ACSSF in the enlarged production and achieve SSF continuity.

6.5.3.1 Fermentation and Analysis of Clavulanic Acid

Design of ACSSF Bioreactor

As shown in Fig. 6.10, the ACSSF bioreactor is made from Plexiglass; it is transparent so the interior can be observed. The bioreactor is 48.5 cm high. The inner diameter is 12 cm. The tank cover is connected with the tank body by a flange. There are an input hole for fermentation broth and seed liquid in the tank cover and an output hole for air. The input hole for the fermentation broth is connected with the fermentation broth reservoir (1) through a pipeline. A peristaltic pump is connected to the pipeline. The air outlet is opened in the air directly. An input hole for air and an output hole for fermentation broth are at the bottom of the tank body. The air input hole is connected with the gas reservoir by a pipeline.

An air flowmeter is connected with the pipeline. The output hole for the fermentation broth is connected with the fermentation broth reservoir, which is used to store fermentation products.

Preparation Before Fermentation

The PUF was cut into small cubes, washed, and dried. The depth of PUF in the bioreactor was 25–35 cm. The material was sterilized at 121 °C for 20 min.

The fermentation medium was stored in the reservoir after sterilization. The seed was cultivated for 3 days and stored in the seed tank.

Inoculation Process

Before fermentation, seed broth was added to the bioreactor using a peristaltic pump from the seed tank. About 10 cm above the inert carrier layer, there is a distribution plate covered with 1-mm holes. In the tank, there is a distribution plate at about 10 cm from the fill layer, covered with 1-mm holes. The seed broth was dispersed evenly through the distribution plate and then slowly dripped onto the carrier, gradually penetrating into the interior of the inert carrier. When the fermentation broth discharged from the output hole at the bottom of the bioreactor, it indicated that the seed broth had been covered on the surface of the carrier, and the inoculation procedure was complete.

Fermentation Process

After inoculation, fermentation broth was added to the bioreactor evenly by a peristaltic pump. Fermentation broth was distributed on the surface of the carrier by the distribution plate. Fermentation broth was absorbed on the surface of the carrier and formed a liquid film. Microorganisms grew in the liquid film and produced products. With the continuous addition of fermentation broth, the absorbed medium increased and exceeded the capacity for water retention. Finally, the medium was discharged into the bottom by squeezing and was collected in the reservoir. Fermentation broth flowed from the top to the bottom, and when the products were discharged along with fermentation broth, a fermentation period was completed.

In the fermentation process, sterilized air flowed from the bottom to the top and provided oxygen to the microorganisms.

In the fermentation process, fermentation and air were both continuous. Fermentation broth was added continuously to the bioreactor and discharged continuously. Inoculation was required only once.

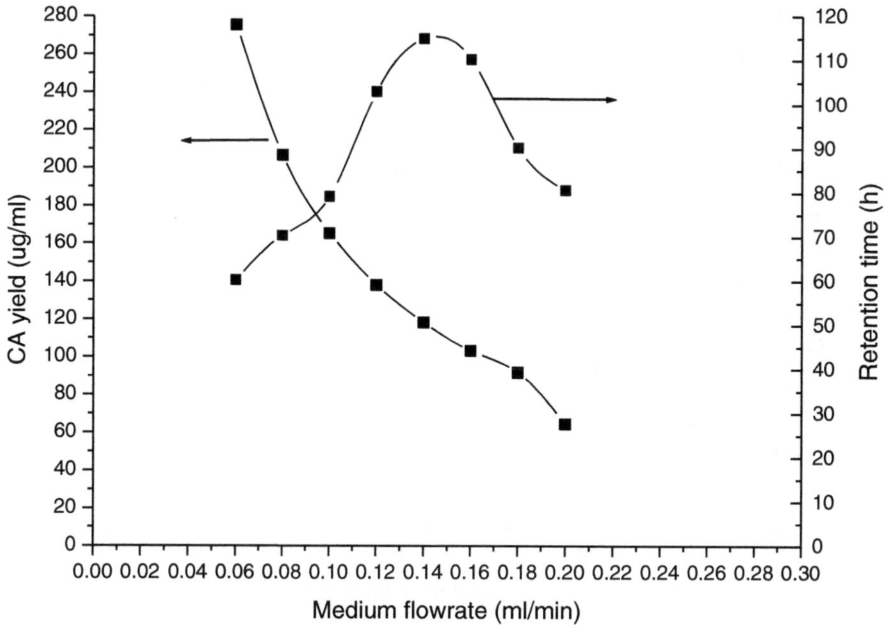

Fig. 6.34 Effect of flow of fermentation broth on clavulanic acid production at 29 °C; airflow was 10 ml/min

6.5.3.2 Results and Discussion: Effects on Clavulanic Acid Production

Effect of Fermentation Broth Flow

As shown in Fig. 6.34, with the increase in flow, the fermentation broth retention time in the bioreactor decreased. When the flow was in the range of 0.06–0.14 ml/min, with the increase in flow, the yield of clavulanic acid increased, and the yield reached the maximum of 268 μg/ml at a flow of 0.14 ml/min. When the flow increased further, the yield decreased.

The carrier liquid adsorption amount was limited. With the continuous addition of fermentation broth, the liquid amount increased and reached the maximum adsorption amount. In this condition, the added fermentation broth flowed downstream gradually and finally separated with carrier. The retention time was from one drop of medium added to one drop of medium discharged; it was also the fermentation time of clavulanic acid. The effect of fermentation flow on the yield of clavulanic acid was achieved by influencing the fermentation time.

As shown in Fig. 6.35, the fermentation broth retention time was inversely proportional to the amount of fermentation broth. With the increase in flow, the retention time decreased, and the fermentation time also decreased. From the results of flask fermentation of clavulanic acid discussed in Sect. 6.5.1, the maximum yield appeared at 48 h. This continuous fermentation experiment confirmed those results.

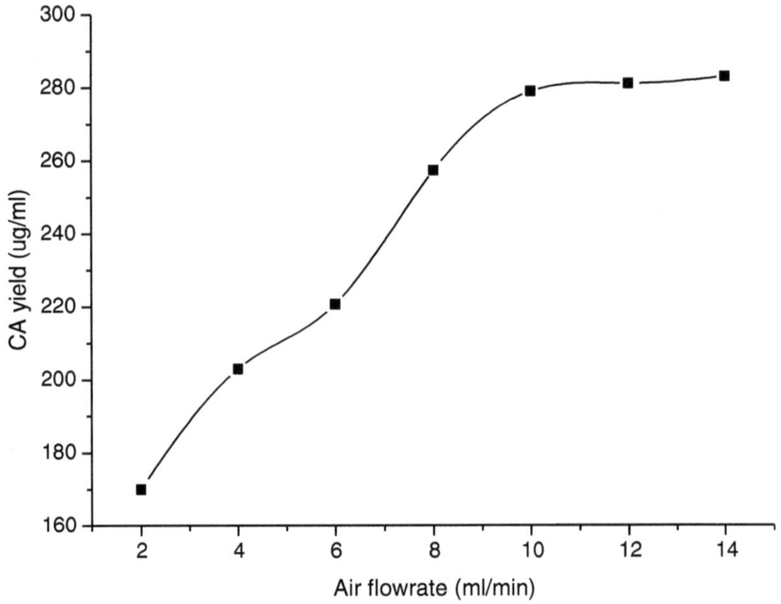

Fig. 6.35 Effect of airflow on clavulanic acid production

When the flow was 0.5 ml/min, the retention time was 47.8 h, which is close to 48 h. Under this flow, the yield of clavulanic acid reached the maximum. When the flow was too fast, the microorganism did not have sufficient time to produce products. In addition, the amount of fermentation broth accumulated in the carrier and built up the pore; air transfer was limited. The microorganism could not obtain adequate oxygen. When the flow was too slow, the retention time was too long, and the amount of clavulanic acid was degraded.

Effect of Airflow

A prominent advantage of ACSSF is good aeration. Microorganisms can obtain sufficient oxygen without mechanical agitation. As shown in Fig. 6.35, when the airflow increased from 2 to 14 ml/min, the yield increased throughout. When the airflow reached 10 ml/min, the yield increased slowly.

Because the carrier was deformed by extrusion, there were dead areas in the bioreactor, which resulted in pores becoming smaller, and the effective absorption amount decreased. The fermentation broth may stagnate in these areas. These all influenced the yield of clavulanic acid. This could improve by changing the airflow.

When the airflow was slow, the air flowed slowly among the pores of the carriers and could not reach the dead areas, which resulted in insufficient oxygen transfer and a decrease in fermentation efficiency. With the increase in airflow, the air flowed fast and loosened the squeezed areas, which reduced the diverse effect of the

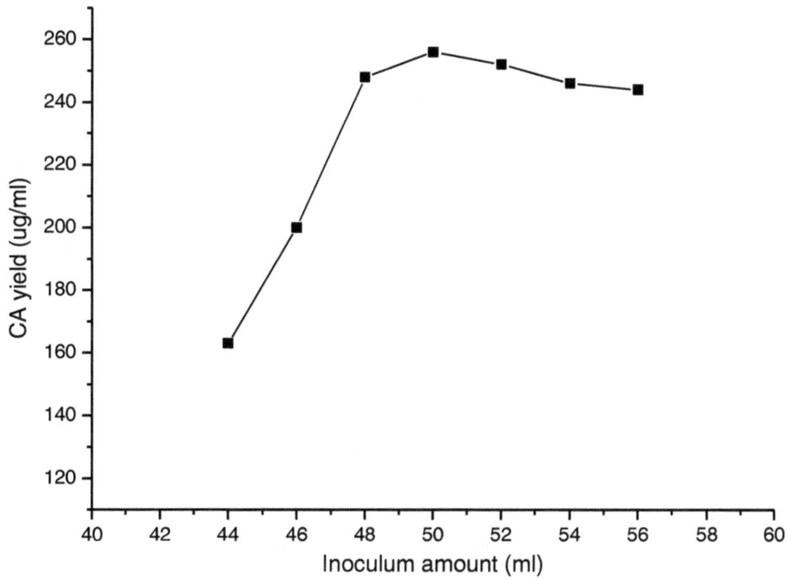

Fig. 6.36 Effect of inoculum amount on clavulanic acid production

carrier depth. The increase in airflow was not unlimited; higher flow might result in diverse effects. If the airflow were extremely high and the ability to penetrate were high, short-circuited channels might appear. In these channel areas, the microorganisms could obtain sufficient oxygen, and microorganisms could not obtain sufficient oxygen in regions other than the channels.

Effect of Inoculum Amount

In the results presented in Sect. 6.5.1, the best inoculum amount was 0.5 ml seed broth per 1 g carrier, and the solid/liquid ratio was 1:5. In the scale-up experiment, this result may be different from the flask experiment. In this experiment, the carrier loading was 100 g; the seed broth amount added was about 50 ml. As shown in Fig. 6.36, when there was little inoculum, as the amount increased, the yield of clavulanic acid increased and reached a maximum at an inoculation amount of 52 ml. There was no significant increase in the yield with the further increase in the inoculation amount.

In the continuous ACSSF process, part of the microorganisms could be absorbed on the surface of the carriers. These immobilized cells could not flow out along the fermentation broth. When the fresh fermentation broth passed through these areas, the immobilized cells could grow in the fresh fermentation broth. Therefore, this fermentation method only needed one inoculation, which overcame the problem of repeated inoculation in traditional SSF.

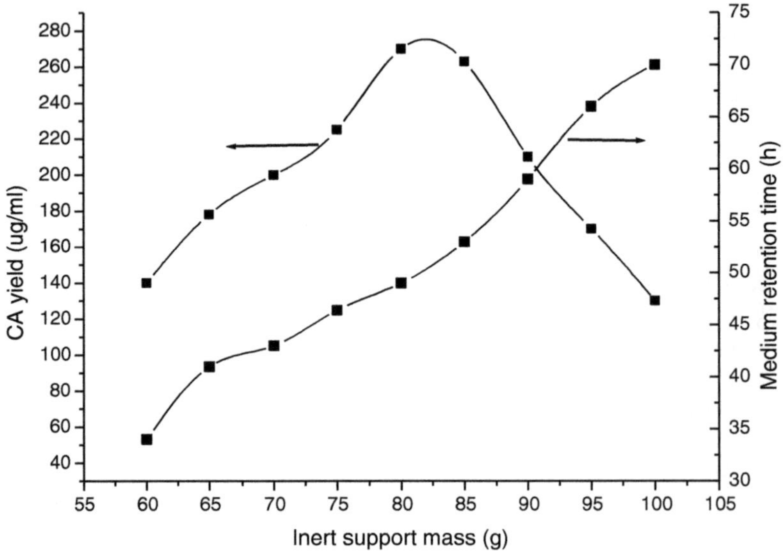

Fig. 6.37 Effect of carrier loading on clavulanic acid production

Effect of Carrier Loading

In continuous ACSSF, the carrier loading could influence the fermentation broth retention time, the effective volume of carrier, and the aeration condition. In this bioreactor, the height was 48.5 cm; the inner diameter was 12 cm. It could be filled with 100 g carrier. As shown in Fig. 6.37, with the increase in carrier loading, the retention time of the fermentation broth was prolonged; the yield of clavulanic acid increased first and then decreased. The maximum yield was 263 μg/ml when the carrier loading was 85 g.

Effect of Carrier Size

As shown in Fig. 6.38, with the increase in carrier size, as the fermentation broth retention time shortened, the yield of clavulanic acid increased first and then decreased. The maximum yield was obtained when the carrier size was $1 \times 1 \times 1$ cm.

Effect of pH

The same time as clavulanic acid was produced by *Streptomyces clavuligerus*, clavulanic acid was degraded. Although this was a means of microbial protection, it was unfavorable for industrial production. The pH is an important factor affecting

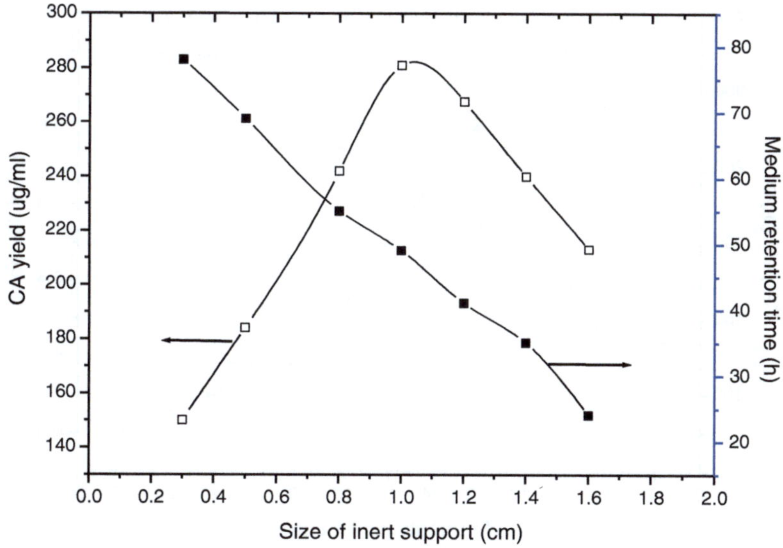

Fig. 6.38 Effect of carrier size on clavulanic acid production

clavulanic acid decomposition. At the beginning of fermentation, the time of microorganism growth and reproduction, the pH was near 7.0. At the later stage, the cells began to produce clavulanic acid, and the pH of the fermentation broth at that time also began to increase. It was easy to degrade clavulanic acid in the alkaline condition. It can be seen that the optimum growth pH and the optimum production of clavulanic acid pH were different. To obtain the maximum yield, the pH should be adjusted at different stages. At the beginning of fermentation, the pH should be adjusted to 7.0 for suitable growth. In the late stage, clavulanic acid is produced at a high rate, and the pH should be adjusted to 6.0 to slow decomposition. As shown in Fig. 6.39, by adjusting the pH, the yield of clavulanic acid increased.

6.5.3.3 Conclusion

The purpose of this study was to design a bioreactor and techniques to produce clavulanic acid and evaluate the feasibility of ACSSF in scaled-up production. The conclusions are as follows:

1. The fermentation broth flow influenced the yield of clavulanic acid by changing the fermentation broth retention time.
2. Increase in the airflow improved the aeration condition and benefitted microorganism growth.
3. The carrier could only provide limited space for microorganism growth. The inoculation amount may influence the yield of clavulanic acid significantly.

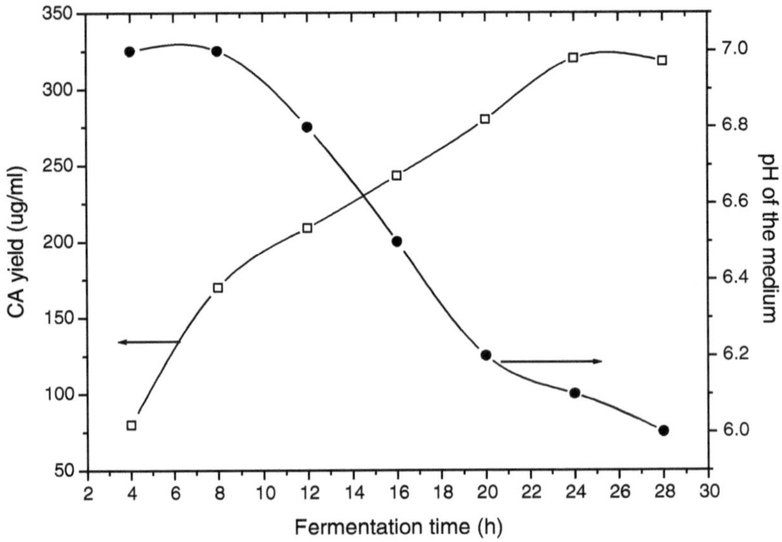

Fig. 6.39 Effect of pH on clavulanic acid production

4. The carrier loading influenced the yield of clavulanic acid by affecting the fermentation broth retention time.
5. The carrier size determined the porosity of the carrier and influenced the yield of clavulanic acid by affecting the retention time.
6. The optimum growth pH and the optimum production of clavulanic acid pH were different. To obtain the maximum yield, the pH should be adjusted at different stages.

6.5.4 *Xanthan Production by SSF on PUF*

Xanthan gum is an extracellular heteropolysaccharide produced by *Xanthomonas campestris*. Because of its unique rheological properties, such as high viscosity and pseudoplasticity, xanthan has wide applications in industries such as those for food, pharmaceuticals, and oil and in other industries.

Xanthan is usually produced industrially by SmF. Because of the high viscosity of solutions that contain xanthan gum, there is stagnant fluid in the fermentor when the gum concentration is high. The oxygen transfer resistance caused by these stagnant zones is a primary restriction in xanthan production (SmF).

Compared with SmF, SSF has several advantages: good aeration, low energy consumption, and high yields. Compared with other materials, such as spent malt grains and ion exchange resins, PUF has a more homogeneous porous structure, and PUF can be reused. More important, posttreatment is simpler as the broth in the PUF cubes can be collected easily without dilution.

Zhang (2010) studied the suitability of SSF on PUF for xanthan preparation and evaluated the effects of inoculum amount, the size of the PUF cube, solid/liquid ratio, and bed depth on xanthan production.

6.5.4.1 Fermentation and Analysis of Xanthan in SSF on PUF

Xanthan Production in SSF on PUF

For xanthan SSF production on PUF, the culture medium was absorbed on PUF. The culture medium consisted of 30 g/L glucose, 1 g/L NH_4NO_3, 2 g/L yeast extract powder, 0.25 g/L $MgSO_4 \cdot 7H_2O$, 0.1 g/L Na_2HPO_4, and 3 g/L $CaCO_3$.

The PUF, which was used as an inert support, was cut into cubes of desired side lengths (including 0.5, 1.0, and 1.5 cm), washed twice with tap water, and dried in an oven before use. Media and PUF cubes in all experiments were sterilized in an autoclave at 121 °C for 15 min before use.

The experiments to determine the suitable SSF culture conditions were carried out in 150-ml beakers (6 cm in diameter) containing PUF cubes, and the beakers were cultured in an incubator. After the beakers containing PUF cubes were sterilized and cooled to room temperature, culture media and inoculum were added. Then, the PUF cubes in these beakers were stirred using a glass stick under strict aseptic conditions. The process parameters were varied: inoculum size (5–25 %), side length of PUF cubes (0.5–1.5 cm), volume of added moistening medium per gram PUF cubes (5–25 ml), and bed depth (1.5–6 cm). The procedure adopted for the determination of xanthan production was to evaluate the effects of individual parameters while keeping all other parameters constant. All the beakers were cultured at 30 °C for 72 h.

Biomass, Glucose, and Xanthan Concentration Determination

In liquid cultures, the pH was adjusted to 3 with an HCl solution (5 mol/L) at the end of fermentation to remove the residual $CaCO_3$ in the broth. Then, 2 ml of the broth were used to determine the biomass and sugar concentrations. The sample was diluted by a factor of 10 with deionized water. A 10-ml portion of the diluted sample was centrifuged at 14,000 *g* for 30 min at 5 °C. The pellet was resuspended in deionized water and analyzed with a spectrophotometer at 560 nm to obtain the biomass concentration. The rest of the diluted sample was used to determine the glucose concentration using the dinitrosalicylic acid reagent method. After the biomass and sugar concentrations were determined, the rest of the broth was precipitated with three volumes of 95 % ethanol to determine the xanthan concentration. The precipitation was filtered and dried at 60 °C for 24 h. The dried sample was then weighed to calculate the raw polysaccharide concentration. The xanthan concentration was determined as the difference between the concentration of raw polysaccharide and the total biomass.

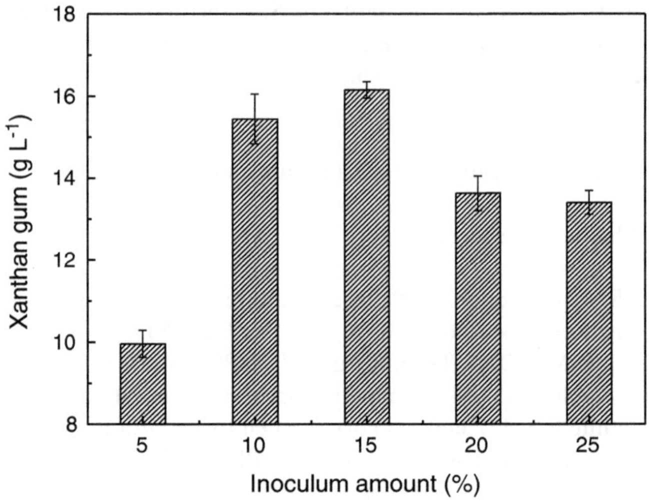

Fig. 6.40 Influence of inoculum amount on xanthan yield. Side length of polyurethane foam (*PUF*) cubes, 0.5 cm; bed depth, 3 cm; moistening medium volume, 10 ml/g PUF cubes; fermentation time, 72 h

In solid-state cultivation, biomass, sugar, and xanthan concentrations were determined after the broth in PUF cubes was collected using a plastic 60-ml syringe, at the front end of which several 1-mm diameter holes were punched. To obtain accurate values for the xanthan concentrations, three volumes of deionized water were added to dilute the broth in the PUF before it was collected. The biomass, glucose, and xanthan concentrations in the broth were calculated using the dilution proportion.

6.5.4.2 Results and Discussion

To determine suitable culture conditions for static SSF, four factors that affected xanthan yields were investigated: the amount of inoculum, the size of the PUF cubes, the ratio of media to cubes, and the bed depth.

Effect of Inoculum Amount on Xanthan Yields

The effect of the inoculum amount on xanthan yields is shown in Fig. 6.40. The xanthan concentration increased as the amount of inoculum increased up to 15 % (v/v), and the maximum gum concentration was 16.1 g/L. When the amount of inoculum was over 15 %, the xanthan concentration decreased with increases in the inoculum.

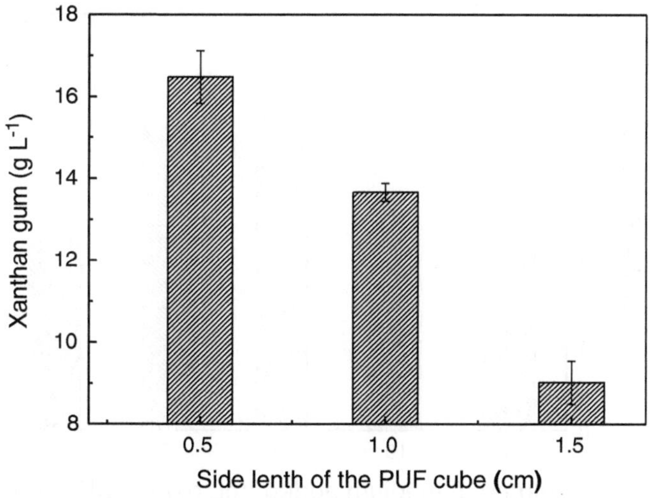

Fig. 6.41 Effect of inert support (polyurethane foam [*PUF*] cubes) size on xanthan gum yield. Inoculum amount, 15 % (v/v); bed depth, 3 cm; moistening medium volume, 10 ml/g PUF cubes; fermentation time, 72 h

Effect of Inert Support (PUF Cubes) Size on Xanthan Gum Yields

Figure 6.41 shows the effect of the size of PUF cubes on xanthan yields. In the present study, the PUF was cut into small cubes with side lengths of 0.5, 1.0, and 1.5 cm. The final concentration of xanthan gum decreased as the size of the cube increased. When cubes with a side length of 0.5 cm were used, 16.5 g/L xanthan was obtained. When the side length was 1.5 cm, the xanthan concentration in the broth was only 9.0 g/L. The results suggest that, under static conditions, smaller PUF cubes are better for xanthan production. Cubes with sizes smaller than $0.5 \times 0.5 \times 0.5$ cm were not convenient to prepare, so we did not culture the *X. campestris* on smaller cubes.

Effect of the Ratio of Medium Volume to PUF Cubes on Xanthan Yields

Figure 6.42 shows the effect of the ratio of medium volume to PUF cubes on xanthan yield. The xanthan yield appeared to decrease when the ratio of medium volume to PUF cubes increased, and the maximum xanthan amount obtained was 15.9 g/L at the ratio of 15:1. The observation that too much medium reduced xanthan gum yield may be because the amount of medium exceeding the dispersion capacity of the PUF restricted oxygen transfer from the environment to the interior of the PUF cubes.

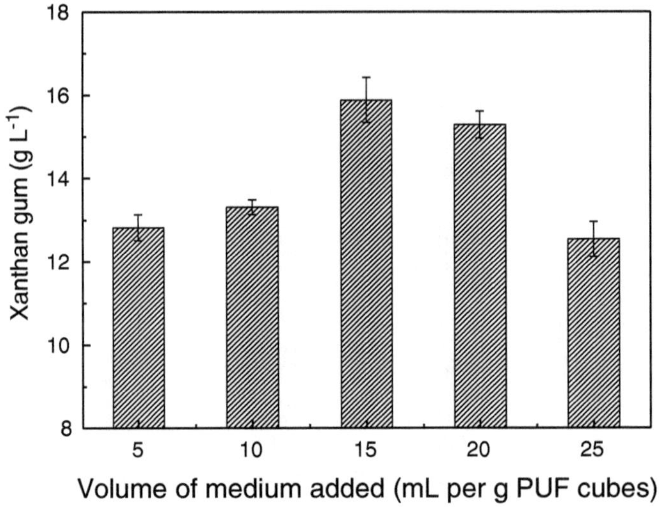

Fig. 6.42 Effect of moistening medium volume on xanthan yield. Inoculum amount, 10 % (v/v); side length of polyurethane foam (*PUF*) cubes, 0.5 cm; bed depth, 4.5 cm; fermentation time, 72 h

Effect of Bed Depth on Xanthan Gum Yields

The effect of bed depth on xanthan gum yields is shown in Fig. 6.43. Xanthan yield increased as the bed depth increased up to 4.5 cm, and the maximum concentration obtained was 17.3 g/L. When the bed depth was over 4.5 cm, the gum yield decreased with increasing bed depth. The bed depth of the substrate was an important parameter. In the present study, oxygen was supplied by air infiltration. Bed depths that were too high negatively affected the respiration of the bacteria.

From these results, the suitable conditions were 15 % inoculum, 0.5-cm (side length) PUF cubes, 15 ml medium per gram PUF cube, and a 4.5-cm bed depth.

6.5.4.3 Conclusion

The use of SSF on PUF is suitable for xanthan preparation, especially when the initial glucose concentration is between 60 and 80 g/L. Hence, SSF on PUF has the potential to be used in xanthan production.

6.5.5 Bacterial Cellulose Production by SSF on PUF

Bacterial cellulose is an extracellular product of vinegar bacteria. It is almost pure cellulose and contains no lignin or other foreign substances. It is the same as cell wall cellulose according to its chemical composition and reactivity.

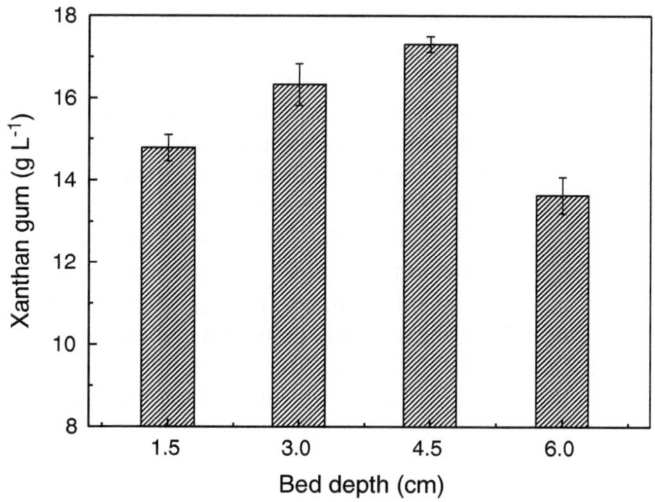

Fig. 6.43 Effect of bed depth on xanthan yield. Inoculum amount, 15 % (v/v); side length of polyurethane foam (*PUF*) cubes, 0.5 cm; moistening medium volume, 10 ml/g PUF cubes; fermentation time, 72 h

Bacterial cellulose is less than 130 nm wide; it is composed of a bundle of much finer microfibrils, less than 10-nm diameter. Bacterial cellulose belongs crystallographically to cellulose I, commonly with natural cellulose of vegetable origin, in which two cellobiose units are arranged parallel in a unit cell, and those cellulose molecules tend to have a specific planar orientation in dried film. The long-chain molecules are aligned parallel in the extended form.

Compared with plant cellulose, bacterial cellulose has some advantages, such as high purity, degree of polymerization, crystallinity, hydrophilicity, Young's modulus, strength, and so on. Bacterial cellulose also has high biocompatibility and good biodegradability. Therefore, it has extensive applications. Among various applications studied so far, it has reached the level of practical use for acoustic diaphragms as bacterial cellulose has been found to bear two essential properties: high sonic velocity and low dynamic loss. The use of films as the raw material for conductive carbon film was investigated and found excellent, although this remains in the laboratory. The use of fragmented bacterial cellulose for papermaking is promising, and test pieces of flexure-durable papers and papers with high filler content, ideal for banknotes and bibles, have been prepared. Other ideas include the use of sheet or film as a temporary skin for medical care and as a separation membrane, the use of fragmented suspension as a viscosity-enhancing agent for various purposes, and others.

The species of bacteria that produces cellulose is generally called *Acetobacter xylinum*. Culture is carried out normally in a static condition at around 28–30 °C. The system becomes turbid, and after a while, a white pellicle appears on the surface; its thickness increases steadily with time, reaching over 25 mm in 4 weeks.

It is important to note that in the process of gel growth, the aerobic bacteria generate cellulose only in the vicinity of the surface, so that productivity depends primarily on surface area, not on vessel volume. With the aim of enhancing productivity, culture in agitated conditions has been studied, although a flat gel is no longer obtained, and use has to be limited to such applications as papermaking.

Chen and Weng tried bacterial cellulose production in SSF on PUF by *Acetobacter xylinum* (Weng 2010). Growth and bacterial cellulose production of *Acetobacter xylinum* on PUF (PUF), which serves as an inert support, were studied under different conditions, including the ratio of culture medium to PUF cubes, bed depth, and initial glucose concentration. Maximal productivity of bacterial cellulose was obtained with 16 ml medium per gram PUF cubes, 3-cm bed depth, and a glucose concentration of 20 g/L. After 72-h fermentation, the concentration of bacterial cellulose was 4.86 g/L, which amounts to 6.65 times that of SmF.

6.6 Perspective on ACSSF

6.6.1 Economy of Commercial Systems

The realization of the potential of SSF on impregnated inert supports as a commercial system depends on its feasibility (Ooijkaas et al. 2000). Aspects to consider include (1) costs of the medium, (2) fermentor costs, (3) costs of downstream processing and formulation, (4) registration costs, and (5) product value. The higher costs of the defined media make natural media the preferred choice, especially for low-cost products. Natural media make use of agricultural materials that are usually inexpensive and widely available. However, for products with high added value, such as metabolites, specific enzymes, or biologically active products, SSF on inert supports can be used because the media costs are normally a fraction of the overall production costs. For example, the preliminary cost calculations for spore production of *C. minitans* on chemically defined media have shown that fermentation costs account for more than 80 % and substrate costs account for less than 20 % of the production costs. In these calculations, only the costs for the defined medium and the fermentation were taken into account. Therefore, if downstream processing, formulation, and registration are also considered, the media costs will become even less important.

It is known that downstream processing can markedly affect the overall production costs owing to recovery losses. Improving the recovery yield and decreasing the number of steps in downstream processing can enhance this situation. Because downstream processing is simplified and improved with inert supports compared with natural substrates, it is anticipated that this will affect the overall production costs in a positive way and consequently outweigh the higher media costs.

6.6.2 *Challenges in ACSSF*

6.6.2.1 Selection of Inert Carriers

At present, the inert carriers used in SSF can be divided into two types. One type is natural materials, such as lignocelluloses (sugarcane bagasse, cassava residue, cocoa coconut, and agricultural residues) and rock (vermiculite, perlite). Another type is synthetic carriers, such as ion exchange resin, polystyrene, PUF, and SiO_2 xerogel. Although these carriers are effective and appear better application foreground compared with nutrient supports, there are still many problems in the selection of inert carriers.

Existing carriers, whether natural or synthetic, are readily available material, and selection of a particular carrier according to special microorganisms has not developed. The living environment and the characteristics of each microorganism are different; it is impossible that there is a carrier suitable for all microorganisms.

So far, there are no criteria for inert carrier selection, and only selecting the most suitable carrier according to particular microbial characteristics can obtain the highest product yields. The choice of carriers needs to consider various factors, such as porous structure, pore size, water absorbency, water retention capacity, mechanical strength, toxicity to microorganisms, stability, whether reused repeatedly, price, and so on.

Currently, PUF and polystyrene are the two carriers studied in depth, but there are many deficiencies in these two carriers. The materials themselves are hydrophobic; the water absorption rate is low. The mechanical strength is low and easy to deform by squeezing. These all reduce the carrier's service life.

Some carriers may obtain a higher yield at the laboratory scale, but the same carrier may not be able to obtain the desired results at a larger scale. In addition, the choice of the carriers depends on the fermentor and the need to consider some other features of the carriers. The carrier cannot form lumps in the fermentor, which will cause difficulty in oxygen transfer and stirring. The carrier should be able to withstand the shearing force generated by stirring to maintain its original structure.

There are two methods to solve these problems. One is to continue to look for suitable carriers by expanding the selection range. Another is to synthesize suitable carriers or improve the existing carriers according to particular microorganisms.

6.6.2.2 Reactor Design

For a fermentation method, whether it is effective or has a development prospect, the key is its actual effect on the industry, that is, whether the reactor is suitable for industrial production. In theory, ACSSF achieves continuous SSF. In the entire process, only one inoculation occurs, and the process parameters are adjustable. But so far, few fermentors are designed for ACSSF. As a potential fermentation device, only combination with advanced technologies can enable its strengths.

Designing a suitable fermentor and combining automated control and online monitoring techniques should be considered. They are the future developmental direction of ACSSF.

References

Acuña-Argüelles M, Gutierrez-Rojas M, Viniegra-Gonzalez G, Favela-Torres E. Effect of water activity on exo-pectinase production by *Aspergillus niger* CH4 on solid state fermentation. Biotechnol Lett. 1994;16:23–8.

Acuña-Argüelles M, Gutierrez-Rojas M, Viniegra-González G, Favela-Torres E. Production and properties of three pectinolytic activities produced by *Aspergillus niger* in submerged and solid-state fermentation. Appl Microbiol Biotechnol. 1995;43:808–14.

Baños JG, Tomasini A, Szakács G, Barrios-González J. High lovastatin production by *Aspergillus terreus* in solid-state fermentation on polyurethane foam: an artificial inert support. J Biosci Bioeng. 2009;108:105–10.

Barrios-Gonzalez J, Tomasini A, Viniegra-Gonzalez G, Lopez L. Penicillin production by solid state fermentation. Biotechnol Lett. 1988;10:793–8.

Barrios-Gonzalez J, González H, Mejia A. Effect of particle size, packing density and agitation on penicillin production in solid state fermentation. Biotechnol Adv. 1993;11:539–47.

Cahn F. Citric acid fermentation on solid materials. Ind Eng Chem. 1935;27:201–4.

Chen HZ. An apparatus and method for continuous solid state fermentation on inert carriers. Chinese Patent 200410101617.9; 2004.

Chen HZ, Xu J. Principle and application of current solid state fermentation. Beijing: Chemical Industry Press; 2004.

Chen HZ, Wang H, Zhang AJ, Li ZH. Alkaline protease production by solid state fermentation on polyurethane foam. Chem Biochem Eng Q. 2006;20:93–7.

Chen HZ, Li C, Li ZH. Device for solid state fermentation on porous carriers. Chinese Patent 02146763.3; 2002.

Christen P, Auria R, Vega C, Villegas E, Revah S. Growth of *Candida utilis* in solid state fermentation. Biotechnol Adv. 1993;11:549–57.

Gautam P, Sabu A, Pandey A, Szakacs G, Soccol CR. Microbial production of extra-cellular phytase using polystyrene as inert solid support. Bioresour Technol. 2002;83:229–33.

Hernandez M, Lonsane B, Raimbault M, Roussos S. Spectra of ergot alkaloids produced by *Claviceps purpurea* in solid state system-influence of the composition of liquid medium used for impregnating sugar cane bagasse. Process Biochem. 1993;28:23–7.

Hrubant G, Rhodes R, Orton W. Continuous solid-substrate fermentation of feedlot waste with grain. Biol Waste. 1989;28:277–91.

Hu T, Zhou Y, Dai L, Wang Y, Liu D, Zhang J, et al. Enhanced cellulase production by solid state fermentation with polyurethane foam as inert supports. Procedia Eng. 2011;18:335–40.

Huerta S, Favela E, Lopez-Ulibarri R, Fonseca A, Viniegra-Gonzalez G, Gutierrez-Rojas M. Absorbed substrate fermentation for pectinase production with *Aspergillus niger*. Biotechnol Technol. 1994;8:837–42.

Lareo C, Sposito AF, Bossio AL, Volpe DC. Characterization of growth and sporulation of *Mucor bacilliformis* in solid state fermentation on an inert support. Enzyme Microbiol Technol. 2006;38:391–9.

Larroche C, Gros J. Strategies for spore production by *Penicillium roquefortii* using solid state fermentation techniques. Process Biochem. 1989;24:97–103.

Marin-Cervantes MC, Matsumoto Y, Ramírez-Coutiño L, Rocha-Pino Z, Viniegra G, Shirai K. Effect of moisture content in polyurethane foams as support for solid-substrate fermentation of *Lecanicillium lecanii* on the production profiles of chitinases. Process Biochem. 2008;43:24–32.

Meyrath J. Production of amylase on vermiculite by *Aspergillus oryzae*. Sci Food Agric. 1965;16:14–8.

Munoz G, Agosin E, Cotoras M, Martin RS, Volpe D. Comparison of aerial and submerged spore properties for *Trichoderma harzianum*. FEMS Microbiol Lett. 1995;125:63–9.

Nagendra Prabhu G, Chandrasekaran M. Polystyrene—an inert carrier for LXXX-glutaminase production by marine *Vibrio costicola* under solid-state fermentation. World J Microbiol Biotechnol. 1995;11:683–4.

Nagendra Prabhu G, Chandrasekaran M. Impact of process parameters on L-glutaminase production by marine *Vibrio costicola* in solid state fermentation using polystyrene as an inert support. Process Biochem. 1997;32:285–9.

Nampoothiri KM, Pandey A. Solid state fermentation for L-glutamic acid production using *Brevibacterium* sp. Biotechnol Lett. 1996;18:199–204.

Ooijkaas LP, Weber FJ, Buitelaar RM, Tramper J, Rinzema A. Defined media and inert supports: their potential as solid-state fermentation production systems. Trends Biotechnol. 2000;18:356–60.

Oriol E, Schettino B, Viniegra-Gonzales G, Raimbault M. Solid-state culture of *Aspergillus niger* on support. J Ferment Technol. 1988;66:57–62.

Ozawa S, Sato K, Endo I. Repeated batch production of alkaline protease by solid-state fermentation using urethane foam as carriers. Bioprocess Biosyst Eng. 1996;14:63–8.

Pandey A. Solid-state fermentation. Biochem Eng J. 2003;13:81–4.

Pandey A, Soccol CR, Nigam P, Soccol VT. Biotechnological potential of agro-industrial residues. I: sugarcane bagasse. Bioresour Technol. 2000;74:69–80.

Peralta-Perez M, Saucedo-Castaneda G, Gutierrez-Rojas M, Campero A. SiO_2 xerogel: a suitable inert support for microbial growth. J Sol-Gel Sci Technol. 2001;20:105–10.

Richard A, Sergio H, Maurice R, Sergio R. Ion exchange resin: a model support for solid state growth fermentation of *Aspergillus niger*. Biotechnol Tech. 1990;4:391–6.

Soccol C, Marin B, Raimbault M, Lebeault JM. Potential of solid state fermentation for production of L(+)-lactic acid by *Rhizopus oryzae*. Appl Microbiol Biotechnol. 1994;41:286–90.

Tomasini A, Fajardo C, Barrios-González J. Gibberellic acid production using different solid-state fermentation systems. World J Microbiol Biotechnol. 1997;13:203–6.

Viniegra-González G, Favela-Torres E, Aguilar CN, Rómero-Gomez SJ, Díaz-Godínez G, Augur C. Advantages of fungal enzyme production in solid state over liquid fermentation systems. Biochem Eng J. 2003;13:157–67.

Wang H. Preparation of adsorptive material based on liquefying steam-exploded wheat straw and its application in solid state fermentation. Doctoral dissertation, Institute of Process Engineering, Chinese Academy of Sciences, Beijing; 2006.

Weber FJ, Tramper J, Rinzema A. A simplified material and energy balance approach for process development and scale-up of *Coniothyrium minitans conidia* production by solid-state cultivation in a packed-bed reactor. Biotechnol Bioeng. 1999;65:447–58.

Weng YY. Production of bacterial cellulose by solid state fermentation on inert support. Master's dissertation, Institute of Process Engineering, Chinese Academy of Sciences, Beijing; 2010.

Weng YY, Chen HZ. Production of bacterial cellulose under solid state fermentation on polyurethane foam. J Cell Sci Technol. 2010;18:1–7.

Zhang ZG. Straw fraction by steam explosion and xanthan production. Doctoral dissertation, Institute of Process Engineering, Chinese Academy of Sciences, Beijing; 2010.

Zhang ZG, Chen HZ. Xanthan production on polyurethane foam and its enhancement by air pressure pulsation. Appl Biochem Biotechnol. 2010;162:2244–58.

Zhu Y, Smits J, Knol W, Bol J. A novel solid-state fermentation system using polyurethane foam as inert carrier. Biotechnol Lett. 1994;16:643–8.

Chapter 7
Development Trends and Application Prospects for Modern Solid-State Fermentation

Abstract Solid-state fermentation (SSF) has unique water- and energy-saving advantages. Some problems related to it need to be solved before effective industrial production. Theoretical and technical breakthroughs are needed to boost the progress of SSF technology and engineering. In-depth knowledge of the nature of SSF is necessary. Research and application of online monitoring technology are key points in regulation of the SSF process. A new solid-state bioreactor was designed and manufactured to meet the needs of current industrial production. It is important to study optimization of the mixed fermentation process. Multidisciplinary research on basic theory and industrial applications is needed, and breakthroughs in SSF for solution of problems should continue. An overview of biomass bioconversion by SSF and corresponding advances achieved in recent years are discussed, especially progress achieved by our group based on the characteristics of solid biomass.

Keywords Solid-state fermentation • Biomass • Value-added bioconversion • Porous media • Gas double dynamic solid-state fermentation

7.1 Development Trends for Modern Solid-State Fermentation

Solid-state fermentation (SSF) is an important part of the fermentation industry. It has become the core of the modern fermentation industry because of its environmental and economic advantages. Solid-state fermentation originated in China. Traditional SSF provides the flavor of fermented foods (e.g., cheese, tempeh, alcohol, etc.) that people consume each day. Composting of agriculture waste is a mixed SSF process that uses a variety of microorganisms. Solid-state fermentation is a cleaner production technology that has unique water- and energy-saving advantages. Since the mid-twentieth century, the SSF production application range has been extended to antibiotics, enzymes, organic acids, alcohol, biopesticides, and recombinant bacterial drugs. Some progress in modern SSF research has been made, such as large-scale, closed pure cultivation and online detection and control of fermentation environmental

DOI 10.1007/978-94-007-6043-1_7, © Springer Science+Business Media Dordrecht 2013

parameters (temperature, humidity, oxygen, carbon dioxide). The SSF operation mode has achieved mechanization and automation. Furthermore, there has been continuous invention of new solid-state fermentors, from tray bioreactor to packed bed bioreactor, rotary bioreactor, fluidized bed bioreactor, mixed bioreactor, and pressure pulsating fermentor. However, SSF production technology and equipment are not yet fully operational for actual production. Many parts need further research, such as the SSF process, production equipment design, automatic control system design, product separation and purification method optimization, the processing of production waste, and so on.

To develop SSF, several new measures should be studied. First, we should include accurate and fast methods in the SSF process for complex parameter characterization, detection, and regulation. Second, we must establish detection technology and platforms to comprehensively track and analyze the function of microorganisms, the dynamic change of functional protein, and the evolution of microbial populations. Molecular ecology can be used to analyze the evolution of microbial populations in the fermentation process. Biochips can be used for analysis of gene expression regulation. Functional proteomics technology can analyze the dynamic variation of active components. All of these are used to establish an effective model for fermentation parameter control and system optimization. Industrialization of SSF is difficult to achieve effectively mainly because of the lack of knowledge concerning transfer in the fermentation process, which makes the process hard to control and amplify.

Solid-state fermentation is similar to the fermentation process in nature, which is not uniform for cell growth, absorption of nutrients, or secretion of metabolites. The fermentation parameters are more difficult to detect and control, and biosensors from liquid fermentation cannot be used in SSF. So far, there is no comprehensive SSF mathematical model in the literature, and its research is still at the experience level (Huang et al. 2003).

Solid-state fermentation for industrial production needs to solve the following problems: (1) The parameters of SSF cannot be accurately regulated, and the automation production process is difficult to achieve. (2) Because of the difficulty in heat and mass transfer within particles and in removing the generated metabolic heat, a temperature gradient results between the material layers; microbial growth is inhibited when the fermentation proceeds at a large industrial scale. (3) The microbial strains used in SSF processes are only suitable for a low-water environment, so the range of products is limited. (4) There is easy contamination of microbes. (5) The design of the SSF bioreactor needs much attention.

7.1.1 In-Depth Knowledge of the Nature of Solid-State Fermentation

Solid-state fermentation refers to a system with little or no free water in the process of growing microorganisms on a solid substance. The process of maintaining microbial activity that requires water mainly uses bound water. Most researchers believe that SSF and solid substrate fermentation are the same concept. Here, they are unified and called SSF. We believe that the nature of the SSF process is microbial degradation of organic matter in the gas-liquid-solid three-phase environment. Media regularity in SSF needs to be characterized from the point of view of porous media, eventually establishing a dynamic model to explore the process of SSF reactions and to make theoretical SSF breakthroughs to boost the progress of technology and engineering.

7.1.2 Research and Application of Online Monitoring Technology in Solid-State Fermentation

The control of parameters is the key point of the regulation of the SSF process; it is also a core issue that needs to be solved in the industrialization process. However, there is little research on the monitoring and control technologies in SSF at home and abroad. The lack is mainly caused by difficulties in sampling and poor representativeness of sampling because of the characteristics of the solid substrate and the mobile gas phase. The heterogeneity of the solid substrate causes complexity in the evolution of microbial populations as well as their biochemical metabolism in the fermentation process. Therefore, the slow development of automated measurement has become an important factor restricting the SSF industrialization process.

At present, studies of SSF process parameters mainly focus on the environmental parameters (e.g., temperature, humidity, pressure, water activity, etc.), as well as the related metabolism substances (O_2, CO_2, and biomass). For the former, online detection is relatively simple and can be achieved utilizing a probe or instrument. For the latter, online detection is often difficult because of the complexity and heterogeneity of the fermentation substrate. On the other hand, the physical parameters of the substrate (such as particle size, substrate density, porosity, substrate heat capacity, solid-phase filled density, etc.) cannot be ignored, yet research in this area is poor. Therefore, it is a top priority of the current study to establish a comprehensive analysis of the fermentation process, such as functional genomics, functional proteomics, and system biology analytical methods, based on the laws of growth and metabolism of microorganisms and to establish accurate and fast SSF technology for characterization, detection, and regulation of the key parameters.

Duan and Chen (2012) established a new type of fractal dimension model combined with high-resolution digital image acquisition technology. The model can be used to characterize changes of cell growth in the SSF process online, as well as changes in the nutritional substrate morphology, providing a new way to optimize the SSF process.

7.1.3 Solid-State Fermentation Equipment Design and Manufacture

Currently, SSF equipment lacks automation. Compared to traditional open or semiopen fermentation equipment, the current solid-state fermentor mainly focuses on the tray bioreactor or ventilation pool, which shows poor fermentation controllability and low efficiency. The modern pure culture SSF equipment has achieved the basic requirements of large-scale pure culture, yet still lacks understanding of the factors that influence the transfer of heat and mass during the bioreactor amplification process (Hölker and Lenz 2005). On the other hand, at present, the existing fermentation industry engineering design mainly focuses on liquid fermentation technology, so there is seldom research in SSF, especially the engineering design and integration optimization of pure culture SSF, which results in the low energy efficiency and low reactor utilization. A series of problems must be solved in the SSF process (Krishna 2005), such as reactor amplification, prevention of contamination, and process monitoring.

There is a great gap between domestic and foreign fermentation equipment. Our requirements are for urgent design of new bioreactors and establishment of new technology, such as forced ventilation, temperature control, ease of operation, low investment, and so on.

7.1.4 Application of Mixed Culture in Solid-State Fermentation

Strains with excellent performance are required for microbial conversion. It was discovered from long-term experiments and production practice that many important biochemical processes must be accomplished by coculturing two or more microorganisms, which is called mixed fermentation (Chen et al. 2006). Mixed SSF is similar to the natural fermentation process, which has been widely used in production practice (Singhania et al. 2009). Compared to pure culture fermentation, mixed SSF has some absolute advantages, and some new substances will be produced during mixed SSF, so it can replace the production of pure culture SSF in many fields (Yang et al. 2011). This is mainly because of the synergistic effect of mixed microorganisms. Among microbes, there are various kinds of synergistic effect, such as those between bacteria and bacteria, bacteria and fungi, and fungi and fungi. Compared with mixed fermentation, the equipment and energy utilization efficiencies in pure culture fermentation are low, which results in a substantial increase in production costs. Multiple strains in mixed SSF can overcome these shortcomings (Yi et al. 2011). However, there are few studies of mixed fermentation optimization processes.

The production of cellulase by microbial mixed SSF has been an effective way to make full use of natural cellulose resources (Zhao et al. 2010). The cellulase produced by fungi was extracellular enzymes. The enzymes are generally secreted into the medium and then obtained by filtration and centrifugation. In practice, cellulase production from SSF has proved superior to liquid fermentation in such areas as stable quality, low investment costs, and low environmental pollution (Zhang 2012). The mixed culture takes advantage of the synergies of the different strains to expand the substrate range and improve product conversion efficiency. However, research needs to be more in depth and extensive because of significant complexity. From the industrial viewpoint, fermentations of mixed cultures are different from pure culture fermentation and its conditions, which are simple and have easy operation. In practice, it has been proved that the microbial growth can be detected during the various expansion stages of the strains by microscopic examination; the fermentation temperature, humidity, inoculum size, fermentation process, and their interaction relationship can be determined based on the number of microbes. SSF can produce high-quality product and avoid large-scale investment, serious environmental pollution, and other negative effects.

7.1.5 Multidisciplinary Crossing in Solid-State Fermentation

Solid-state fermentation originated from ancient food brewing, leading to two major research branches: aerobic and anaerobic SSF. After the historical development of the fermentation industry, SSF has become an important part of modern fermentation and engendered a research center. Obviously, because of the continuous development of science and technology, SSF needs to be developed in a multidisciplinary range, such as fluid dynamics, computational mathematics, porous medium theory, and the mathematical model. Consequently, comprehensive solution of the key technical issues, such as strain screening parameter detection, process control, equipment research and development, process integration, and large-scale engineering, can be obtained.

The world is experiencing a new technological revolution. Many disciplines are forming a multidisciplinary crossing that forces people to continue to develop and expand their knowledge. It is possible to solve critical problems in scientific fields in an interdisciplinary manner. For this reason, we should study SSF from the views of a large multidisciplinary field rather than the biological scope. Multidisciplinary research is needed for concerted efforts on the basic theory and industrial applications and for continued breakthroughs in SSF for solution of the problems.

7.2 Application Prospects for Modern Solid-State Fermentation

7.2.1 Introduction

Conversion of solid biomass into value-added products is desirable because of recent high crude oil prices and increasing concerns over global warming. Because of its abundance, renewability, and cost-effective characteristics, biomass (especially lignocellulosic feedstock) is therefore recognized as an effective alternative substrate. It is reported that annual yields of biomass residues exceed about 220 billion tons worldwide (Zhang et al. 2007). Straw makes up most of the abundant renewable resources; the world's annual output is over 2.9 billion tons. Of this amount, the proportions of cornstalks, wheat straw, rice straw, and barley straw are around 35, 21, 19, and 10 %, respectively. Biomass has always been a major source of energy for humankind and is presently estimated to supply 14 % of worldwide energy consumption. However, most biomass is not utilized because annual production of biomass is equivalent to ten times the world's annual energy consumption (McKendry 2002).

Bioconversion of biomass (i.e., disposal and utilization of biomass with modern biological technology) is crucial and significant for value-added utilization of solids because of the high-efficiency, low energy consumption, and nonpolluting characteristics of biomass. During the bioconversion process, solid biomass is mostly biodegraded into its constituent monosaccharides by microorganisms or enzymes and further produces high value-added chemicals and biologicals (Chen 2008). Unfortunately, the complex structure of solid biomass and high cost of key enzymes cause a dilemma. Thus, a number of breakthroughs are needed before value-added utilization of solids is achieved.

Solid-state fermentation has emerged as a potential technology that could employ solid material as either a natural substrate or an inert support. As a fermentation process occurring in the absence or near absence of free water, it can be especially attractive for the production of microbial products such as feed, fuel, food, industrial chemicals, and pharmaceutical products because SSF holds tremendous potential for metabolite production using solid wastes (Holker et al. 2004). In addition, superior productivity of SSF over submerged fermentation (SmF) has been confirmed in a number of products at the laboratory scale, although there are still many technological and economic challenges to be addressed in the development of SSF. Owing to its remarkable ability agroindustrial residue disposal, during past decades many patents and publications have appeared on fundamental and applied aspects of SSF to achieve industrial SSF.

This chapter attempts to prove the good applicability of SSF technology for solid biomass bioconversion. Based on characteristics of solid biomass and the purposes of SSF technology operated on biomass, a systematic framework for biomass bioconversion by SSF technology has been established. Corresponding advances achieved recently are introduced based on this framework.

7.2.2 *Applicability of Solid Biomass Used as Substrate*

7.2.2.1 Characteristics of Solid Substrates Used

Two types of SSF systems can be distinguished by the nature of the solid phase used. The system most commonly used involves cultivation on a natural material; the second most common system involves cultivation on an inert support impregnated with a liquid medium. Both of these substrates are insoluble in water but could absorb water onto their matrix, which provides the required moisture in the SSF system for microorganism growth and metabolic activities. This requires the SSF solid substrate to have water-holding capacity.

In addition, in the commonly used system, the main roles of substrate are as a nutrient source and supporting carrier. Polysaccharides (cellulose, hemicellulose, pectins, starch, etc.), lignin, protein, and other items often can be metabolized by different microorganisms as a source of carbon and energy. Furthermore, the solids could supply vitamins, minerals, and other nutrients. In a supporting role, substrate supplies a biological interface for microorganisms; generally, bacterial and yeast cultures could grow on the surface of substrate fibrils and particles, and fungal mycelia could penetrate into the substrate particles for nutrition (Ooijkaas et al. 2000).

Mass and heat transfer are more important for SSF because the continuous phase is air, compared with water in SmF (Chen and Xu 2004). The supporting role and the size of the substrate determine the void space air occupies. Thus, the porosity of materials, which governs the accessible surface area for microbial growth, is another important characteristic for solid substrate utilization (Krishna 2005).

7.2.2.2 Solid Biomass Bioconversion Characteristics

Biomass is plant material derived from the reaction between CO_2 in the air, water, and sunlight via photosynthesis to produce carbohydrates and other organics that build blocks of biomass. The particular physical and chemical features of biomass make its bioconversion process have some special characteristics.

Biomass Recalcitrance

The nature of biomass is polymeric. Biomass can be separated into chemical component parts: sugars (as cellulose, hemicellulose, or starch), lignin, protein, glycerides, terpenes, and others (Ragauskas et al. 2006). Natural resistance of these polymers to microbial and enzymatic deconstruction collectively is known as *biomass recalcitrance*. Himmel et al. (2007) summarized a series of corresponding factors that contribute to this recalcitrance. Also, they pointed out that chemical and structural features of biomass resulting from hydrogen bonding, including intra- and intermolecular hydrogen bonding and hydrogen bonding between cellulose and water (Himmel et al. 2007; Xiros et al. 2012; Ye and Farriol 2005), greatly affect liquid penetration or enzyme accessibility and activity. Generally, lignocellulosic material lacks solubility in most conventional solvents, although it has significant swelling ability in water. Thus, although solvent-free operation is the best option for "green" processes, at least a dispersant is required as a reaction medium because most biomass comprises insoluble and structurally complex solids.

Inhomogeneity

The inhomogeneity of bioconversion results from the types, batches, sources, morphologies, structures, and chemical compositions of biomass. Furthermore, the sluggish solubility of biomass components often makes the reactions concerning the substrate heterogeneous (Rinaldi and Schüth 2009). Our preliminary studies were of the inhomogeneity of straw and its effects on bioconversion; the heterogeneity of physical characteristics, such as particle size, water retention ability, volumetric heat capacity, and thermal conductivity, have influence on SSF transfer properties (Duan and Chen 2012). The inhomogeneity of chemical compositions and cell types of corn stover has been proved to affect its enzymatic hydrolysis and fermentation performance significantly (Chen et al. 2011a, b). All the results showed the importance of inhomogeneity in biomass bioconversion.

Substrate Specificity

Although some microorganisms and enzymes display relatively broad substrate specificity in biodegradation, the characteristics of a multicomponent compound mean that separate utilization of a single microorganism or enzyme cannot result in total utilization of biomass and high value-added application. In this respect, synergistic interaction of different enzymes and mixed fermentation are necessary for fractional conversion and refinement in value-added bioconversion.

Porosity

Porosity is one of the most important properties of solid materials; the capillary voids in lignocellulosic fibers include (1) gross capillaries such as the cell lumina, pit apertures, and pit membrane pores, which are visible via light microscopy with diameters in the range between 20 nm and 10 μm; and (2) cell wall capillaries, such as the spaces among microfibrils and cellulose molecules in the amorphous regions, which are about 0.5–7.5 nm in diameter (Fan et al. 1980).

Cellulytic enzyme molecules commonly range from 1.3 to 7.9 nm in diameter and have accessible pores equal to or larger than a nominal diameter of 5.1 nm (Sangseethong et al. 2007); β-glucose has a length of 0.58 nm along axis C-1 and C-4 (Rinaldi and Schüth 2009). The difference of pore sizes guarantees successful enzymatic hydrolysis of lignocellulose. All of these show clearly the importance of capillary structure, porosity, and pore size distribution for bioconversion of biomass (Grethlein 1985).

According to the International Union of Pure and Applied Chemistry (IUPAC) recommendation, porous materials are classified in three groups: microporous (pore size <2 nm), mesoporous (2–50 nm), and macroporous (>50 nm) (Wu et al. 2012). Solid biomass pore sizes range from microporous to mesoporous, and the different pores provide passage for mass and heat transfer in SSF.

Based on the porous structures of biomass, the bioconversion characteristics can be grouped into intraparticle properties (thermal properties, moisture content, particle size, porosity, kinetics and energetics of biological processes, etc.) and extraparticle properties (heat transfer from reactor to particle, permeability and mass transfer conditions, etc.). One of the most important is the capillary water-holding capacity of biomass caused by capillary structure. This bound water is crucial for microbial growth and fermentation; at the same time, the changing pores caused by water-swollen solids are favorable for biodegradation and bioconversion of biomass.

Essentially, the SSF system is a combination of solid, liquid, and gas phases. The solid matrix itself is the solid phase; the water held in the matrix is the liquid phase; and the gas in the large pores makes up the gas phase, which is the continuous phase in the SSF system (Chen and Xu 2004). That is, the substrate is a complex medium combined with a solid skeleton and fluid; it possesses the basic characteristics of porous medium suggested by Bear in 1972 (Bear 1972). Thus, we could analyze and study the SSF mass and heat transfer process using the corresponding theory of porous media.

In the three-phase structure of SSF substrates, the volume occupied by the continuous air phase within the substrate is dependent on the substrate characteristics and the water holding ability. The amount of holding water should be optimal. Too much water compacts the substrate, impedes oxygen transfer, and favors contamination. On the other hand, too little water creates adverse conditions that inhibit growth and enzyme activity and limit the transfer of nutrients (Raghavarao et al. 2003).

In terms of pore size distribution measurement, specific surface, or volume, the classic method is the Brunauer-Emmett-Teller (BET) method using nitrogen adsorption. However, this method measures the pores available to a nitrogen molecule, which is different from a cellulase molecule in size. Another problem with this technique involves dried substrate, which causes different pore characteristics compared with the substrate in its swollen state (Zhao et al. 2012). There are several methods to measure the pore size, such as mercury porosimetry, small-angle x-ray scattering, water vapor sorption, size exclusion, differential scanning calorimetry, and nuclear magnetic resonance. Hong et al. developed a novel and more accurate approach for quantitative determination of cellulose accessibility based on the adsorption of a nonhydrolytic fusion protein containing cellulose-binding module and a green fluorescent protein (Hong et al. 2007). But, all of these methods are not suitable for porosity detection in industrial SSF.

In short, the bioconversion characteristics of solid biomass mean that it is difficult to utilize these solid materials directly, simultaneously, or economically. However, solids are generally considered the best substrates for SSF processes and enzyme production because their given physical and chemical characteristic advantages despite the existent shortcoming such as biomass recalcitrance. This is because solids could be used as an SSF nutrient and support carrier, could supply space for heat and mass transfer, and could provide water-holding capacity.

I believe that, to achieve key breakthroughs in biomass bioconversion technology, researchers need to work from the perspectives of pretreatment, component fractionation, and fractional conversion to increase the porosity and solve the problems of biomass recalcitrance, inhomogeneity, and substrate specificity.

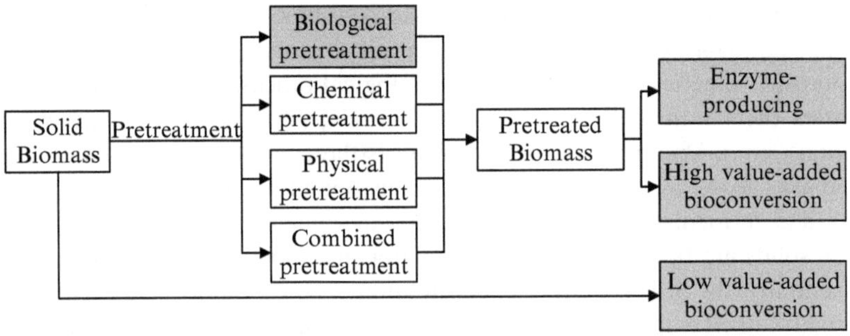

Fig. 7.1 Bioconversion technology for biomass based on solid-state fermentation (*SSF*)

7.2.3 Biomass Bioconversion Technology Based on Solid-State Fermentation

Based on the analysis of solid material characteristics, the applicability of solid biomass in SSF, and the research study of pretreatment, component fractionation, and fractional conversion, I propose a value-added bioconversion system for biomass focused on SSF in this chapter (Fig. 7.1). This system includes four parts according to the purpose of SSF technology used: biological pretreatment technology, enzyme production for biomass bioconversion, high value-added bioconversion of biomass, and low value-added bioconversion. This section introduces this system and corresponding advances achieved in recent years, especially progress achieved in my work group.

7.2.3.1 Biological Pretreatment Technology

In most cases, pretreatment of lignocellulosic biomass or the addition of available carbon sources is needed before fermentation because of the difficulty in direct use of polymers by microorganisms. This operation increases substrate porosity because of lignin biodegradation and redistribution. Therefore, it enables greater exposure of cellulose surface area for effective hydrolysis or fermentation with minimal energy consumption. Among various preprocessing technologies, biological pretreatment is desirable because it requires relatively low energy and mild environmental conditions (Sun and Cheng 2002). The progressive effect of biological pretreatment on enzymatic hydrolysis has been demonstrated previously (Xu et al. 2010).

In biological pretreatment processes, microorganisms such as brown, white, and soft rot fungi are mostly used to degrade lignocellulosic biomass. Brown rots mainly attack cellulose; white and soft rots attack both cellulose and lignin. Traditionally, white rot fungi are the most effective basidiomycetes for biological pretreatment of lignocellulosic materials (Sun and Cheng 2002).

This method is considered to be environmentally friendly and energy saving as it is performed at a low temperature and needs no chemicals. However, the rate of hydrolysis in most biological pretreatment processes is low, and the short processing period makes practical application a future possibility (Wan and Li 2012).

7.2.3.2 Enzyme Production for Biomass Bioconversion

Enzymes for biomass bioconversion are a series of hydrolases induced by biomass and could utilize biomass as a substrate. Cellulase, ligninase, xylanase, pectinase, and the like are the most common enzymes. Filamentous fungi, *Trichoderma*, *Penicillium*, *Aspergillus*, *Rhizopus*, and others are involved in the majority of enzyme-producing processes. Especially, *Trichoderma* and *Aspergillus* species are known for their capacity to secrete high levels of enzymes. Here, the example of cellulase production is used to introduce the gas double dynamic SSF (GDD-SSF) technology, which represents major progress in modern SSF. In addition, a corresponding comparison of cellulose production with other SSF technologies is made in this section.

Traditional SSF operates under static conditions, resulting in poor heat and mass transfer effects of steep gaseous concentration gradients and heat gradients, which may adversely affect solid-state fermentor performances. Much research attention is directed toward design development; for example, agitation and rotation in SSF are often carried out to improve mass and heat transfer (Mitchell et al. 2000). Large-scale fermentation has been carried out in tray-, drum-, or deep-trough-type fermentors; the shearing force caused by agitation and rotation has adverse effects on media porosity and disrupts fungal mycelia (Marsh et al. 2000).

The gas double dynamic SSF bioreactor (GDSFB) devised by my group has internal air circulation and periodic pulsation of air pressure (Fig. 7.2) (Zeng and Chen 2009; Chen et al. 2002). Pressure pulsation occurs through feeding and exhaling sterile air in the reactor; it can replace traditional mechanical mixing not only because it promotes evaporation and cooling (Yang et al. 2002; Liu et al. 2007; Liu and Yang 2006), but also because it is consistent with the principle of artificial periodic stimulation (Li and Chen 2005). Internal air circulation is another dynamic process. Its main purpose is to strengthen internal heat transfer. At different fermentation stages, the rate of air circulation in the bioreactor changes according to the varying speed of metabolic heat production. Thus, the GDSFB not only could provide sufficient O_2 but also could provide more room for fungal propagation and improve heat transfer within the substrate (Chen and Qiu 2010). The GDD-SSF technology has the potential to realize industrial production of enzymes.

A laboratory study showed that the average cellulase activity in a GDSFB could increase from 10.8 IU/g dry substrate (DS) to 20.4 IU/g DS (Xu et al. 2002). An industrial-scale 100-m^3 GDSFB was designed and manufactured to produce cellulase; it not only increased cellulase activity two to three times but also shortened the required fermentation period (Chen and Qiu 2010).

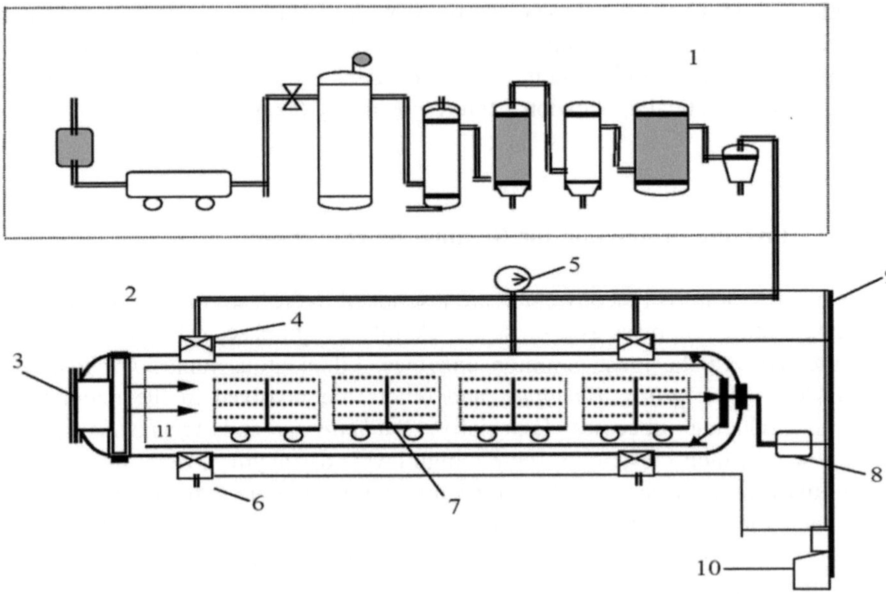

Fig. 7.2 Schematic of a gas double dynamic solid-state fermentation bioreactor (*GDSFB*) (Chen and Xu 2004). *1* Aseptic air system. *2* The GDSFB vessel. *3* Quick-opening mechanism. *4* Air inlet valve. *5* Electric contact pressure gauge. *6* Air outlet valve. *7* Media trays. *8* Circulation blower. *9* Controlling line. *10* Industrial computer. *11* Airflow circulation

Table 7.1 shows the productivity of cellulase in pilot- and large-scale solid-state fermentors, indicating that industrial-scale GDD-SSF had clear advantages in both enzyme activities and fermentor scale. Table 7.2 shows a comparative study of production of other enzymes and metabolites by GDD-SSF and static SSF at different scales. All of these results illustrate the great potential of GDD-SSF in industrial production of enzymes.

7.2.3.3 High Value-Added Bioconversion of Biomass

Solid-state fermentation may be utilized to obtain value-added compounds such as enzymes, mushrooms, amino acids, biopesticides, biofuels, biosurfactants, organic acids, flavors, colorants, aromatic compounds, biologically active secondary metabolites, and other substances of interest in the food industry (Chandel et al. 2012). However, high value-added bioconversion of biomass refers to the anaerobic SSF process of higher carbon utilization; there is no need for an oxygen supply, and there is less carbon dioxide emission and lower water and energy consumption in solid state biodegradation. Here I give a brief review of their independent development of the solid-state enzymatic saccharification, fermentation, and separation system and continuous anaerobic SSF coupled with CO_2 gas-stripping technology. Both technologies are introduced by using ethanol production as an example.

Table 7.1 Production of cellulase under solid-state fermentation in pilot and large scale (Chen and Qiu 2010)

Strains	Fermentor type	Mass and heat transfer mode	Cellulase activity (U/g dry substrate)	Reference
Trichoderma viride	100-m^3 gas double dynamic fermentor	Circulation and periodic air pulsation pressure	60	Unpublished results
Aspergillus niger and *Trichoderma reseei*	50-L tray fermentor	Laminar airflow	24.17	Dhillon et al. (2011)
Trichoderma harzianum	130-L packed bed	Without agitation	18.26	Roussos et al. (1993)
Trichoderma viride SL-1	4-m^3 rotary drum fermentor	Forced aeration	14	Sun et al. (1999)
Trichoderma harzianum	50-L rotary drum solid-state bioreactor	Rotating and air circulation	10.1	Alam (2009)
Thermoascus aurantiacus	10-L rotating drum-type bioreactor	Intermittent agitation	5.48	Kalogeris et al. (2003)

Table 7.2 Comparative study of enzyme production by gas double dynamic solid-state fermentation (GDD-SSF) and static solid-state fermentation (SSF)

Strains	Products	Static SSF	GDD-SSF	Scale of fermentor	Reference
Mycelia sterilia YY-5	Laccase	1,433 IU/g, 4 day	1,855.8 IU/g, 3 day	50 L	Unpublished results
Bacillus thuringiensis CM-1	Bacillas thuringiensis (Bt) wet powder		Above 18,000 IU/mg, 48 h	70 m^3	Chen et al. (2002)
Penicillium decumbens JUA10	Cellulase	10.8 IU/g, 84 h	20.4 IU/g, 60 h	50 L	Li and Chen (2008)
Aspergillus niger WH-4	Feruloyl esterase	557 mU/g, 72 h	881 mU/g, 48 h	25 L	Zeng and Chen (2009)

Solid-State Enzymatic Saccharification, Fermentation, and Separation System

To alleviate the problems of low substrate loading, nonisothermality, and product inhibition during simultaneous saccharification and fermentation, a nonisothermal simultaneous solid-state saccharification, fermentation, and separation process was investigated. This process involved solid-state enzymatic saccharification and fermentation coupled with a CO_2 gas-stripping loop system (Chen et al. 2004, 2007).

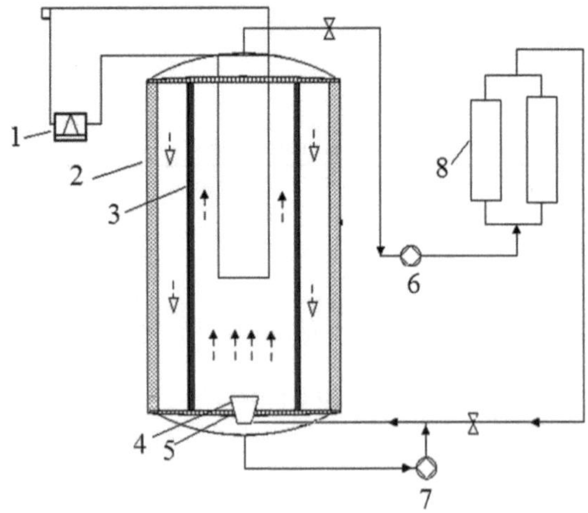

Fig. 7.3 Flowchart of coupling of divisional cycle-enzymatic hydrolysis, fermentation, and gas-stripping separation of ethanol (Chen et al. 2004). *1* Condensing water circulation device. *2* Outer cylinder with heating jacket. *3* Thermal insulation inner column. *4* Float valve. *5* Filter plate. *6* CO_2 circulation pump. *7* Hydrolyzate circulation pump. *8* Absorbed column

This system consists of a fermentor and an online separation device for the product (Fig. 7.3). A thermal insulation inner column divides the fermentor into an enzymatic saccharification zone and a fermentation zone. The temperatures of the separate zones are maintained by a condensing water circulation device and a heating jacket device. The divisional saccharification and fermentation are coupled by hydrolysate circulation. The online separation device consists of at least two parallel adsorbed columns and a CO_2 circulation pump. The CO_2 circulation pump is activated when the ethanol concentration is higher than 5 %. The mixture of CO_2 and carried ethanol enter one of the columns for ethanol adsorption and then switch to another column when the first is saturated. After saturation, ethanol is recovered by heating the column.

This novel industrial-level, 110-m^3 bioreactor was successfully put into operation in 2006. The operating results indicated an appropriate dosage of cellulase decreased to 20 IU/g dry straw at a solid substrate concentration of 150 g/L (i.e., the solid/liquid ratio in the system was higher than 1:7). The average yield of ethanol was about 15–16 % at 72 h, and the cellulose conversion ratio could reach 71.2 %. The final ethanol concentration desorbed from activated carbon could reach 30–40 % (unpublished results).

According to a variety of reports on ethanol production by a simultaneous saccharification and fermentation process, most raw materials used were cellulose powder, paper pulp wastewater, treated wheat straw, bagasse, and so on; the substrate concentration was about 50–150 g/L. Enzyme usage was 7–28 IU/g dry substrate; the cellulose conversion ratio was 48–85 %, and the ethanol concentration could reach 14–47 g/L with a fermentation period of 1–7 days. Thus, the novel industrial-level bioreactor could successfully provide better results than presently provided in large-scale production.

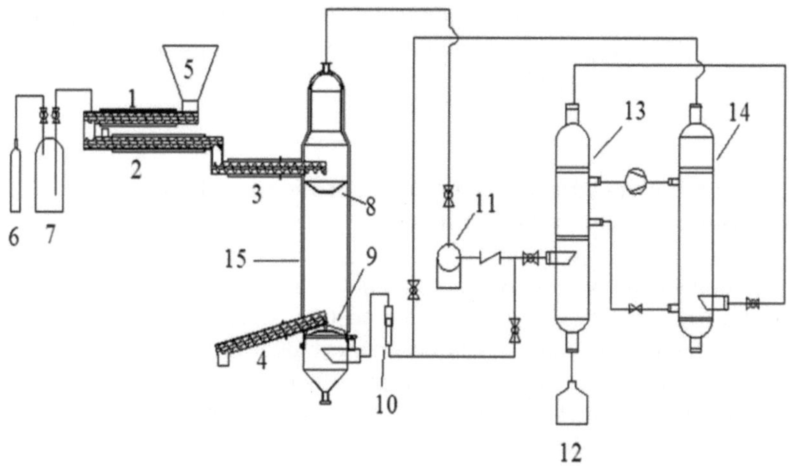

Fig. 7.4 A 150-L reactor of coupling of continuous solid-state fermentation of ethanol, gas stripping, heat pump separation (Chen et al. 2008). *1–4* Screw transport unit. *5* Feed tank. *6* Nutrition salt tank. *7* Inoculum tank. *8* Feed distributed unit. *9* Discharged unit. *10* Flow meter. *11* Fans. *12* Ethanol receiver. *13* Heat pump evaporator. *14* Heat pump condenser. *15* Fermentation reactor

Continuous Anaerobic Solid-State Fermentation Coupled with CO_2 Gas-Stripping Technology

Currently, various types of SSF bioreactors have common problems (e.g., batch fermentation without effective process integration). If continuous SSF and the separation process are integrated, it will greatly enhance production efficiency and decrease corresponding labor.

To my knowledge, no other report exists of bioreactors that could achieve pure culture and continuous anaerobic SSF simultaneously. Continuous SSF of ethanol coupled with product separation technology is proposed based on the current urgent need for biofuel and its problems of SSF (Chen et al. 2008). A sealed multistage screw feeding and charging machine coupled with an inoculating process is used to actualize aseptic culture and continuous anaerobic fermentation; a heat pump is imported to ensure heat and moisture insulation and provide a SSF gas-stripping and condensate separation system to realize stable control of continuous SSF and the effective separation of fermented productions. Guided by the ideas discussed, a continuous anaerobic SSF bioreactor has been established with an effective volume of 150 L (Fig. 7.4), and the technology has succeeded in ethanol production from sweet sorghum. Using this bioreactor, the final average ethanol concentration was above 10.8 %, and the yield of ethanol was about 19.49 % at 40 h. It was about 89 % of the maximum theoretical ethanol yield (unpublished data).

The coupling of heat pumps and gas stripping could provide constant temperature fermentation during the entire process, settle the problem of heat and mass transfer, and control the relative content of water and gas to provide optimal conditions. Furthermore, the recovered metabolic heat was used for gas stripping to make full use of energy. It is an economical SSF technology. A pilot-scale (6-m^3) bioreactor is being debugged.

7.2.3.4 Low Value-Added Bioconversion of Biomass: Direct Bioconversion of Solid Biomass

Direct bioconversion of solid biomass refers to direct conversion of lignocellulose by microbes to a bio-based product. Composting, silage, single-cell protein production, and feed fermentation are typical direct bioconversion processes. The main raw materials for direct bioconversion are generally organic solid wastes. The essential purpose of direct bioconversion of biomass is to actualize waste treatment rather than achieve product. It is a comparatively low value-added bioconversion technology.

References

Alam M. Solid state bioconversion of oil palm empty fruit bunches for cellulase enzyme production using a rotary drum bioreactor. Biochem Eng J. 2009;46:61–4.

Bear J. Dynamics of fluids in porous media. New York: American Elsevier; 1972.

Chandel AK, da Silva SS, Carvalho W, Singh OV. Sugarcane bagasse and leaves: foreseeable biomass of biofuel and bio-products. J Chem Technol Biotechnol. 2012;87:11–20.

Chen HZ. Biomass science and engineering. Beijing: Chemical Industrial Press; 2008.

Chen HZ, Li ZH. Ethanol preparation by solid-state enzymatic saccharification solid-state enzymatic, fermentation and separation technique and apparatus. Chinese Patent No. 1493694; 2004.

Chen HZ, Qiu WH. Key technologies for bioethanol production from lignocellulose. Biotechnol Adv. 2010;28:556–62.

Chen HZ, Xu J. Principle and application of modern solid-state fermentation. Beijing: Chemical Industry Press; 2004.

Chen HZ, Xu FJ, Tian ZH, Li ZH. A novel industrial-level reactor with two dynamic changes of air for solid-state fermentation. J Biosci Bioeng. 2002;93:211–4.

Chen HW, Ye SH, Ji H. Production of protein feed from a mixture of lees fermented by multi-strains. China Brew. 2006;6:74–7.

Chen HZ, Xu J, Li ZH. Temperature cycling to improve the ethanol production with solid state simultaneous saccharification and fermentation. Appl Biochem Microbiol. 2007;43:57–60.

Chen HZ, Ding WY, Dai SH, Li J, Wang L. Continuous anaerobic SSF coupled with CO_2 gas stripping technique and apparatus. Chinese Patent No. 101633940; 2008.

Chen HZ, Li HQ, Liu LY. The inhomogeneity of corn stover and its effects on bioconversion. Biomass Bioenerg. 2011a;35:1940–5.

Chen HZ, He Q, Liu LY. Cellulase production from the corn stover fraction based on the organ and tissue. Biotechnol Bioprocess Eng. 2011b;16:867–74.

Dhillon GS, Oberoi HS, Kaur S, Bansal S, Brar SK. Value-addition of agricultural wastes for augmented cellulase and xylanase production through solid-state tray fermentation employing mixed-culture of fungi. Ind Crops Prod. 2011;34:1160–7.

Duan YY, Chen HZ. Effect of three-phase structure of solid-state fermentation substrates on its transfer properties. CIESC J. 2012;63:1204–10.

Fan L, Lee YH, Beardmore D. Major chemical and physical features of cellulosic materials as substrates for enzymatic hydrolysis. Adv Biochem Eng. 1980;14:101–17.

Grethlein HE. The effect of pore size distribution on the rate of enzymatic hydrolysis of cellulosic substrates. Nat Biotechnol. 1985;3:155–60.

Himmel ME, Ding SY, Johnson DK, Adney WS, Nimlos MR, Brady JW, et al. Biomass recalcitrance: engineering plants and enzymes for biofuels production. Science. 2007;315:804–7.

Hölker H, Lenz J. Solid-state fermentation—are there any biotechnological advantages? Curr Opin Microbiol. 2005;8:301–6.

Holker U, Hofer M, Lenz J. Biotechnological advantages of laboratory-scale solid-state fermentation with fungi. Appl Microbiol Biotechnol. 2004;64:175–86.

Hong J, Ye XH, Zhang YHP. Quantitative determination of cellulose accessibility to cellulase based on adsorption of a nonhydrolytic fusion protein containing CBM and GFP with its applications. Langmuir. 2007;23:12535–40.

Huang DM, Wu QF, Lu JM, et al. Current research progress of solid state fermentation technology and equipment. Food Ferment Ind. 2003;6:87–91.

Kalogeris E, Iniotaki F, Topakas E, Christakopoulos P, Kekos D, Macris B. Performance of an intermittent agitation rotating drum type bioreactor for solid-state fermentation of wheat straw. Bioresour Technol. 2003;86:207–13.

Krishna C. Solid-state fermentation systems—an overview. Crit Rev Biotechnol. 2005;25:1–30.

Li HQ, Chen HZ. The periodic change of environment factors in solid state fermentation and effect on microorganism fermentation. Chin J Biotechnol. 2005;21:440–5.

Li HQ, Chen HZ. Detoxification of steam-exploded corn straw produced by an industrial-scale reactor. Process Biochem. 2008;43:1447–51.

Liu J, Yang J. Process calorimetry on solid-state fermentation of vinegar wastes in bioreactor with air pressure pulsation. Chem Biochem Eng Q. 2006;20:449–55.

Liu J, Li D, Yang J. Operating characteristics of solid-state fermentation bioreactor with air pressure pulsation. Appl Biochem Microbiol. 2007;43:211–6.

Marsh A, Stuart D, Mitchell D, Howes T. Characterizing mixing in a rotating drum bioreactor for solid-state fermentation. Biotechnol Lett. 2000;22:473–7.

McKendry P. Energy production from biomass (part 1): overview of biomass. Bioresour Technol. 2002;83:37–46.

Mitchell DA, Krieger N, Stuart DM, Pandey A. New developments in solid-state fermentation: II. Rational approaches to the design, operation and scale-up of bioreactors. Process Biochem. 2000;35:1211–25.

Ooijkaas LP, Weber FJ, Buitelaar RM, Tramper J, Rinzema A. Defined media and inert supports: their potential as solid-state fermentation production systems. Trends Biotechnol. 2000;18:356–60.

Ragauskas AJ, Williams CK, Davison BH, Britovsek G, Cairney J, Eckert CA, et al. The path forward for biofuels and biomaterials. Science. 2006;311:484–9.

Raghavarao KSMS, Ranganathan TV, Karanth NG. Some engineering aspects of solid-state fermentation. Biochem Eng J. 2003;13:127–35.

Rinaldi R, Schüth F. Design of solid catalysts for the conversion of biomass. Energy Environ Sci. 2009;2:610–26.

Roussos S, Raimbault M, Prebois JP, Lonsane B. Zymotis, a large scale solid state fermenter design and evaluation. Appl Biochem Biotechnol. 1993;42:37–52.

Sangseethong K, Meunier-Goddik L, Tantasucharit U, Liaw ET, Penner M. Rationale for particle size effect on rates of enzymatic saccharification of microcrystalline cellulose. J Food Biochem. 2007;22:321–30.

Singhania RR, Patel AK, Soccol CR. Recent advances in solid-state fermentation. Biochem Eng J. 2009;44:13–8.

Sun Y, Cheng JY. Hydrolysis of lignocellulosic materials for ethanol production: a review. Bioresour Technol. 2002;83:1–11.

Sun T, Liu BH, Li ZH, Liu DM. Effects of air pressure amplitude on cellulase productivity by *Trichoderma viride* SL-1 in periodic pressure solid state fermenter. Process Biochem. 1999;34:25–9.

Wan CX, Li YB. Fungal pretreatment of lignocellulosic biomass. Biotechnol Adv. 2012. doi:10.1016.

Wu DC, Xu F, Sun B, Fu RW, He HK, Matyjaszewski K. Design and preparation of porous polymers. Chem Rev. 2012. doi:10.1021/cr200440z.

Xiros C, Vafiadi C, Topakas E, Christakopoulos P. Decrement of cellulose recalcitrance by treatment with ionic liquid (1-ethyl-3- methylimidazolium acetate) as a strategy to enhance enzymatic hydrolysis. J Chem Technol Biotechnol. 2012;87:629–34.

Xu FJ, Chen HZ, Li ZH. Effect of periodically dynamic changes of air on cellulase production in solid-state fermentation. Enzyme Microb Technol. 2002;30:45–8.

Xu CY, Ma FY, Zhang XY, Chen SL. Biological pretreatment of corn stover by *Irpex lacteus* for enzymatic hydrolysis. J Agric Food Chem. 2010;58:10893–8.

Yang X, Huang T, Tsao GT. Pressure pulsation in solid-phase fermentation. Appl Biochem Biotechnol. 2002;98:599–610.

Yang QJ, Liu ZW, Xiong RQ. Study on flavor fermentation optimization of soy sauce by multi-strains fermentation. Guangdong Agric Sci. 2011;21:106–9.

Ye D, Farriol X. Improving accessibility and reactivity of celluloses of annual plants for the synthesis of methylcellulose. Cellulose. 2005;12:507–15.

Yi AN, Jiang SS, Liang E. Reduction of corn stalk with solid fermentation by multi-strains. J Jilin Agric Sci. 2011;36:47–50.

Zeng W, Chen HZ. Air pressure pulsation solid state fermentation of feruloyl esterase by *Aspergillus niger*. Bioresour Technol. 2009;100:1371–5.

Zhang JJ. The technology of degradant for roughage treatment of multiple strain combination of microorganism fermentation and their innovative usage. China Dairy Cattle. 2012;6:12–6.

Zhang ML, Fan YT, Xing Y, Pan CM, Zhang GS, Lay JJ. Enhanced biohydrogen production from cornstalk wastes with acidification pretreatment by mixed anaerobic cultures. Biomass Bioenerg. 2007;31:250–4.

Zhao FY, Fan NJ, Zhu JC. Isolation and characterization of an efficient cellulose-decomposing strain YN1. Microbiol China. 2010;4:496–502.

Zhao XB, Zhang LH, Liu DH. Biomass recalcitrance. Part I: the chemical compositions and physical structures affecting the enzymatic hydrolysis of lignocellulose. Biofuels Bioprod Biorefin. 2012;6(4):465–82. doi:10.1002/bbb.1331.

MIX
Papier aus verantwortungsvollen Quellen
Paper from responsible sources
FSC® C105338

If you have any concerns about our products,
you can contact us on
ProductSafety@springernature.com

In case Publisher is established outside the EU,
the EU authorized representative is:
Springer Nature Customer Service Center GmbH
Europaplatz 3, 69115 Heidelberg, Germany

Printed by Libri Plureos GmbH
in Hamburg, Germany